T0175125

Essential
STATISTICAL
CONCEPTS
for the Quality
Professional

Essential
STATISTICAL
CONCEPTS
for the Quality
Professional

D.H. Stamatis

CRC Press
Taylor & Francis Group
Boca Raton London New York

CRC Press is an imprint of the
Taylor & Francis Group, an **informa** business

CRC Press
Taylor & Francis Group
6000 Broken Sound Parkway NW, Suite 300
Boca Raton, FL 33487-2742

First issued in paperback 2019

© 2012 by Taylor & Francis Group, LLC
CRC Press is an imprint of Taylor & Francis Group, an Informa business

No claim to original U.S. Government works

ISBN-13: 978-1-4398-9457-6 (hbk)
ISBN-13: 978-0-367-38142-4 (pbk)

Visit the Taylor & Francis Web site at
http://www.taylorandfrancis.com

and the CRC Press Web site at
http://www.crcpress.com

To my grandchildren:

Jamey, Stacey, Caitlyn, and Dean

Contents

List of Illustrations

List of Tables

Preface

Many books and articles have been written on how to identify the root cause of a problem. However, the essence of any root cause analysis in our modern quality thinking is to go beyond the actual problem. This means that not only do we have to fix the problem at hand but that we also have to identify why the failure occurred and what was the opportunity to apply the appropriate knowledge so that avoidance of the problem would be the case.

This approach is somewhat new and unique; however, there are tools and methodologies that can help one evaluate the system for prevention. These tools and methodologies focus on structured, repeatable processes that can be instrumental in finding real, fixable causes of human errors and equipment failures that lead to most quality issues. Traditional approaches to this are the Failure Mode and Effect Analysis (FMEA), Advanced Product Quality Planning (APQP), and others.

As good and effective as these methodologies are, we as quality professionals need to go beyond these. For example,

1. Use as much evidence of the failure (problem) to understand what happened before trying to decide why it happened.
2. Always look to identify multiple opportunities to stop the problem (multiple causal factors).
3. When possible, have built-in expert systems that can be used by problem solvers to find the root causes of each of the causal factors.
4. Encourage problem solvers to look beyond the immediate causes and find correctable systemic issues.
5. Encourage problem solvers to find effective actions to prevent the problem's recurrence when all the root and generic causes have been identified. (There is a risk here because of possible interpolation; however, it is worthwhile.)

One will notice (Table P.1) that all these are based on four things: (1) strategy; (2) governance; (3) integrated business processes; and (4) methods with appropriate data. In this book, we focus on the methods, especially statistical ways of solving issues and problems.

However, before we begin our discussion, we must understand some very important reasons why we pursue problem solving and why we use statistics. Generally speaking, when someone asks the question "why problem solving?" it is not unusual to have a response like

- We do not need a problem-solving approach because we have our own approach to solving problems.
- We really do not have any idea about problem solving. This may come in different flavors such as
  - We do not need to get data; we have a feeling of what the problem is.
  - We have a set way of doing this and it is a repeatable process.
  - The process is very complicated and we cannot define it—so we have to use trial and error.
  - We have tried a systematic approach before but it does not work for us.
  - We are doing our best—there is nothing else that can be done.

Obviously, this kind of thinking is flawed. All problems must be approached in a systematic way to make sure that the appropriate problem is investigated and resolved in the most efficient way. Here, we must also recognize that there may be many ways to solve a particular problem, but there is only one approach that will work all the time. That approach is statistical thinking and the use of statistics as much as possible.

TABLE P.1

The Four Basic Items for Viewing the New Approach to Solving Root Causes

Strategy
Vision, value, and culture

Management
Executive ownership, decision making, accountability, and appropriate compensation

Integrated Business Processes
Automated, closed loop, customer relationship management and business intelligence integration, life-cycle management, and communication

Method	Reporting	Research
Data collection, contact management, sampling method, and survey questions	Selection of methodology for data analysis, benchmarking, micro and macro improvements, and dissemination practices	Business impact: financial, operational, constituency links, and use of customer information

Micro Approach	Macro Approach
Identify and resolve special causes	Identify and resolve common causes
Focus on changing individual-specific issues	Focus on solving systematic concerns
Make specific improvements that affect the customer	Make the effort to improve the entire organization
Focus on short-term solutions	Find long-term solutions

To illustrate this approach, let me use an analogy using the justice system as it is supposed to work:

- A crime is committed—a rule or a specification has been broken.
- Investigation of the crime and the crime scene is initiated—check out the problem, determine what is appropriate and applicable, and gather clues that will help in prosecution.
- Issue a warrant for the criminal(s) after you have identified the significant few, which you believe affect the output. To do this, a clear definition of the crime must be stated, otherwise the case may be questionable at best.
- Build the case. Use appropriate clues and circumstances to evaluate cause and effect. Statistically speaking you may use design of experiments (DOE), t-tests, F-tests, chi-square, analysis of variance (ANOVA), and anything else that you deem necessary and appropriate.
- Present the case and state the facts—statistical analysis.
- Present the closing arguments—practical significance.
- Lessons learned—what did you learn and what do you still need to know?

This analogy is, of course, fundamental in understanding that problem solving and the use of statistics are inherently build into the continual improvement attitude of any quality professional in any environment.

In the language of quality, this is, in fact, the classic approach of the traditional **P**lan, **D**o, **C**heck/**S**tudy, **A**ct (PDC/SA) with an additional component which is the **I**ngrain (some call it Infusion and some call it Institutionalization) of the gain into the organization. This approach is summarized in Table P.2.

Specifically, this book will cover the following:

Introduction: *Some Issues and Concerns about Using Statistics.* An overview of some pitfalls that should be considered when undertaking an experiment for improvement.

Chapter 1: *What Is Statistics?* explains what statistics is and how it is used in improvement endeavors.

Chapter 2: *Data* explains the lifeblood of all experiments. It identifies data and how they are used in an experimentation process.

Chapter 3: *Summarizing Data* emphasizes the need for summarizing data and some techniques to facilitate a decision based on sound interpretation of data.

TABLE P.2

A Summary of the PDC/SA Approach

Plan	Identify the problem Identify the tools useful for identifying potential causes	Define the problem in practical terms. Usually, the problems are due to centering or spread Use graphs to illustrate the problem. Remember that a picture is worth 1000 words. Also, do not underestimate the value of observing the process yourself
Do	Gather evidence (data) to narrow down the number of causes	Use historical and applicable data; use comparative and associative tests such as graphs; *t*-tests; correlation, and regression. Above all, do not underestimate your own experience or of those involved with the process—including operators
Check/ Study	Check: Verify problem and appropriate tools Study: Use appropriate tools to confirm causation	Use graphs when appropriate and applicable Use screen designs then follow up with fractional and full factorial analysis depending on time, cost, and significance of interactions. Do not be intimidated by central composite designs and surface methodology. Use them; they can help you optimize your results. Whenever possible, use cube plots, main effects plots, interaction plots, and Pareto diagram to see the effects of key factors
Act	Make the necessary changes	Document what you have done. This may be in a written form or via illustrations; implement necessary training; implement better controls; define the settings of your process control based on the analysis. Above all, make sure that you develop a new maintenance schedule for the new system
Ingrain	Demonstrate "ongoing" improvement changes	Use control charts, graphs, and capability analysis to demonstrate the "effective change" that has taken place. Once the sustainability is demonstrated, the new process should only be checked periodically or after a new change in the process

Special note: For those who use the Six Sigma methodology, the PDCA model is also applicable. Thus: Plan in Quality equates to Define in Six Sigma; Do equates to Six Sigma's Measure phase; Check equates to Analyze; and Act equates to Improve and Control.

Chapter 4: *Tests and Confidence Intervals for Means* introduces the reader to some common tests and confidence intervals for means.

Chapter 5: *Tests and Confidence Intervals for Standard Deviations* introduces the reader to some common tests and confidence intervals for the standard deviation.

Chapter 6: *Tests of Distributions and Means* introduces the reader to distributions and tests for normality and means.

Chapter 7: *Understanding and Planning the Experiment and Analysis of Variance* focuses on specific plans for any experiment for improvement and introduces the reader to ANOVA.

Chapter 8: *Fitting Functions* expands the notion of experimentation by dealing with mathematical models such as regression to optimize improvement and understand the relationship among several factors.

Chapter 9: *Typical Sampling Techniques* emphasizes the need for sampling and introduces some specific techniques in order to make sure that accuracy and precision of the data are appropriate and applicable for the study at hand.

Chapter 10: *Understanding Computer Programs for Design and Estimating Design Power* gives an anthology of designs for ANOVA and estimating power with each approach.

Appendix A: *Minitab Computer Usage* explains the most intricate part of the software (the work screen) with specific examples and graphics. It also addresses some of the actual screens for basic statistics and MSA analysis.

Appendix B: *Formulae Based on Statistical Categories* provides selected formulas for several categories; for example, parameter, regression, ANOVA, and others.

Appendix C: *General Statistical Formulae* provides many common and powerful formulas that may be used in the course of implementing statistics in an improvement environment.

Appendix D: *Hypothesis Testing Roadmap* provides a cursory roadmap for selecting specific statistical tests depending on the sample.

Appendix E: *Test Indicators* provides indicators for selecting appropriate basic statistic tests, statistical process control charts, and multivariate tests.

Appendix F: *Hypothesis Testing—Selected Explanations and Examples* provides the reader with selected additional background information on hypothesis testing along with examples.

Appendix G: *When to Use Quality Tools—A Selected List* provides a selected list of tools and their application.

Glossary provides a lexicon of statistical terms and concepts that the modern quality professional should be aware of and use in his or her application of statistics, by no means is it an exhaustive list.

Selected Bibliography provides readers with some additional current sources for pursuing their own research on the specific tests and methodologies covered in this book.

Acknowledgments

Being in the field of quality for over 30 years, I have come to realize that most people like quality, and they want quality. However, when a professional practitioner tries to explain quality quantitatively, he or she generally falls short of convincing the other person who may be a supervisor, management, customer, supplier, or whoever of how he or she has arrived at the specific decision and how sure that decision makes him or her feel.

Over the years, this was very difficult indeed; however, with the introduction of computers and their accessibility on the work floor and office, the issue of what is common and easy and understandable, as well as applicable, has become a concern for many. The concern is that they have an answer but they are not sure what it means.

To identify all situations and all applications is not possible. That is why I have written this book to help professionals in the application of statistics for improvement no matter what the situation. I had much help in formalizing what is indeed very important and useful for a professional using statistics for improvement. Obviously, I have not covered all issues and all tests available, but with the help of many I believe that what is covered in this book will indeed be very helpful.

As I said, many individuals have helped in crystallizing some of these issues, and I want to thank them for sharing their thoughts. I cannot thank everyone but rest assured that I appreciate all their inputs.

High on the list is Dr. M. Lindsay who introduced me to DOE in manufacturing. Also Dr. J. Framenco who gave me the first opportunity to run DOE in both classical and Taguchi methods at Ford Motor Company.

Dr. J. Kapur and Dr. R. Roy helped in applying DOE in many situations within the automotive industry.

Dr. L. Williams early on was patient enough to answer all my questions about SAS and the specific applications that it offered.

The late Dr. H. Bajaria for his thoughtful comments and interaction. His discussion points are still memorable due to the vivid analogies he used. His contribution to statistical problem-solving discussions as well as the long discussions about the benefits of classical DOE and Taguchi were very helpful.

Dr. R. Munro asked thoughtful and always provocative questions about applicability and appropriateness as well as helping in the selection process of the flow of the book.

Dr. S. Stamatis, my son, who even though had his hands full finishing his own dissertation, was always available to help with computer work but also—and more importantly—to debate and discuss the content of the book.

He always wanted a little more theory, while I was more concerned about the practical side of the content. Thanks, son. I wish you the best in Iowa.

C. Wong, a long-time friend, kept pushing me to write this book. His enthusiasm and belief in me are greatly appreciated.

I also want to thank K. Wong for the long hours of typing previous drafts, especially the statistical formulas. Thank you very much. You were a tremendous help in bringing this work to the finish line.

I also want to thank Minitab, Inc. for giving me permission to use the screens of Minitab® Statistical Software.

A special thanks to Random House, Inc., for allowing me to reproduce a portion of the *Iacocca: An Autobiography* by Lee Iacocca pages 50–52.

I want to thank the producers of this book for making it not only pleasant to the eye but also for improving the graphics and layout of the text to make for better flow and easier understanding.

Finally, I thank my wife Carla for always encouraging me and supporting me in everything I do. Of course, I cannot overlook her general comments and editing throughout the writing of this work. Thanks, Carla J.

Introduction

Some Issues and Concerns about Using Statistics

There is a great misconception about quality practitioners and the use of statistics in general. There is a prevalent idea that the higher the level of statistical analysis performed on a process, given a set of data, the more positive the results and understanding. That is not always the case. To avoid erroneous results, the following must be addressed at the start of any statistical analysis.

1. **Avoid Falling Victims to the "Law of the Instrument":** The Law of the Instrument states: "Give a small boy a hammer and everything he encounters needs pounding." Some experimenters seem to "suffer" from familiarity of a specialized forecasting technique and seem to use it on every occasion. However, there is no such thing as the ideal method. The "right" statistical technique depends on a number of conditions:

   - Is the purpose of the study clear and practical?
   - What is the accuracy required? Is the acceptable range of the statistical analysis appropriate? What is the level of risk?
   - What about the personnel doing the study—are they available? Do they have the appropriate background? What is their level of expertise and how much time do they have to spend on the study?
   - What is the working budget? Is it flexible? For example: Can you hire outside experts and/or consultants?
   - What kind of data is available?
   - What are management's inclinations? Do they want a highly scientific study or a quick answer?

2. **Use Multimethods:** Gunther (2010) made a study of lucky people. He reported his findings in the *Luck Factor*. One of his findings was that lucky people seemed to base their actions on forecasts derived from both hard and soft data (Corollary 2 of "The Hunching Skill"— Collect "soft" facts along with the "hard"). In statistical analysis, it's a good idea to follow Gunther's Corollary 2. Use a collage of techniques, some quantitative and some qualitative. This approach helps

the study to stand the test of common sense. Too often experimenters get so enamored with a technique, they accept the results without question. This danger is perhaps especially prevalent when using "esoteric" quantitative techniques. (In manufacturing this is very dangerous when one evaluates the process controls for the failures. We do have a tendency to repeat past controls even though they may have worked marginally or not at all, just because they have become comfortable. To avoid this, *one of the techniques you should always use—even though you may not incorporate its results in your findings—is a trend line.*)

3. **Be wary of Transition Periods:** There are certain stages of the product/service life-cycle curve when statistical analysis is most difficult. Transition periods are those periods where the product/service is changing from one stage to another (i.e., from introduction to growth) or when it is undergoing rapid growth or decline or when structural changes are taking place. If you use statistical techniques during these periods, make sure that you adjust the coefficients/formulas with qualitative judgments.

4. **Search for Causal Relationships:** There is the story about an Indian lying down with his ear pressed to the ground. A cowboy rode up, saw the Indian and asked, "What is it?" The Indian replied, "Half hour from here. Wagon. Four wheels. Pulled by two horses. One beige. One roan. Driven by white man in black suit. Black hat. Woman riding next." The cowboy cut him off. "Why that's tremendous. To be able to hear all that." The Indian replied, "Not hear—see. Wagon run over me one half hour ago!"

   Sometimes things aren't what they seem. Make sure you understand the correct causal relationships; if you haven't done so, put together a demand formula (transformation function) (multiple regression) for your specific product or process (sometimes even for the industry). You may not have all the data necessary to "run" the formulas, but the mere attempt of formulation will help you better understand the causal relationships within your company or even your industry.

5. **Consider Potential Competitive Actions:** The classic battle situation between a mongoose and a cobra describes this notion. The two are almost equally matched: the mongoose with its lightning speed and the cobra with its deadly, swift strike. Yet the mongoose almost invariably wins because of its "competitive strategy."

   The two meet—the mongoose feigns an attack. The cobra strikes, fully extending itself. And thus begins the dance of death. But the mongoose, perhaps smarter, has a strategy. After each successive feign attack, the mongoose shortens its distance. The cobra, however, continues to fully extend itself trying to bite the mongoose.

Soon, however, the mongoose's feigned attack is nothing more than a movement of the head; the cobra's strike, however, leaves the cobra fully extended. While the cobra is thus extended and its mobility greatly reduced, the mongoose attacks, crushing the cobra's head with its jaws.

Too many experimenters just do not take into account potential competitive actions. Such forecasts usually are based on the assumption that competitors will be benign—they will continue to do the same thing that they have been doing in the past; for example, when you do benchmarking make sure you take this competitive action into account.

6. **Get Management Involvement:** Management must believe in the statistical analysis if it is to be used. They must be convinced of the "data-driven" solutions. All statistical analyses will always be questionable because of the different motives of management personnel.

    The statistical analysis will be especially suspect if it is not "surprise free," that is, if it differs from what management expects to happen—"conventional wisdom." Therefore, believability will be highly questionable if the analysis differs from the past trend or the firm is operating in a constant change environment where structural stability is under stress. Thus, the less the analysis is surprise free, the greater the need for involvement. Two techniques to get management involved are

    - Have management involved at the outset
    - Communicate as often as possible via formal and informal means

7. **Avoid the Dowser-Rod Syndrome:** Some people claim that people who "witch" for water are using the dowser rod as an excuse for drilling a well in a location where they would like the well to be. In statistical analysis, make sure that you do not arrive at a specific analytical tool that will give you an excuse for pursuing a course of action you would like to follow. Make sure that the statistic is more than a methodological excuse. Avoid wishful thinking coloring your analysis.

8. **Be Careful of Intuition:** Be careful of intuition, especially in areas about which you know little. Occasionally, you may run into a person who has a sixth sense—able to predict what is going to happen. But you find these people only *rarely*. The last person I heard of who could do this was, unfortunately, crucified over 2000 years ago. More common is the type of situation where a person knows an industry so thoroughly that he develops what he thinks is an intuitive sense. But what he does is subconsciously draw trend lines and adjust them for seasonal and competitive behavior, and the like. Many times he

is quite accurate. But then, the so-called intuitive forecasting can get one into trouble especially where a person believes he has a sixth sense, but actually has been good at forecasting because of his intimate knowledge of a particular type of business. When a person believes that he has a sixth sense and starts shooting from the hip, trouble may ensue.

9. **Seek Out—But Scrutinize—Secondary Data:** Much has been written about existing data. Unfortunately, it's hard to find. Although there are a number of sources, here are some of the major ones you shouldn't overlook:

   - Your industry association
   - Libraries
   - Syndicated data sources, such as Mead Data Control. (Spend noon hours talking with representatives.)

   Make sure, however, that you check the authenticity of the secondary data. First, what is the timeliness of the data? Then, what was the character of the authors/agency who wrote the article/conducted the research? Did they have a vested interest in the point of view that they were trying to make?

   Then, if the data was based on a survey, examine these three levels:

   - Level 1. Was it the right universe?
   - Level 2. Was the sample representative of that universe?
   - Level 3. What about the survey instrument and execution itself? Did they seem appropriate? Or were they likely to create bias?

10. **Watch Out for Target Fixation:** One of the dangers of low-level fighter strafing is that pilots tend to become so involved in destroying a truck, tank, etc. that they lose their sense of closure. "That truck, I've hit it a number of times. Why doesn't it blow up?" Just a few more cannon bursts and I'll have him ... And so on. But unfortunately, the pilot gets so wrapped up that he gets too close to the ground. Sometimes you'll get so involved in trying to find a certain bit of information on a particular customer and or product or service that you'll spend a disproportionate amount of time in the search.

    Avoid target fixation by planning and budgeting the statistical analysis process. This is just common, managerial sense. But a short review won't hurt. Understand (or determine):

    - The accuracy required. What is the magnitude of the decision resting on this forecast? What is the acceptable range of the forecast accuracy in terms of dollars and/or units?
    - When must the forecast be completed?

- What kind of budget do you have to work with in terms of personnel time and money?

Then lay out the steps that you will have to complete in order to finish the forecast. Why not use an arrow diagram to help organize your thinking? On each activity specify the time and resources required. This will help you avoid spending excessive time on noncritical areas.

If you're having someone else do the forecast, make sure that the person knows, in specific terms, what you want accomplished (see above). Avoid such loose terms as "determine potential." Rather, let that person know the accuracy required, how the tasks fits into the forecasting process, and what actions will be taken as a result of the forecast. It might not be a bad idea to also have the person lay out the planned procedure on an arrow diagram.

11. **Don't Wait for All the Facts:** Here's an excerpt from Lee Iacocca's book, *Iacocca*, New York: Bantam Books, 1984, pp. 50–52. Although he's referring specifically to decision making, the general principle applies to "putting the forecast to bed."

> If I had to sum up in one word the qualities that make a good manager, I'd say that it all comes down to decisiveness. You can use the fanciest computers in the world and you can gather all the charts and numbers, but in the end you have to bring all your information together, set up a timetable, and act.
>
> And I don't mean act rashly. In the press, I'm sometimes described as a flamboyant leader and a hip shooter, a kind of fly by the seat of the pants operator. I may occasionally give that impression, but if that image were really true, I could never have been successful in this business.
>
> Actually, my management style has always been pretty conservative. Whenever I've taken risks, it's been after satisfying myself that the research and the market studies supported my instincts. I may act on my intuition—but only if my hunches are supported by the facts.
>
> Too many managers let themselves get weighted down in their decision making, especially those with too much education. I once said to Philip Caldwell, who became the top man at Ford after I left: "The trouble with you, Phil, is that you went to Harvard, where they taught you not to take any action until you've got all the facts. You've got ninety-five percent of them, but it's going to take you another six months to get that last five percent. And by the time you do, your facts will be out of date because the market has moved on you. That's what life is all about—timing."
>
> A good business leader can't operate that way. It's perfectly natural to want all the facts and to hold out for the research that guarantees a particular program will work. After all, if you're about to spend $300

million on a new product, you want to be absolutely sure you're on the right track. That's fine in theory, but real life just doesn't work that way. Obviously, you're responsible for gathering as many relevant facts and projections as you possibly can. But at some point you've got to take that leap of faith. First, because even the right decision is wrong if it's made too late. Second, because in most cases there's no such thing as certainty. There are times when even the best manager is like the little boy with the big dog waiting to see where the dog wants to go so he can take him there.

What constitutes enough information for the decision maker? It's impossible to put a number on it, but clearly when you move ahead with only 50 percent of the facts, the odds are stacked against you. If that's the case, you had better be very lucky—or else come up with some terrific hunches. There are times when that kind of gamble is called for, but it's certainly no way to run a railroad. At the same time, you'll never know 100 percent of what you need. Like many industries these days, the car business is constantly changing. For us in Detroit, the great challenge is always to figure out what's going to appeal to customers three years down the road, I'm writing these words in 1984, and we're already planning our models for 1987 and 1988. Somehow I have to try to predict what's going to sell three and four years from now, even though I can't say with any certainty what the public will want next month.

When you don't have all the facts, you sometimes have to draw on your experience. Whenever I read in a newspaper that Lee Iacocca likes to shoot from the hip, I say to myself: "Well, maybe he's been shooting for so long that by this time he has a pretty good idea of how to hit the target." To a certain extent, I've always operated by gut feeling. I like to be in the trenches. I was never one of those guys who could just sit around and strategize endlessly.

But there's a new breed of businessmen, mostly people with M.B.A.'s, who are wary of intuitive decisions. In part, they're right. Normally, intuition is not a good enough basis for making a move. But many of these guys go to the opposite extreme. They seem to think that every business problem can be structured and reduced to a case study. That may be true in school, but in business there has to be somebody around who will say: "Okay, folks, it's time. Be ready to go in one hour." When I read historical accounts of World War II and D-day, I'm always struck by the same thought: Eisenhower almost blew it because he kept vacillating. But finally he said: "No matter what the weather looks like, we have to go ahead now. Waiting any longer could be even more dangerous. So let's move it!"

The same lesson applies to corporate life. There will always be those who will want to take an extra month or two to do further research on the shape of the roof on a new car. While that research may be helpful, it can wreak havoc on your production plans. After a certain point, when most of the relevant facts are in, you find yourself at the mercy of the law of diminishing returns. That's why a certain amount of risk taking is essential. I realize it's not for everybody. There are some people who won't leave home in the morning without an umbrella even if the sun is shining. Unfortunately, the world doesn't always wait for you while you try to anticipate your losses. Sometimes you just have to take a chance and correct your mistakes as you go along.

Back in the 1960s and through most of the 1970s, these things didn't matter as much as they do now. In those days the car industry was like a golden goose. We were making money almost without trying. But today, few businesses can afford the luxury of slow decision making, whether it involves a guy who's in the wrong job or the planning of a whole new line of cars five years down the road.

Despite what the textbooks say, most important decisions in corporate life are made by individuals, not by committees. My policy has always been to be democratic all the way to the point of decision. Then I become the ruthless commander. "Okay, I've heard everybody," I say, "Now here's what we're going to do. You always need committees, because that's where people share their knowledge and intentions. But when committees replace individuals—and Ford these days has more committees than General Motors—then productivity begins to decline.

To sum up: nothing stands still in this world. I like to go duck hunting, where constant movement and change are facts of life. You can aim at a duck and get it in your sights, but the duck is always moving. In order to hit the duck, you have to move your gun. But a committee faced with a major decision can't always move as quickly as the events it's trying to respond to. By the time the committee is ready to shoot, the duck has flown away.

References

Gunther, M. (2010). The Luck Factor: *Why Some People Are Luckier Than Others and How You Can Become One of Them*. New York: Harriman House.
Iacocca, L. with W. Novak. (1984). *Iacocca: An Autobiography*. New York: Random House, Inc.

1

What Is Statistics?

There is a tremendous confusion about statistics and what it means. When one speaks about statistics, the listener gets frustrated and the common response is, "It is all Greek to me." Hopefully, this chapter explains what statistics is and how it is used in improvement endeavors.

The word *statistics* has two different meanings. More commonly, statistics means a collection of numerical facts or data, such as stock prices, profits of firms, annual incomes of college graduates, and so on. In its second meaning, it refers to an academic discipline, similar to physics, biology, and computer science.

The field of statistics can be divided into two parts: *descriptive* statistics and *inferential* statistics. Some years ago, the study of statistics consisted mainly of the study of methods for summarizing and describing numerical data. This study has become known as descriptive statistics because it primarily describes the characteristics of large masses of data. (Formally, descriptive statistics consists of procedures for tabulating or graphing the general characteristics of a set of data and describing some characteristics of this set, such as measures of *central tendency* or *dispersion*.) In many such statistical studies, data are presented in a convenient, easy-to-interpret form, usually as a table or graph. Such studies usually describe certain characteristics of the data, such as the *center* and the *spread* of the data.

The methods used to collect statistical data often require knowledge of sampling theory and experimental design. Once the data have been collected, descriptive statistics is used to summarize the raw data in a simple, meaningful way. Often, this is done by grouping the data into classes, thus forming what are called *frequency distributions*. When data are grouped into classes or categories, it becomes easy to see where most of the values are concentrated, but some information or detail may be lost due to the grouping. This approach of summarizing data by using tables and graphs is utilized heavily throughout the Six Sigma methodology.

The second branch of statistics is called inferential statistics. While descriptive statistics describe characteristics of the observed data, inferential statistics provide methods for making generalizations about the population based on the sample of observed data. (Formally, inferential statistics consists of a set of procedures that help in making inferences and predictions about a whole population based on information from a sample of the population.) For example, before building a large shopping center in a certain location, a real estate developer would want to know about the local community: the income and age distribution of residents in the area; their annual expenditures on items such as jewelry, clothing, sporting goods, and books; what proportion of them dines in expensive restaurants; and so on.

To get this information, the developer would probably rely on a sample of data obtained by interviewing some residents. The developer would then organize the data so that general characteristics would become evident. Using the methods of descriptive statistics, the developer would construct frequency distributions, which will show how many of the observations lie in a specific interval or category. Sample means could be calculated, which would show, the average annual income in the community, the average age, the average annual expenditure on sporting goods, and so on.

Deductive and Inductive Statistics

The field of statistics can also be divided into deductive statistics and inductive statistics. In *deductive statistics,* we deduce the properties of a sample of observations from the known characteristics of the population. In *inductive statistics,* we reverse the procedure—we start with a sample of observations and try to draw general conclusions about the properties of the population based on the characteristics of the sample. (Formally, in deductive statistics, we try to deduce properties of a given sample on the basis of known characteristics of the population. On the other hand, in inductive statistics, we infer properties of a population on the basis of observations of a selected sample of the population.)

The field of deductive statistics is often referred to as *probability theory.* As an example of a problem involving probability theory, suppose that an insurance agent has data that show that about 1% of the population will be involved in an accident this year. Thus, the agent knows something about the entire population. Now suppose that the agent insures a sample of 50 people, the agent might want to know the probability that no people in this particular sample will be involved in a serious accident. In this example, the agent is trying to use probability theory to deduce what is likely to be observed in a sample when given information about the characteristics of the entire population.

In inductive statistics, inferences about populations are made based on information contained in a sample of data. In inductive statistics, only the sample is known, and we try to determine the characteristics of the population from the information contained in the sample.

Stages of a Sample Study

A sample study has three major stages: (1) the *plan* or *sampling design*, (2) the actual *data collection*, and (3) the *analysis of the data* and the *statement of conclusions*. Each of these stages consists of several activities. The following list describes some of the choices that must be made in creating the sampling design:

1. Identifying the target population and, if necessary, constructing a frame. A frame is a listing of all units in a given problem.
2. If necessary, designing a questionnaire that is simple to fill out and that elicits honest answers.
3. Choosing an appropriate sampling technique.
4. Deciding what estimates are needed, what tests are to be performed, or what forecasts have to be made. Selecting the appropriate statistical procedures to fulfill these goals.
5. Determining the sample size. In doing so, the gains in accuracy or reliability that a large sample provides must be balanced against the increased costs of getting more information.

It has been said many times that "garbage in garbage out." In reference to data, this is of paramount importance because it is the data that will drive the decision, and the outcome of the analysis to come up with a solution to a problem. Therefore, the data collection phase is a priority for honesty, integrity, and generally, it requires the most time and cost. The major steps in data collection are as follows:

1. Selecting the sample of elementary units. An elementary unit is a person or object on which a measurement is taken.
2. Obtaining the sample of observations from the sample of elementary units.

The statistical analysis phase of a sampling study consists of some of the following activities:

- Describing the data using graphs or tables and constructing frequency distributions
- Calculating various sample statistics, such as the sample mean, the sample variance, or the sample proportion
- Estimating the appropriate population parameters and constructing point estimates and interval estimates

- Testing hypotheses and making forecasts
- Stating the conclusion

The most common sample procedures used are as follows:

1. Simple random samples—It is a method that a sample is obtained, in which (a) every possible set of n units has the same probability of being selected and (b) the selection of any one unit in no way affects the chance of selecting any other unit.
2. Systematic random samples—It is a method that a sample is obtained in the following way:
   a. Number the units in the frame from 1 to N
   b. Calculate the ratio N/n, where N is the population size and n is the desired sample size
   c. Let k be the largest integer less than or equal to N/n
   d. Randomly select an elementary unit from the first k units in the frame and then select every kth unit thereafter.
3. Stratified random samples—It is a method that a sample is obtained by separating the population into a number of nonoverlapping subpopulations, called strata, and then selecting a simple random sample from each stratum.
4. Cluster samples—It is a method of obtaining a sample by separating the population into subpopulations called clusters and then selecting clusters by simple random sampling. After the clusters have been selected, a complete census is taken of every unit in each selected cluster.
5. Randomized response sampling—It is a sampling that encourages subjects to provide truthful answers to sensitive questions.
6. Sequential sampling—The sample size is not determined or fixed before the sample is taken. The final size of the sample depends on the results obtained.

Warning!!!

One can hardly list all possible types of nonsampling error, all the ways that a sample can yield misleading data, and all sources of invalid information about a target process. Only a selected few are listed here:

1. In the planning stage:
   a. Selection bias is a systematic tendency to favor the inclusion in a sample of selected basic units with particular characteristics, while excluding other units with other characteristics.

b. Response bias is a tendency for selection of a sample to be wrong in some systematic way.

2. In the collection stage:

a. Selection bias is apt to enter the sample when experimenters are instructed to select within broad guidelines, the particular characteristics that they will sample.

b. Response bias can arise for a number of reasons during data collection. Both the experimenter and the process may be at fault.

c. Nonresponse bias may arise when no data (legitimate) is available from the sample.

3. In the processing stage:

a. The emergence of bias during data collection can conceivably be minimized by the careful design of the sample. Nevertheless, bias can enter even at the data-processing stage. People who code, edit, keypunch, or type the data into a computer database, tabulate, print, and otherwise manipulate data have many opportunities for making noncanceling errors. One of the areas of major concern in the quality area is the issue of data "outliers" or "wild values." We have a tendency to eliminate unbelievable data (high, low, or just different from the majority) and/or to substitute zero for a "no value" and vice versa.

Control the Data

To optimize the results of your data, the following may be considered.

1. **Weighting sample data:** It is a technique nothing more than the multiplication of sample observations by one or more factors to increase or decrease the emphasis that will be given to the observation. The troublesome aspect of weighting is related to the selection or calculation of the weighting factors. The specifications of the weighting scheme must be defined in terms of our overall objective: What is the purpose of weighting? In most cases, the obvious answer is that we would like our sample data to be representative of the population. The immediate follow-up to the first question is another: In what ways are the data to be representative of the population? The answer to this question should lead us to select an appropriate technique.

2. **Beware of the homing pigeon syndrome:** This is where you become completely dependent for data on the incoming paper flow; you lose the interactive process and find out only what the sender wants you to know.

3. **Reports/data required from the bottom up must be balanced by data interchange from the top down:** Asking the same old questions gets the same old answers. If the system does not allow for an interchange in the data-flow process, you will soon find yourself asking the wrong questions, at which point the answers do not matter.

4. **Have appropriate sample for the specific project:** It is imperative that we have the correct sample plan figured out before we begin experimenting. It is beyond the focus of this book to discuss the mathematical calculations for sampling. However, any statistics book may be of help.

5. **Missing data:** It is one of the most common problems in the field of quality. We all have a tendency to fill in the blanks of missing data. When that happens, it is the worst possible alternative that we as experimenters can do. The reason for this is that (1) we do not really know the outcome—therefore, we are guessing and (2) we are forcing a distribution of our data that are not really representative of what we are studying. If we do have missing data—for whatever reason—treat it as such. In case of computer utilization, most software programs have a built-in command for handling the missing data. The identification of missing data conventionally is noted as a dot ("."). In Minitab, the * is used to identify any missing data.

Many times we run a specific statistical technique using off-the-shelf software programs. However, when the analysis is all done we notice that our results are based on much fewer responses than anticipated. Why? Because, generally many of the software packages assume a multivariate procedure and drop any response from the analysis that fails to answer even a single item from your set of independent or dependent variables. What do you do? The question is very important; however, there is no definitive answer. Some approaches of resolving this dilemma are as follows:

1. Use only the responses with complete information. This has the effect of you basing your study on fewer respondents than planned for. As a consequence, the validity of the study may be questioned.

2. Use the highest or lower value of your sample. This is the least preferred method as it will skew the distribution in the direction of the selected value.

3. Use the middle point between the neighbor values. This will also force the distribution toward a leptokurtic shape.

4. Use a zero (0) value. This will also force the distribution toward a leptokurtic shape. It is not recommended to use a zero value.

5. Use the mean as a substitute for any value that is missing. Most software packages will do this if you tell them to; however, there are some repercussions if you do this:

   a. If there is a given response that, say, 90% is not available, the mean answer of the remaining 10% is substituted and used as if that is what the other 90% responded with. In effect, your response distribution is forced to be more leptokurtic.

   b. A response with few or no data is included in the analysis, unless the experimenter overrides the substitution of means. In either case, the response distribution is forced to be more leptokurtic.

An alternative that seems to be gaining favor is to use only those responses that are corresponding to the dependent variable and then substitute the response's own mean on the ones for which there is no response. For more information on missing data see Cohen and Cohen (1983), Cox and Snell (1981), and Milliken and Johnson (1984).

6. **Types of errors**: When you have collected the data, there have been several assumptions about the data and its significance. One of the important factors in deciding on the appropriateness is the issue of error. There are two basic errors for most of the applications in the quality world. They are as follows:

   a. Type alpha (α) or Type I error. This is the producer's error. It means that the producer is rejecting a good item.

   b. Type beta (β) or Type II error. This is the consumer's error. It means that the consumer is getting better quality than what he or she is paying for. (The way you or an organization define the error has a tremendous impact on whether the appropriate sample has been selected, the analysis is proper, the experiment was conducted based on sound design, and so on.)

7. **Power analysis**: A hypothesis test tells us the probability of our result (or a more extreme result) occurring, if the null hypothesis is true. If the probability is lower than a prespecified value (α, usually 0.05), it is rejected. If not, it is accepted. The ability to reject the null hypothesis depends on the following:

   a. **Alpha (α):** Usually is set to be 0.05, although this is somewhat arbitrary. This is the probability of a Type I error, that is, the probability of rejecting the null hypothesis, given that the null hypothesis is true. To use the search analogy, it is the probability of thinking we have found something when it is not really there.

    b. **Sample size:** A larger sample leads to more accurate parameter estimates, which lead to a greater ability to find what we were looking for. The harder we look, the more likely we are to find it.

    c. **Effect size:** The size of the effect in the population. The bigger it is, the easier it will be to find.

However, the above is not strictly correct. Cohen (1989) has pointed out, "all null hypotheses, at least in their 2-tailed forms, are false." Whatever we are looking for is always going to be there—it might just be there in such small quantities that we are not bothered about finding it. Therefore, to make sure that we are testing appropriately a power analysis is necessary.

Power analysis allows us to make sure that we have looked hard enough to find it and whether there is enough of it to bother us. The size of the thing we are looking for is known as the *effect size*. Several methods exist for deciding what effect size we would be interested in. Different statistical tests have different effect sizes developed for them; however, the general principle is the same. That is:

- **Base it on substantive knowledge:** Kraemer and Thiemann (1987) provide the following example. It is hypothesized that 40-year-old men who drink more than three cups of coffee per day will score more highly on the Cornell Medical Index (CMI) than men who do not drink coffee. The CMI ranges from 0 to 195, and previous research has shown that scores on the CMI increase by about 3.5 points for every decade of life. It is decided that an increase caused by drinking coffee, which was equivalent to about 10 years of age, would be enough to warrant concern; thus, an effect size can be calculated based on that assumption.

- **Base it on previous research:** See what effect sizes other researchers studying similar fields have found. Use this as an estimate of the sample size.

- **Use conventions:** Cohen (1989) has defined small, medium, and large effect sizes for many types of tests. These form useful conventions and can guide you, if you know approximately how strong the effect is likely to be.

Performing Power Analysis

Three types of power analysis exist, *a priori*, *post hoc*, and *compromise*. The first two are common, whereas the third one is quite complex and rarely used. It is also controversial.

- *A priori* **power analysis:** Ideally, power analysis is carried out *a priori*, that is, during the design stage of the study. A study can conclude whether a null hypothesis was true or false. The real answer, that is, the state of the world, can be that the hypothesis is true or false. Given the three factors—α, sample size, and effect size—a fourth variable, called β, can be calculated. Where α is the probability of a Type I error (i.e., rejection of a correct null hypothesis) and β is the probability of a Type II error (i.e., acceptance of a false null hypothesis).

The probability of correctly accepting the null hypothesis is equal to $1 - \alpha$, which is fixed, whereas the probability of incorrectly rejecting the null hypothesis is β. The probability of correctly rejecting the null hypothesis is equal to $1 - \beta$, which is called power. The power of a test refers to its ability to detect what it is looking for. To return to the search analogy, the power of a test is our probability of finding what we were looking for, given its size.

A power analysis program can be used to determine power, given the values of a sample size and effect size. If the power is deemed to be insufficient, steps can be taken to increase the power (most commonly, but not exclusively, by increasing the sample size).

A priori power analysis can ensure that you do not waste time and resources carrying out a study that has very little chance of finding a significant effect and can also ensure that you do not waste time and resources testing more subjects than are necessary to detect an effect.

- *Post hoc* **analysis:** Whereas *a priori* analysis is done before a study has been carried out, *post hoc* analysis is done after a study has been carried out to help explain the results of a study that did not find any significant effects. Imagine that a study had been carried out to see whether CMI scores were significantly correlated with coffee consumption (measured in average number of cups per day). A researcher carries out a study using 40 subjects, fails to find a significant correlation, and, therefore, concludes that coffee consumption does not alter CMI score.

  The effect size for correlation coefficients is simply r, the correlation coefficient. A *power analysis* can be carried out to find what effect size it would have been likely to detect. This is most easily done by examining a graph of power as a function of effect size. Usually, the x-axis goes from 0.1 to 0.5. Here, we must point out that Cohen (1989) defines a small effect size to be $r = 0.1$, a medium effect size to be $r = 0.3$, and a large effect size to be $r = 0.5$.

  In the example used by Kraemer and Thiemann (1987), the power to detect a large effect size is very high, above 0.95. It could, therefore, be safely concluded from this study that coffee consumption does not have a large effect on CMI score. At the medium effect size ($r = 0.3$), the power to detect a significant result is slightly above 0.6. Although this is reasonable power, it is not sufficient to safely assume that there is not a medium effect. Finally, at a small effect size, the power is very low—around 0.15. This is certainly not sufficient to conclude that there is not a small effect.

Whether the power of the study was sufficient to decide that there is no effect in the population would depend on the degree of correlation that would be determined to be large enough to be important.

Of special concern here is that when one calculates the power using a statistical software package, he or she must bear in mind that slightly different definition of *post hoc* power may be used. Make sure that you understand what and how the software package responds to the power effect. The confusion in software packages comes from the fact that the power given is the power that you had to detect the effect size that you found, not the effect size that might have been. This means that the power is a function of the *p*-value. If $p = 0.05$, then the power $= 0.50$. (Note that it's not necessarily the case when you have a multivariate outcome.)

- **Increasing power:** Although studies with excessive power exist, they tend to be few and far between. The main problem in designing and carrying out studies is to gain sufficient power. The greater the power the better it is. The rule of thumb for a minimum acceptable power is 80%—many large "definitive" studies have power around 99.9%. However, some have argued that Type I error should be equal to Type II error, and some have argued for high power because Type II error is more reliable and advantageous to have. The way to increase power in any study is by the following four ways:

  1. Increase sample size: As we have seen, the main way of increasing power is to increase sample size. Also, power is greater in one-tailed tests than in comparable two-tailed tests.

  2. Increase α: As we increase the α, we also increase the power.

  3. Shrink standard deviations: By using more homogenous groups (in an experimental study), the relative effect size increases. Similarly, increasing the reliability of the measures will have the same effect.

  4. Use analysis of covariance (ANCOVA): Adding covariates to an experimental study statistically reduces the error variance, and therefore, increases the relative effect size. Two rules of thumb to remember are as follows:

     a. Addition of covariates with an R^2 of 0.49 (i.e., a correlation of 0.7) increases power to the same extent as a doubling in sample size.

     b. Addition of covariates with an R^2 of 0.25 (correlation of 0.5) increases power to the same extent as an increase in sample size of one third.

Significance Testing

One of the most troubling issues in quality today is the issue of significance. Software packages generally test two types of statistical hypotheses simultaneously. The first type has to do with all of the independent variables as a group. It can be worded in several equivalent ways: Is the percentage of variance of the dependent variable that is explained by all of the independent variables (taken as a bunch) greater than zero or can I do a better job of predicting/explaining the dependent variable using all of these independent variables than not using any of them? Anyway, you will generally see an *F*-test and its attendant significance, which help you make a decision about whether or not all of the variables, as a group, help out. This, by the way, is a one-sided alternative hypothesis, since you cannot explain a negative portion of variance.

Next, there is usually a table that shows a regression coefficient for each variable in the model, a *t*-test for each, and a two-sided significance level. (This later can be converted to a one-sided significance level dividing it by two, which you will need to do if you have posited a particular direction or sign, *a priori*, for a given regression coefficient.) Now here is the interesting thing. You will sometimes see regression results in which the overall regression (the *F*-statistic) is significant, but none of the individual coefficients are accompanied by a *t*-statistic that is even remotely significant. This is especially common when you are not using stepwise regression and are forcing the entire set of independent variables into the equation, which can result from correlation between the independent variables. If they are highly correlated, then as a set they can have a significant effect on the dependent variable, whereas individually they may not.

One can see that the issue of significance may indeed play a very important role in the analysis. Let us then examine this little more closely.

1. In everyday language, the term *significance* means *importance*, while in statistics—this is very important—it means PROBABLY TRUE. The implication of these simple definitions is the fact that a particular study may be true without being important. As a result, when a quality engineer says that a particular result is highly significant he or she means that the result is very probably true. Under no circumstances, he or she means that the result is highly important.

2. Significance levels show you how probably true a result is. The most common level is 95% with no special significance; in other words, the result has a 95% chance of being true. This is also misleading, however, because most software packages show you "0.05," meaning that the finding has a 5% chance of not being true, which is the same as a 95%

chance of being true. The 95% level comes from academic publications, where theory usually has to have at least 95% chance of being true to be considered worth reporting. However, under no circumstances the 95% level of significance is sacred. It may be set at any level by the experimenter. In fact, in the business world if something has a 90% chance of being true ($p = 0.1$), it certainly can't be considered proven, but it may be much better to act as if it were true rather than false.

3. To find the exact significance level, subtract the number shown from one. For example, a value shown in the computer output as "0.01" means there is a 99% ($1 - 0.01 = 99$) chance of being true. A strong warning about this significance is the fact that IT MUST BE SET *a priori*—before the experiment and analysis have taken place. Do not fall into the trap when you see the significance on the computer print-out and get tempted to change the parameter of the significance. For more information on significance see Lehman (1986), Fuller (1987), and Rousseeuw and Leroy (1987).

Wrong Signs

This happens all the time. For example, you know that overall satisfaction and convenience are positively correlated. Yet, in multiple regression, the sign of the coefficient for convenience is negative. Why?

There are a couple of things that can be going on. First the *t*-statistic for the coefficient may not be statistically significant. We interpret this as an indication that the coefficient is not significantly different from zero and, hence, the sign (and magnitude for that matter) of the coefficient are spurious. Almost half the time and for a truly nonsignificant effect, the sign will be wrong.

The second thing that can be happening is the partialling effect noted earlier. It could be that the slope is negative given the effect of the other variables in the regression (partial) even though all by itself the variable shows a positive correlation and slope (total). For more information on signs see Hays (1981), Snell (1987), Box and Draper (1987), and Gibbons (1985).

Summary

This chapter tries to demystify statistics for the general reader. It identifies the key things that are important in addressing statistics for improvement and provides some guidelines to keep away from the traditional traps. In the next chapter, we will address the lifeblood of all statistics and experimentation, which of course are the data.

References

Box, G. E. P. and Draper, N. R. (1987). *Empirical Model Building and Response Surfaces.* John Wiley & Sons, New York.

Cohen, J. (1989). *Statistical Power Analysis for the Behavioural Sciences,* 2nd ed. Erlbaum, Hillsdale, NJ.

Cohen, J. and Cohen, P. (1983). *Applied Regression/Correlation Analysis for the Behavior Sciences,* 2nd ed. Erlbaum, Hillsdale, NJ.

Cox, D. R. and Snell, E. J. (1981). *Applied Statistics: Principles and Examples.* Chapman & Hall, New York.

Fuller, W. A. (1987). *Measurement Error Models.* John Wiley & Sons, New York.

Gibbons, J. D. (1985). *Nonparametric Statistical Inference,* 2nd ed. (Revised and expanded). Marcel Dekker, New York.

Hays, W. L. (1981). *Statistics,* 3rd ed. Holt, Rinehart & Winston, New York.

Kraemer, H. C. and Thiemann, S. (1987). *How Many Subjects? Statistical Power Analysis in Research.* Sage, Newbury Park, CA.

Lehman, E. L. (1986). *Testing Statistical Hypothesis,* 2nd ed. John Wiley & Sons, New York.

Milliken, G. A. and Johnson, D. E. (1984). *Analysis of Messy Data.* Van Nostrand Reinhold, New York.

Rousseeuw, P. J. and Leroy, A. M. (1987). *Robust Regression and Outlier Detection.* John Wiley & Sons, New York.

Snell, E. J. (1987). *Applied Statistics: A Handbook of BMDP Analyses.* Chapman & Hall, New York.

Selected Bibliography

Abelson, R. P. (1995). *Statistics As Principled Argument.* Erlbaum, Hillsdale, NJ.

Chow, S. L. (1996). *Statistical Significance: Rationale, Validity and Utility.* Sage, London.

Keren, G. and Lewis, C. (1993). *A Handbook for Data Analysis in the Behavioral Sciences.* Erlbaum, Hillsdale, NJ.

Murphy, K. R. and Myors, B. (2003). *Statistical Power Analysis: A Simple and General Model for Traditional and Modern Hypothesis Tests,* 2nd ed. Erlbaum, Hillsdale, NJ.

2

Data

In the previous chapter, we discussed the meaning of statistics and how one can use that information to facilitate a study with good results. In this chapter, we define data and categorize them in such a way that appropriate and applicable statistical analyses may be performed given a specific category of data. We also address some issues of sampling and the significance of normality.

Everything we do is based on data. So, the question that arises quite often is, is it datum or data? Grammatically speaking, the singular word is datum and the plural is data. However, because we have more than one word, the convention is that we use data. In common usage, "data" are any *materials* that serve as a basis for drawing conclusions. (Notice that the word we use is material. That is because material may be quantifiable or numerical and measurable, or on the other hand, may be an attribute or qualitative. In either case, it can be used for drawing conclusions.) Drawing conclusions from data is an activity that everyone engages in—bankers, scholars, politicians, doctors, engineers, common individuals in their everyday life, and corporate presidents. In theory, we base our everyday purchasing requirements, foreign policy, methods of treating diseases, corporate marketing strategies, process efficiency, and quality on "data."

There are many sources of data. We can conduct our own surveys or experiments (primary), look at information from surveys other people have conducted (secondary), or examine data from all sorts of existing records—such as stock transactions, election tallies, or inspection records. But acquiring data is not enough. We must determine what conclusions are justified based on the data. That's known as "data analysis." People and organizations deal with data in many different ways. Some people accumulate data but don't bother to evaluate it objectively. They think they know the answers before they start. Others want to examine the data but don't know where to begin. Sometimes people carefully analyze data, but the data are inappropriate for the conclusions that they want to draw. Unless the data are correctly analyzed, the "conclusions" based on them may be erroneous. A superior treatment for a disease may be dismissed as ineffectual; you may purchase stocks that don't perform well and lose your life savings, or you may target your marketing campaign to the wrong audience, costing your company millions of dollars, or worse yet, you may adjust the wrong item in the process and, as a consequence, you may affect the response of the customer in a very unexpected way. The consequences of bad data analysis can be severe and far reaching. That's why we want to know how to analyze data well.

One can analyze data in many different ways. Sometimes all one needs to do is describe the data. For example, how many people say they are going to buy a new product you're introducing? What proportion of them are men and what proportion are women? What is their average income? What product characteristic is the customer delighted with? In other situations, you want to draw more far-reaching conclusions on the basis of data you have at hand. You want to know whether your candidate stands a chance of being elected, whether a new drug is better than the one usually used, or you want to improve a design of a product that the customer is really excited about. You don't have all of the information you would like. You have data from some people or samples, but you would like to draw conclusions about a much larger audience and/or samples.

At this juncture your answer may be that "I do not have to worry about all this because the computer will do it for me." That is not an absolute truth. Computers simplify many tasks, including data analysis. By using a computer to analyze your data, you greatly reduce both the possibility of computational error and the time required. Learning about computers and preparing data for analysis by computer does require time, but in the long run it substantially decreases the time and effort required. Using a computer also makes learning about data analysis much easier. You don't have to spend time learning formulas. The computer can do the calculating for you. Instead, your effort can go into the more interesting components of data analysis—generating ideas, choosing analyses, and interpreting their results.

Because of this characteristic, this book doesn't emphasize formulas. It emphasizes understanding the concepts underlying data analysis. The computer can be used to calculate results. You need to learn how to interpret them. Selected formulas and selection guidelines have been included in Appendices B through F.

Describing Data

Once you've prepared a data file, you are ready to start analyzing the data. The first step in data analysis is describing the data. You look at the information you have gathered and summarize it in various ways. You count the number of people giving each of the possible responses. You describe the values by calculating averages and seeing how much the responses vary. You look at several characteristics together. How many men and how many women are satisfied with your new product? What are their average ages?

You also identify values that appear to be unusual—such as ages in the 100s, incomes in the millions—and check the original records to ensure that these values were picked up correctly. You don't want to waste time analyzing incorrect data.

Data Orientation

Before we do anything about the process we must understand it and know the characteristics of what we are about to do. We must understand the differences between variable and attribute data because the definition of the data will guide us in the appropriate sampling as well as the appropriate selection of the appropriate and applicable tool.

There are two kinds of data: variable (measurable) and attribute (go/no go). With variable data one has maximum opportunity and flexibility to study the process with a variety of methods and approaches. On the other hand, with attribute data both the opportunity and the flexibility are much less. In addition, if you deal with attribute data, the power of the data is increased with many samples, whereas with the variable data that power is sufficient with much less sampling.

Nominal, Ordinal, Interval, and Ratio

Variables can be classified into different groups based on how they are measured. Machine number, cost, and weight are all different types of variables. Machine number is called a *nominal* variable since the numerical code assigned to the possible responses conveys no information. They are merely labels or names. (That is why the level of measurement is called nominal—from the Latin word for "name.") Codes assigned to possible responses merely identify the response. The actual code number means nothing. For typical statistics used with this kind of data see Figure 2.1.

If the possible responses can be arranged in order, or identified as a degree of something or a product quality characteristic variable, the variable is called *ordinal*: Its codes have an order, nothing more. (*Ordinal* is from—you guessed it—a Latin word meaning "order.") Variables such as dollars, job satisfaction, condition of health, and happiness with one's social life, all of which are usually measured on a scale going from much too little, are ordinal variables. The numbers assigned to the responses allow you to put the responses in order. But the actual distances between the numeric codes

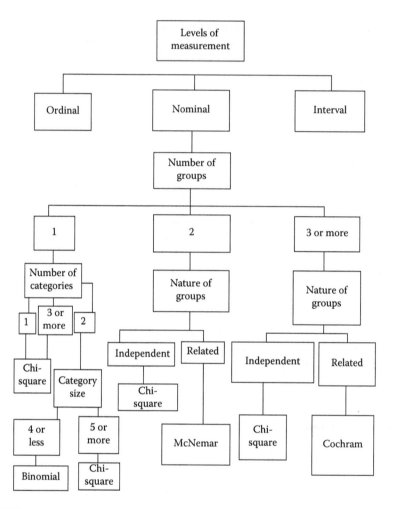

FIGURE 2.1
Nominal data statistics.

mean nothing. For typical statistics used with this kind of data see Figure 2.2.

Temperature can be measured and recorded on a scale that is much more precise than job satisfaction. The interval, or distance, between values is meaningful everywhere on the scale. The difference between 100°F and 101°F is the same as the difference between 102° and 103°. Since temperature measured on the Fahrenheit scale does not have a true zero, however, you can't say that an 80° day is twice as hot as a 40° day. A temperature of zero does not mean there is no heat. The zero point is determined by convention. (If you insist, I'll admit that temperatures

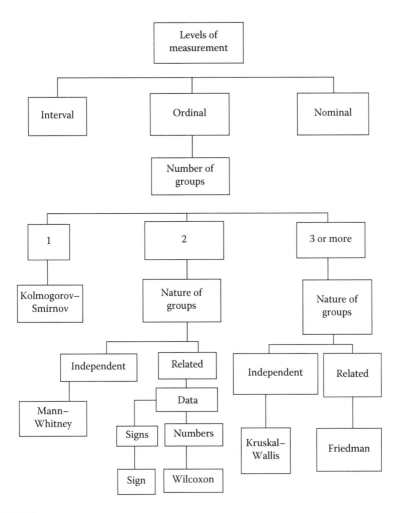

FIGURE 2.2
Ordinal data statistics.

do have an absolute zero point, but that has very little to do with, say, the measurement of body temperatures.) Thus, body temperature can be called an *interval* variable.

The last type of measurement scale is called a *ratio or interval* scale. The only difference between a ratio scale and an interval scale is that the ratio scale has an absolute zero. Zero means *zero*. It's not just an arbitrary point on the scale that somebody happened to label with zero. Height, weight, distance, age, and education can all be measured on a ratio scale. Zero education means no education at all, just like zero weight means no weight at all. On a ratio scale, the proportions, or ratios, between items are meaningful. A 200-lb person is

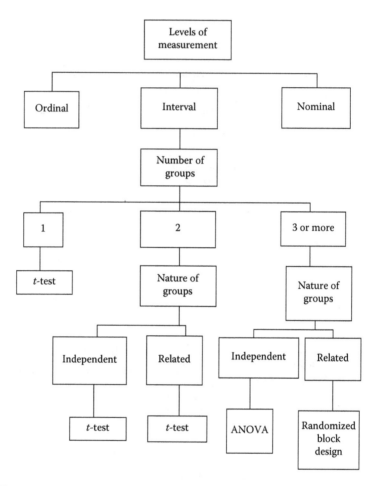

FIGURE 2.3
Ratio (interval) data statistics.

twice as heavy as a 100-lb person. A 1000-m race is twice as long as a 500-m race. I suppose there's no need to tell you what language the words *interval* and *ratio* come from.

Why all the fuss? Why have we spent all this time describing these "levels of measurement"? The reason is straightforward—the way in which you analyze your data depends on how you have measured it. Certain analyses make sense with certain types of data. Even something as simple as interpreting cumulative percentages requires you to know what scale your data are measured on. For example, cumulative percentages don't make much sense for variables measured on a nominal scale. So, depending on the level of measurement, the appropriate technique, test, and analysis must be selected; otherwise, the results will be meaningless. For typical statistics used with this kind of data see Figure 2.3.

Sampling Considerations

Suppose we are making simple pens. We make about 10,000 per hour. In these pens, we are also responsible for many dimensions and most—if not all—must be controlled. Some examples of our controlling characteristics are as follows: length of the pen, length of the cap, diameter of the hole, tapering, no fill, thickness, and many more.

The point here is that no one will be able to check every characteristic of all the 10,000 pens produced per hour. We would try to measure just the important ones, but even then if we decide that the diameter of the hole is very important, we still have to contend with 10,000 measurements. If we assume that each measurement takes 10 sec to measure and record each diameter of the hole we will need more than 27 ½ {[(10 sec × 10,000)]/3,600} h to check every pen produced. Of course, this is for only one characteristic and for only 1 h of production. Obviously, something is not right. How can we possibly tell whether our pens are good or bad without looking at all of them? How can we tell whether other characteristics are changing without seeing and or measuring all of them?

We can use a sampling plan. In a sampling plan, the selection of a few parts out of the operation is checked and then the results are projected into the entire population. It is not necessary to look at every pen in order to see what is happening to the whole process.

There are several types of sampling plans. It is very important for the analyst to select the type that will result with the most and best information of our needs. For example, in manufacturing more often than not we choose the consecutive sampling plan.

What is a consecutive plan? It is a plan that is taking a few parts in the order in which they are made so we tell how things should be done. Why would we choose such a plan? Because using a consecutive sampling plan will allow us to tell exactly when something starts to change. We can then pinpoint the reason for the change. This is very important information, because if we find that nothing has changed—which is good—we must find out what has kept the process to stay the same. Either way, this information can be very helpful for our understanding of how to improve the process.

Additional considerations for sampling are rational samples and rational subsamples.

Rational Samples

What are they?

- Groups of measurements subjected only to common cause variation
- Collections of individual measurements, the variation among which is attributable only to a constant system of chance causes

How should they be chosen?

- Strive to choose samples to minimize the occurrence of special causes within the sample.
- Strive to maximize the opportunity to detect special causes when they occur between samples.

Rational Subsamples

The small number of observations taken periodically should be rational sub-samples. This means that they should be taken in such a way that only common cause variability can be attributed to the points in a particular subsample. There shouldn't be any assignable causes of variability that affect some of the points in the subsample and not others. Typically, rational subsamples are obtained by taking observations nearby in time. For example, every half hour we might examine five consecutive soda cans coming off the production line. However, if we are not careful, even with a rational subsample we may end up in trouble. Make sure you review your sampling plan with a statistician or with a quality professional before you start taking samples.

Rational Sampling Pitfalls

To be sure, rational sampling is an important concept and is used very frequently. However, there are some cautions. The three major ones are as follows:

1. Combination: Several machines make the same product. The product is combined and a single control chart is maintained.
2. Stratification: Each machine contributes equally to the sample composition.
3. Mixing: Combine the outputs of several machines into a single stream and then sample.

Frequency of the Sample

No one will deny the fact that sampling is important. However, for that importance to be of significance, the sample must be random and representative of the whole. To make sure that this is actually happening, statisticians have developed many techniques to test the appropriateness of the sample. Here, however, we will address some simple and straightforward approaches.

To begin with, the questions that we should pose are as follows:

- Is the process stable?
- What is the desired confidence level? (90%, 95%, or 99%)?
- Is the size of difference interested in a particular change (Δ) appropriate?
- Is the amount of variation (unexplained) within our boundaries?

If we have answers to these questions, then we may proceed to use one of the following formulas.

Condition 1: When a difference of two averages of size Δ is considered economically significant use

$$n = \frac{2xt^2xs^2}{\Delta^2}$$

Condition 2: When a standard deviation must be known within Δ for economic reasons use

$$n = \frac{t^2xs^2}{2x\Delta^2}$$

where

$\quad t = 1.64 \approx 1.6$ for 90% confidence
$\quad 1.96 \approx 2$ for 95% confidence
$\quad 2.58 \approx 3$ for 99% confidence

s = the standard deviation and Δ is the desired difference
x = the multiplication operation

Degrees of Freedom

In statistics, the number of degrees of freedom (dof) is the number of independent pieces of data being used to make a calculation. Another way of saying it is: the number of values in the final calculation of a statistic that are free to vary. It is usually denoted with the Greek letter *nu*, ν, or with the English *f* or *df*. The number of dof is a measure of how certain we are that our sample population is representative of the entire population—the more dof, usually the more certain we can be that we have accurately sampled the entire population. For statistics in analytical chemistry, this is usually the number of observations or measurements (N) made in a certain experiment.

The dof can be viewed as the number of independent parameters available to fit a model to data. Generally, the more parameters you have, the more accurate your fit will be. However, for each estimate made in a calculation, you remove 1 dof. This is because each assumption or approximation you make puts more restrictions on how many parameters are used to generate the model. Putting in another way, for each estimate you make, your model becomes less accurate.

The most obvious and simple example to show the *df* is in the calculation of the variance for a sample population. The formula for the variance is

$$s^2 = \frac{1}{n-1}\sum_i (x_i - \bar{x})^2$$

The dof is $v = n - 1$, because an estimate was made that the sample mean is a good estimate of the population mean, so we have 1 less dof than the number of independent observations. In other words, we used 1 dof when the average was calculated.

In many statistical calculations, such as regression, ANOVA, outliers, and *t*-tests, you will need to know or calculate the number of dof. If you use a software program, that is calculated for you.

Where Do I Start?

Once you have identified the problem and begin the collection of data, it is time to characterize the problem using basic statistics. This means that the response variable must be understood (preferably as a measurable characteristic) and the identification of as many as possible key input variables must be recognized.

When this is done, the focus is on whether the problem lies within the location of the data or the spread of the data or both. This implies that the investigator has to identify which inputs influence the response or, mathematically speaking, which Xs influence the Ys. This investigation requires adjustments of the factors in such a way that the "right" combinations will produce the desirable results.

It must be recognized here that the desirable results in actuality are differences. These differences are from the old to the new situation of the process. So, in order for anyone to determine whether or not the difference is for real or random we must test that difference statistically or practically. The pivotal point of the decision to determine the real difference from a random one is (a) could I be wrong? and (b) if so, how often? To make this decision, we use tables of the specific test that we are using or we use the probability value. In the past, tables of significance were published in all statistical books. However, with the proliferation of computers, now the probability (*p*) value is used.

What Is a *p*-Value?

Perhaps one of the most confusing items in statistical analysis is the issue of the *p*. What does it mean and how do we apply it? In other words, the problem with the *p* value is the interpretation. I will try to give a simple explanation here, for it is very important that we all understand not only the significance but also the interpretation.

We wish to test a null hypothesis against an alternative hypothesis using a data set. The two hypotheses specify two statistical models for the process that produced the data. The alternative hypothesis is what we expect to be true if the null hypothesis is false. We cannot prove that the alternative hypothesis is true, but we may be able to demonstrate that the alternative is much more plausible than the null hypothesis given the data. This demonstration is usually expressed in terms of a probability (a *p*-value) quantifying the strength of the evidence against the null hypothesis in favor of the alternative.

We ask whether the data appear to be consistent with the null hypothesis or whether it is unlikely that we would obtain data of this kind if the null hypothesis were true, assuming that at least one of the two hypotheses is true. We address this question by calculating the value of a test statistic, that is, a particular real-valued function of the data. To decide whether the value of the test statistic is consistent with the null hypothesis, we need to know what sampling variability to expect in our test statistic if the null hypothesis is true. In other words, we need to know the null distribution, the distribution of the test statistic when the null hypothesis is true. In many applications, the test statistic is defined so that its null distribution is a "named" distribution for which tables are widely accessible, for example, the standard normal distribution, the binomial distribution with $n = 100$ and $p = 1/2$, the t distribution with 4 dof, the chi-square distribution with 23 dof, the F-distribution with 2 and 20 dof, and many more.

Now, given the value of the test statistic (a number), and the null distribution of the test statistic (a theoretical distribution usually represented by a probability density), we want to see whether the test statistic is in the middle of the distribution (consistent with the null hypothesis) or out in a tail of the distribution (making the alternative hypothesis seem more plausible). Sometimes we will want to consider the right-hand tail, sometimes the left-hand tail, and sometimes both tails, depending on how the test statistic and alternative hypothesis are defined. Suppose that large positive values of the test statistic seem more plausible under the alternative hypothesis than under the null hypothesis. Then we want a measure of how far out our test statistic is in the right-hand tail of the null distribution. The *p*-value provides a measure of this distance. The *p*-value (in this situation) is the probability to the right of our test statistic calculated using the null distribution. The further out the test statistic is in the tail, the smaller the p-value, and the stronger the evidence against the null hypothesis in favor of the alternative.

The p-value can be interpreted in terms of a hypothetical repetition of the study. Suppose the null hypothesis is true and a new data set is obtained independently of the first data set but using the same sampling procedure. If the new data set is used to calculate a new value of the test statistic (same formula but new data), what is the probability that the new value will be further out in the tail (assuming a one-tailed test) than the original value? This probability is the p-value.

The p-value is often incorrectly interpreted as the probability that the null hypothesis is true. Try not to make this mistake. In a frequent interpretation of probability, there is nothing random about whether the hypothesis is true; the randomness is in the process generating the data. One can interpret "the probability that the null hypothesis is true" using subjective probability, a measure of one's belief that the null hypothesis is true. One can then calculate this subjective probability by specifying a prior probability (subjective belief before looking at the data) that the null hypothesis is true, and then use the data and the model to update one's subjective probability. This is called the Bayesian approach because Bayes' theorem is used to update subjective probabilities to reflect new information.

When reporting a p-value to persons unfamiliar with statistics, it is often necessary to use descriptive language to indicate the strength of the evidence. I tend to use the following sort of language. Obviously, the cutoffs are somewhat arbitrary and another person might use different language.

- $p > 0.10$: No evidence against the null hypothesis. The data appear to be consistent with the null hypothesis.
- $0.05 < p < 0.10$: Weak evidence against the null hypothesis in favor of the alternative.
- $0.01 < p < 0.05$: Moderate evidence against the null hypothesis in favor of the alternative.
- $0.001 < p < 0.01$: Strong evidence against the null hypothesis in favor of the alternative.
- $p < 0.001$: Very strong evidence against the null hypothesis in favor of the alternative.

While using this kind of language, one should keep in mind the difference between statistical significance and practical significance. In a large study, one may obtain a small p-value even though the magnitude of the effect being tested is too small to be of importance. It is a good idea to support a p-value with a confidence interval for the parameter being tested.

A p-value can also be reported more formally in terms of a fixed level α test. Here, α is a number selected independently of the data, usually 0.05 or

0.01, more rarely 0.10. We reject the null hypothesis at level α, if the p-value is smaller than α; otherwise we fail to reject the null hypothesis at level α. I am not fond of this kind of language because it suggests a more definite, clear-cut answer than is often available. There is essentially no difference between a p-value of 0.051 and 0.049. In some situations, it may be necessary to proceed with some course of action based on our belief in whether the null or alternative hypothesis is true. More often, it seems better to report the p-value as a measure of evidence.

A fixed-level α test can be calculated without first calculating a p-value. This is done by comparing the test statistic with a critical value of the null distribution corresponding to the level α. This is usually the easiest approach when doing hand calculations and using statistical tables, which provide percentiles for a relatively small set of probabilities. Most statistical software packages produce p-values that can be compared directly with α. There is no need to repeat the calculation by hand.

Fixed-level α tests are needed for discussing the power of a test, a useful concept when planning a study. Suppose we are comparing a new medical treatment with a standard treatment, the control. The null hypothesis is that of no treatment effect (no difference between treatment and control). The alternative hypothesis is that the treatment effect (mean difference of treatment minus control using some outcome variable) is positive. We want to have a good chance of reporting a small p-value, assuming, the alternative hypothesis is true and the magnitude of the effect is large enough to be of practical importance. The power of a level α test is defined to be the probability that the null hypothesis will be rejected at level α (i.e., the p-value will be less than α) assuming the alternative hypothesis is true. The power generally depends on the variability of the data (lower variance, higher power), the sample size (higher n, higher power), and the magnitude of the effect (larger effect, higher power).

Assessing Normality

In most situations dealing with quality problems, perhaps one of the most important issues that everyone has to address is the issue of normality. If normality exists, then we can conduct a variety of statistical tests. If not, then we either have to transform the data or conduct very sophisticated tests for that nonnormal data.

There are two ways to test normality: (a) graphical means (probability paper, log paper, Daniel plots, Q-Q plots) or (b) statistical tests using the null hypothesis. With the graphical means we look for a straight line. On the other hand, when we test the null hypothesis we look at some preset probability. For more on this see Chapter 6.

Ryan–Joiner Test

Among the many tests available to assess normality is the Ryan–Joiner test. It uses the principle of null hypothesis as follows:

Null hypothesis (H_0): The data $\{x_1, ..., x_n\}$ are a random sample of size n from a normal distribution.

Alternative hypothesis (H_1): The data are a random sample from some other distribution.

We test this hypothesis with a test statistic r, which is the correlation between the data and the normal scores.

The rationale for this test is that if the data are a sample from a normal distribution, then the normal probability plot (plot of normal scores against the data) will be close to a straight line, and the correlation r will be close to 1. If the data are sampled from a nonnormal distribution, then the plot may show a marked deviation from a straight line, resulting in a smaller correlation r. Smaller values of r are, therefore, regarded as stronger evidence against the null hypothesis.

P-value: The probability is the value to the left of the observed correlation r calculated using the null distribution, that is, the area under the density to the left of r. You do not need to know how to calculate this. Most software programs do the calculation for you.

Summary

In this chapter, we have addressed the lifeblood of all statistics and experimentation: data. Specifically, we defined data and categorized them in such a way that appropriate and applicable statistical analyses may be performed given a specific category of data. We also addressed some issues of sampling and the significance of normality. In the next chapter, we will address how and why we summarize the data.

Selected Bibliography

Casagrande, J. T., Pike, M. C. and Smith, P. G. (1978). An improved approximate formula for calculating sample sizes for comparing two binomial distributions. *Biometrics*. 34:483–486.

Dupont, W. D. (1988). Power calculations for matched case-control studies. *Biometrics*. 44:1157–1168.

Dupont, W. D. (2008). *Statistical Modeling for Biomedical Researchers*. 2nd ed. Cambridge, UK: Cambridge University Press.

Dupont, W. D. and Plummer, W. D. (1990). Power and sample size calculations: A review and computer program. *Controlled Clinical Trials*. 11:116–128.

Dupont, W. D. and Plummer, W. D. (1998). Power and sample size calculations for studies involving linear regression. *Controlled Clinical Trials*. 19:589–601.

Fleiss, J. L. (1981). *Statistical Methods for Rates and Proportions*. 2nd ed. New York: John Wiley.

Mantel, N. and Haenszel, W. (1959). Statistical aspects of the analysis of data from retrospective studies of disease. *Journal of the National Cancer Institute*. 22:719–748.

Pearson, E. S. and Hartley, H. O. (1970). *Biometrika Tables for Statisticians*. Vol. I. 3rd ed. Cambridge, UK: Cambridge University Press.

Schlesselman, J. (1982). *Case-control Studies: Design, Conduct, Analysis*. New York. Oxford University Press.

Wittes, J. and Wallenstein, S. (1987). The power of the Mantel-Haenszel test. *Journal of American Statistical Association*. 82:1104–1109.

3

Summarizing Data

In the previous chapter, we addressed the issues of data and explained the concerns that one may address in gathering this data. In this chapter, we focus on summarizing the data in such a way that we make the data come to life.

The purpose of collecting data is not to categorize everything into neat figures, but to provide a basis for action. The data themselves can be in any form, that is, measurement data—continuous data, length, weight, time and so on, or attribute data—countable data, number of defects, number of defectives, percentage of defectives, and so on.

In addition, there are also data on relative merits, data on sequences, and data on grade points, which are somewhat more complicated but useful to those with the experience to draw appropriate conclusions from them.

After the data are collected, they are analyzed, and information is extracted through the use of statistical methods such as histograms, Pareto diagrams, check sheets, cause-and-effect diagram, scatter diagram, box plots, dot plots, stem and leaf displays, descriptive statistics, control charts, and many other tools and methods.

Therefore, data should be collected and organized in such a way as to simplify later analysis. For example, the nature of the data must always be identified. Time may elapse between the collection and the analysis of the data. Moreover, the data may be used other times for different uses. It is necessary to record not only the purpose of the measurement and data characteristics but also the date, the instrument used, the person doing it, the method, and everything else pertaining to the data collection process.

Some of the key characteristics in the process of summarizing the data to find causal factors are as follows:

- Clarify the purpose of collecting the data. Only when the purpose is clear, then the proper disposition can be made.
- Collect data efficiently. Unless the data are collected within the budget constraints and in such a way that they indeed reflect the process, product, or system, the analysis will fail.
- Take action according to the data. The decision must be made on the basis of the data; otherwise, they will not be collected in a positive manner. Make a habit of discussing a problem on the basis of the data and respecting the facts shown by them.
- Determine the shape of the distribution.
- Determine the relationship with specifications.

- Be prepared to change the histogram with some kind of a transformation function (e.g., logarithm, square root, or some other transformation function). A case can be made when a distribution is of a bimodal nature.

- Prepare a cause-and-effect diagram with as many people as possible. Their input will make it an educational experience. Everyone taking part in making this diagram will gain new knowledge. Even people who do not yet know a great deal about their jobs can learn a lot from making a cause-and-effect diagram or merely studying a complete one. It generates as well as facilitates discussion.

- Seek cause(s) actively.

- The use of a Pareto diagram should be the first step in making improvements. In making these improvements, the following considerations are important:

  - Everyone concerned must cooperate.
  - Item of concern has a strong impact.
  - Selected goal is concrete and doable.
  - Recognition that if all workers try to make improvements individually with no definite basis for their efforts, a lot of energy will produce few results. The Pareto diagram is very useful in drawing the cooperation of all concerned.

- In using a scatter diagram be aware of stratification. Conflicting results in correlation may be indicated. Also look for the peaks and troughs as well as the range where correlation exists.

- Control chart: Even though it is getting replaced by many acceptance sampling plan systems, it is still a graphical aid for the detection of quality variation in output from a given process. It is a summarizing as well as an evaluative method to improve the process on an ongoing basis by which corrective action, when necessary, may take place. (A strong reminder here: The term "control" is used in the title of "control chart," but, in fact, we all must recognize and always remember that the control chart does not control anything.)

Frequency Distributions

The construction of a numerical distribution consists essentially of three steps.

a. Choosing the classes into which the data are to be grouped
b. Sorting or tallying the data into the appropriate classes
c. Counting the number of items in each class

Since the last two of these steps are procedural, we are going to discuss the problem of choosing suitable classifications.

The two things necessary for consideration in the first step are to determine the number of classes into which the data are to be grouped, and the range of values each class is to cover, that is, from where to where each class is to go. The following are some rules of thumb:

1. We seldom use fewer than 6 or more than 15 classes; the exact number used in a given situation depends on the nature, magnitude, and range of the data. A rough rule of thumb is to use the squared root of the sample measurements ($\sqrt{n}$).

2. We always choose such classes that all of the data can be accommodated.

3. We always make sure that each item belongs to only one class; that is, we avoid overlapping classes. Successive classes having one or more values in common are not allowed.

4. Whenever possible, we make the class interval of equal length; that is, we make them cover equal ranges of values. It is also generally desirable to make these ranges multiples of 5, 10, 100, and so on, to facilitate the reading and the use of the resulting table.

Histograms

A histogram is a type of graph that shows the distribution of whatever you are measuring. In our modern world where we design products to be produced to given measurement, a histogram portrays how the actual measurement of different units of a product varies around the desired value. The frequency of occurrence of any given measurement is represented by the height of vertical columns on the graph.

The shape or contour formed by the tops of the columns has a special meaning. This shape can be associated with statistical distribution that in turn can be analyzed with mathematical tools. These various shapes are often given definitions such as normal, bimodal, or skewed, and a special significance can sometimes be attached to the causes of these shapes. A normal distribution causes the distribution to have a "bell" shape and is referred to as a "bell-shaped curve."

Why Histograms Are Used

Histograms are used because they help to summarize data and tell a story that would otherwise be lengthy and less effective in a narrative form. Also, they are standardized in format and thus lend themselves to a high degree of communication between their users.

Based on a long history of experience in analyzing observations made on happenings in nature, activities of people, and machine operations, it has been found that repetitive operations yield slightly different results. Sometimes these differences are significant and sometimes not; a histogram aids in determining the significance. The shape of the diagram gives the user some clues as to the statistical distribution represented and, thereby, a clue as to what might be causing the distribution. For example, if a machine is worn out and will not hold close tolerances, measurement of parts coming off that machine will produce a histogram that is wide and flat.

Sometimes in analyzing experiments it is preferable to present data in what is called a "cumulative frequency distribution," or simply a "cumulative distribution," which shows directly how many of the items are less than or greater than various values. Successively adding the frequencies will produce the cumulative frequency.

Graphical Presentations

Where frequency distributions are constructed primarily to condense large sets of data, display them in an easy-to-digest form; it is usually advisable to present them graphically, that is, in a form that appeals to the human power of visualization. The most common among all graphical presentations of statistical data is the histogram.

A histogram is constructed by representing the measurements of observations that are grouped on the horizontal scale and the class frequency on the vertical scale. Once the scales have been identified, then the drawing of the rectangles whose bases equal the class interval and whose heights are determined by the corresponding class frequencies, the markings on the horizontal scale can be the class limits, as for example, the class boundaries, the class marks, or arbitrary key values. For easy readability, it is generally preferable to indicate the class limits, although the rectangles actually go from one class boundary to the next. Histograms cannot be used in connection with frequency distributions having open classes and they must be used with extreme care if the classes are not all equal.

An alternative, though less widely used, form of graphical presentation is the frequency polygon. Here, the class frequencies are plotted at the class marks and the successive points are connected by means of straight lines. If we apply this same technique to a cumulative distribution, we obtain a so-called ogive. The ogive corresponds to the "or less" principle.

Dot Plots

The dot diagram is a valuable device for displaying the distribution of a small body of data (up to about 20). In particular it shows:

1. The general location of the observations
2. The spread of the observations

A comparison with the histogram will identify the following characteristics:

- Whereas the histogram groups the data into just a few intervals, the dot plot groups the data as little as possible. Ideally, if we had wider paper or printer with higher resolution, we would not group the data at all.
- Whereas the histogram tends to be more useful with large data base, the dot plot is useful with small data base.
- Whereas the histogram shows the shape of the distribution, the dot plot does not.

In the final analysis, dot plots are very useful to compare two or more sets of data for location and spread. To construct a dot plot, draw the *x*-axis with the value scale of your choice and then plot the observations.

Stem and Leaf Displays

A stem and leaf display is a variation of the histogram and the check sheet. Although it was designed by J. Tukey in the late 1960s for variable data, it may be used with any set of numbers. The shape of the stem and leaf display is very similar to the histogram with two major differences.

1. Whereas the histograms deal in intervals, the stem and leaf display depends on the actual data to generate the distribution. The advantage of this is that the experimenter knows exactly where the data are coming from.
2. Whereas the scale for the intervals is on the *x*-axis, in the stem and leaf display, it is on the vertical side.

In evaluating the stem and leaf, the numbers are both to the left and to the right of a vertical line. The numbers to the left are called stem and the numbers to the right are called leafs. To actually do the display, we must split the raw data into two parts (that is what the vertical line is all about). The split is usually between the 10 digits that become the stem and the ones digit that becomes the leaf. (If the data have more than two digits, you still split the numbers according to your requirements. For example, if you have three-digit numbers, say one of the numbers is 302, your stem may be 3 or 30 and the leaf may be 02 or 2, respectively.) The reader will notice that if the stem in this case is identified as 3 the experimenter must be very cautious in the interpretation. We suggest that the stem be identified as the number that can cause the least confusion. In this case, we would recommend that the number 30 be the stem and the number 2 be the leaf. It is important to note here that the readings are the actual measurement and the leaves always correspond to the individual stem that they came from.

Box Plots

Yet another way to summarize data, which is both easy and effective, is through a box plot. Box plots can be used in two different ways: either to describe a single variable in a data set or to compare two or more variables. The key features in understanding a box plot are the following:

- The right and left of the box are at the third and first quartiles. Therefore, the length of the box equals the interquartile range (IQR), and the box itself represents the middle 50% of the observations. The height of the box has no significance.
- The vertical line inside the box indicates the location of the median. The point inside the box indicates the location of the mean.
- Horizontal lines are drawn from each side of the box. They extend to the most extreme observations that are no further than 1.5 IQRs from the box. They are useful for indicating variability and skewness.
- Observations farther than 1.5 IQRs from the box are shown as individual points. If they are between 1.5 IQRs and 3 IQRs from the box, they are called mild outliers and are hollow. Otherwise, they are called extreme outliers and are solid. (This convention was instituted by the statistician J. Tukey in the late 1960s and is still in practice.)

Box plots are not usually used as part of the basic SPC; however, they are used extensively in ANOVA and advanced SPC as well as in the Six Sigma methodology.

Scatter Diagrams

When one is interested in the relationship between the dependent variable and the independent variable, or merely between cause and effect, then one may use a scatter diagram. A scatter diagram is a grouping of plotted points on a two axis graph—usually, x-axis and y-axis. These plotted points form a pattern that indicates whether there is any connection between the two variables. The major objective of making a scatter diagram is to find out whether there is a relationship (visual correlation) between the variables under study, and if there is, how much one variable influences the other.

By finding out whether one variable influences another and the extent of that influence, one can imply that a cause-and-effect relationship exists and take appropriate measures to rectify the casual variable.

To construct a scatter diagram follow these steps:

1. Select two variables that seem to relate in a cause-and-effect rela-tionship. This can be based on experiences or theory. For example, the grind bearing diameter seems to vary as different speeds are used in the grinders. Here, the cause would seem to be the speed and the undesirable effect, a varying diameter.
2. Construct a vertical axis that will encompass all the values that you have obtained for the diameter.
3. Plot all the measured values for the diameters and the speeds.

Reading Scatter Diagrams

The idea of a scatter diagram is to show whether or not a relationship exists between variables. In fact, in some cases, one may be tempted to say that one causes the other to change. Knowing this, one can proceed to control the appeared "casual" variable. (Even though it is tempting to think of causation at this stage, the scatter diagram does not provide us with a strong and irre-futable answer as to the causation. The best it offers is a visual relationship. Period!)

Scatter diagrams can be varied. The way the plotted points fall depends on the collected data. The shape and the direction of that shape, however, determines whether the two variables have a relationship at all, and if so, what the relationship is and how strong it is. The basic shapes of the scatter diagram are as follows:

Positive Correlation

This diagram shows a distinct, positive relationship between X and Y. This means that as X is increased, so does Y. Or, if we controlled one of these vari-ables the other one will also be controlled.

A comment about the scatter diagram is necessary here. That is, the closeness of the plotted points indicates the degree of the relationship between the variables. To facilitate this closeness, imagine a line that is drawn between all the points. Such a line will have half of the points on one side, and the other half on the other side. As the points fall close to the line on either side, it is correspondingly easier to determine the relationship accurately. On the other hand, as the points scatter about, then determination becomes less accurate. This line turns out to be the correlation line or regression line.

No Relationship

This means that given our data, we cannot show a relationship. Points that seem to fall all over the two-axis plot or form a circular pattern indicate that the two variables do not influence one another. That means, as we move or change one variable the change in the other is not predictable. We conclude in such a situation that the two chosen variables are independent of each other and that some other variable (cause) must be found that influences the variable Y (effect).

Negative Relationship

If the points form a tight pattern about an imaginary line and the direction of such a line is from the upper left to the lower right, then we say that there is a definite relationship between X and Y, such that, if one variable changes in one direction, the other one will change in the opposite direction. This kind of relationship between two variables is called a negative correlation, and it implies that indeed there exists a cause-and-effect relationship between the two variables.

Possible Positive and Negative Correlation

When the points are dispersed about the imaginary line, however, that dispersion is in a pattern where a direction is visible but a definite causation is not determined, then we say that either a possible positive relationship or a possible negative relationship exists. This means, depending on the direction of the imaginary line, one can say that changes in X will produce changes in Y; however, these changes will not necessarily be equal. Of importance in these possible relationships is the fact that not only X but also the other factors that are present will also influence changes in Y.

If we are interested in a more sophisticated way to identify the relationship, then we may use a correlation, a multiple correlation analysis, or even a regression analysis. We may want to consider correlation if any of the following conditions is present:

- Many variables changing simultaneously
- Variables have complicated effects
  - Nonlinear
  - Interaction exists between variables

- Unexplained or random variation is high
- Accurate predictions are necessary for optimization
- Need to know which variables are important and which are not
- Fortuitous variables are present
- No information on important variables
- Odd ball results happen

Even though correlation is indeed a very powerful tool, the truth of the matter is that in basic analysis the scatter plot is just as good. For elementary analysis if one was to experiment with "correlation", they should also be aware of its pitfalls and difficulties. Some are identified below:

- One has to deal with many variables
- Variables may change constantly
- Nonlinear effects must be accounted for
- Interaction effects must be accounted for
- Experimental error must be understood and accounted for
- Variable changes are small
- Fortuitous variables present
- No information on important variables

It is for these difficulties that a correlation analysis is usually done by a specialist and, thus, for a basic analysis the scatter plot is preferred.

Descriptive Statistics

Perhaps the most used statistics on a daily basis are the descriptive statistics. They are numerical summaries of data sets (frequency distributions) and are divided into categories. The first is measures of central tendency that describe the location of the data set and are the mean, median, and mode. The second is measures of dispersion that describe the spread (or variation) of a data set and are the range and the standard deviation.

The Mean (Average)

The most commonly used measure of central tendency is the mean. The mean of a set of values is the average or "balance point" of those values. "Xbar," sometimes designated as an X with a line just above it ($\bar{X}$), is a symbol that indicates true sample mean. This statistic is familiar, easy to calculate, and useful for symmetric distributions. However, if there is an extremely high or low value in the sample, the mean can be misleading. The population mean is designated with the Greek letter μ (mu).

The Median (Middle)

The median is the second most commonly used measure of central tendency. The median is the middle value in a data set. X with a ~ above it ($\tilde{X}$) is the symbol used to indicate the median. Half of the measurements lie below the median and half of them lie above the median. The median is not affected by extremely large or small measurements and is useful for describing nonsymmetric distributions.

The Mode (Most)

Another common measure of central tendency is the mode. The mode is the measurement that occurs most often in a sample. The mode is the highest point on a histogram or a frequency polygon. (Special note: In the normal distribution, the mean, median, and the mode are the same.)

Measures of Dispersion

Measures of process dispersion portray the amount of spread or variation in a data set. The range and standard deviation are two measures of dispersion. The range is used for small samples ($n < 8$). The standard deviation is used for large samples ($n \geq 8$).

Measures of dispersion aid in the interpretation of the mean. The mean gives the locations of the sample (as a group), but does not represent each measurement equally. Measures of dispersion describe how well the mean represents each measurement. When the range is large, the mean is a relatively poor estimate for members of the sample whose values are far from the mean. When the range is small, the mean is a relatively good estimate for all members of the sample.

Range

The simplest measure of dispersion is the range (R). The range of a sample is calculated as the difference between the largest and the smallest value. The difference is actually the distance between the largest and the smallest values. The closer these values are, the smaller the range will be. A range can never be a negative number.

Standard Deviation

The standard deviation (s for samples and σ for populations) is a measure of variation that is needed to describe the spread of populations. The range is a statistic that is limited to simple data because it requires a known largest and smallest value. These values are unknown for a population because the population is an estimated distribution.

The normal distribution has a curvature that slopes away from the mean. As the curve extends from the mean, it changes the shape of the distribution. The precise point where the curve changes directions is called the inflection point. The standard deviation is the distance from the mean to the inflection point of the curve. The standard deviation is extremely valuable for describing the spread of a population because as the spread of the distribution increases the distance to the inflection point increases.

Normality

In most applications and studies, the distributions that have a bell shape turn out to be a particular type of distribution, called the *normal distribution*, which describes them well. The normal distribution is very important in data analysis.

A mathematical equation defines the normal distribution exactly. For a particular mean and standard deviation, this equation determines what percentage of the observations falls where. If you folded it in the center, the two sides would match; that is, they're identical. The center of the distribution is at the mean. The mean of a normal distribution is also the frequently occurring value (the mode), and it's the value that splits the distribution into two equal parts (the median). In any normal distribution, the mean, median, and mode all have the same value.

Areas in the Normal Distribution

For a normal distribution, the percentage of values falling within any interval can be calculated exactly. For example, in a normal distribution with a mean of 100 and a standard deviation of 15, 68% of all values fall between 85 (1 standard deviation less than the mean) and 115 (1 standard deviation more than the mean). And 95% of all values fall in the range between 70 and 130, within 2 standard deviations from the mean.

A normal distribution can have any mean and standard deviation. However, the percentage of cases falling within a particular number of standard deviations from the mean is always the same. The shape of a normal distribution doesn't change. Most of the observations are near the average, and a mathematical function describes how many observations are at any given distance (measured in standard deviations) from the mean. Means and standard deviations differ from variable to variable, but the percentage of cases within specific intervals is always the same in a true normal distribution.

It turns out that many variables you can measure have a distribution close to the mathematical ideal of a normal distribution. We say these variables are "normally distributed," even though their distributions are not exactly normal. Usually, when we say this we mean that the histograms show that the actual distribution is indeed pretty close to normal.

Standard Score

This is perhaps one of the most underutilized statistic in the quality professional's repertoire. If I tell you that I own 250 books, you probably won't be able to make very much of this information. You won't know how my library compares to that of the average consultant. And you won't know how unusual I am compared to them (based only on the number of books!). Wouldn't it be much more informative if I told you that I own the average number of books, or that I am 2 standard deviations above the average? Then, if you know that the number of books owned by consultants is normally distributed, you could calculate exactly what percentage of my colleagues have more books than I do.

To describe my library better in this way, you can calculate what's called a *standard score*. It describes the location of a particular case in a distribution: whether it's above average or below average and how much above or below. The computation is simple:

1. Take the value and subtract the mean from it. If the difference is positive, you know the case is above the mean. If it's negative, the case is below the mean.
2. Divide the difference by the standard deviation. This tells you how many standard deviation units a score is above or below the average.

For example, if book ownership among consultants is normally distributed with a mean of 150 and a standard deviation of 50, you can calculate my standard score for the 250 books I own like this:

Step 1:

250 (my books) – 150 (average number of books)
100 (I own 100 books more than the average consultant)

Step 2:

100 (difference from step 1) / 50 (standard deviation of books) = 2 (standard score)

My standard score is 2. Since its sign is positive, it indicates that I have more books than average. The number 2 indicates that I am 2 standard deviation units above the mean. In a normal distribution, 95% of all cases are *within* 2 standard deviations of the mean. Therefore, you know that my library is remarkable.

In a sample, the average of the standard scores for a variable is always 0, and the standard deviation is always 1. Suppose you ask 15 people on the street how many hamburgers they consume in a week. If you calculate the mean and standard deviation for the number of hamburgers eaten by these 15 people and then compute a standard score for each person, you'll get 15 standard scores. The average of the scores will be 0, and their standard deviation will be 1.

When you use standard scores, you can compare values for a case on *different* variables. If you have standard scores of 2.9 for number of books, –1.2 for metabolic rate, and 0.0 for weight, then you know:

- You have many more books than average.
- You have a slower metabolism than average.
- Your weight is exactly the average.

You couldn't meaningfully compare the original numbers since they all have different means and standard deviations. Owning 20 cars is much more extraordinary than owning 20 shirts.

It is important to recognize that only for this example we focused on 2 standard deviations. There is nothing that prevents us from going even further. For example, when we evaluate control charts we are interested in evaluating the process on the basis of three different standard deviations. On the other hand, when we use the Six Sigma methodology, we talk about six different standard deviations.

A Sample from the Normal Distribution

Even if a variable is normally distributed in the population, a sample from the population doesn't necessarily have a distribution that's exactly normal. Samples vary, so the distributions for individual samples vary as well. However, if a sample is reasonably large and it comes from a normal population, its distribution should look more or less normal.

Distributions That Aren't Normal

The normal distribution is often used as a reference for describing other distributions. A distribution is called skewed if it isn't symmetric but has more cases, more of a "tail," toward one end of the distribution than the other. If the long tail is toward larger values, the distribution is called positively skewed, or skewed to the right. If the tail is toward smaller values, the distribution is negatively skewed, or skewed to the left. A variable like income has a positively skewed distribution. That's because some incomes are very much above average and make a long tail to the right. Since incomes are rarely less than zero, the tail to the left is not so long.

If a larger proportion of cases fall into the tails of a distribution than into those of a normal distribution, the distribution has positive kurtosis. If fewer cases fall into the tails, the distribution has negative kurtosis. You can compute statistics that measure how much skewness and kurtosis there is in a distribution, in comparison to a normal distribution. These statistics are zero if the observed distribution is exactly normal. Positive values for kurtosis indicate that a distribution has heavier tails than a normal distribution. Negative values indicate that a distribution has lighter tails than a normal distribution. Of course, the measures of skewness and kurtosis for samples from a normal distribution will not be exactly zero. Because of variation from sample to sample, they will fluctuate around zero. To use the computer for the calculations, one only needs to identify the command with skewness and kurtosis and the rest is done by the computer.

The Standard Error of the Mean

The second most misunderstood and underused statistic in the quality profession is the standard error of the mean. If the mean of the distribution of sample means is the population mean, what's the standard deviation of the distribution of sample means? Is it also just the standard deviation of the population? No! We already have talked about that the standard deviation of the mean depends on two things:

1. How large a sample you take. Larger samples meant a smaller standard deviation for the sample means.
2. How much variability there is in the population. Less variability in the samples also meant a smaller standard deviation for the sample means.

To calculate the exact standard deviation of the distribution of sample means, you must know:

1. The standard deviation in the population
2. The number of cases in the sample

All you have to do is divide the standard deviation by the square root of the sample size. The result, the standard deviation of the distribution of sample means is called the *standard error of the mean*. Although it has an impressive name, it's still just a standard deviation—the standard deviation of the sample means. Think about the formula for computing the standard error of the mean: take the standard deviation of the variable and divide by the square root of the sample size. Suppose the standard deviation of number of books owned is 50, and the sample size is four cases. Then, the standard error is 50 divided by the square root of 4, or 25. If the sample size is *increased* to 9, the standard error decreases to 50 divided by the square root of 9, or 16.7. If the sample size is *increased* to 100, the standard error is only 5. The larger the sample size, the less variability there is in the sample means.

Tests of Distributions as a Whole

So far we have addressed some tests to determine whether a sample came from a population having mean μ. In fact, we showed how some of these tests were designed to clarify whether a sample came from a normal population having standard deviation σ. Now we are going to discuss limited—but common—tests to determine whether a sample comes from a population having a distribution of any specified form. Such information is often valuable. They are valuable because not all distributions are normal and it is helpful to know whether we are dealing with normal or nonnormal distributions.

We have been focusing on the importance of selecting a sample in a random fashion. It is very important that we do that. However, quite often, it is difficult, without examining the sample itself, to be sure that the selection does not involve some hidden trend. It is, therefore, incumbent upon the experimenter to perform some tests. There are many tests available for testing for normality; however, here, we will address the one most frequently used.

The first test is the run test for randomness, which can be applied as a one-sided or an equal-tails test. This test is sometimes called the Bradley Test. A run is defined as a series of increasing values or a series of decreasing values. The number of increasing, or decreasing, values is the length of the run. In a random data set, the probability that the $(i + 1)$th value is larger or smaller than the *i*th value follows a binomial distribution, which forms the basis of the run test.

The second most often used test to identify normality is the chi-square (χ^2). It is generally known as the test for goodness of fit for the normal distribution. It is used when the sample values from an experiment fall into *r* categories. To decide at the significance level *a* whether the data constitute a sample from a population with distribution function $f(x)$, we first compute the expected number of observations that would fall in each category as predicted by $f(x)$. The grouping should be arranged so that this theoretical frequency is at least 5 for each category. If necessary, several of the original groups may be combined in order to assure this. To compare the observed frequencies, n_i for the *i*th category, with the expected (theoretical) frequencies, e_i, we use:

$$\chi^2 = \sum_{i=1}^{r} \frac{(n_i - e_i)^2}{e_i}$$

Once we have the calculated value of the chi-square, we proceed to check the significance by the traditional approach; that is, if the calculated value of χ^2 exceeds the critical value from a table or the probability set before the experiment for the degrees of freedom, we reject the hypothesis.

Yet a third test that one may want to conduct is the Poisson distribution. Here again we do not know which specific Poisson distribution to expect, so we must use *m*, the mean of the sample. The formula that we use is:

$$P(x) = \frac{m^x e^{-m}}{x!}$$

where *m* is the mean of the sample; *x* is (0, 1, 2, 3, ... *n*) the individual defects; *e* is the natural logarithm, which is approximately equal to 2.718; and *x*! is the factorial. When the factorial is used remember always that $0! = 1$.

Significance is determined whether or not the calculated value is greater than the table value or the preset probability given the degrees of freedom.

To test the hypothesis that a random sample comes from a population having a normal distribution, we may fit a normal curve to the data and then test to see whether the hypothesis is justified. The testing may be done by means of the chi-square test for goodness of fit, in which case the mean and standard deviation for the fitted normal curve should be estimated from the grouped sample data.

An obvious but rough check on normality can also be made by plotting the fitted normal curve to the same scale as the histogram of the grouped data. A more convenient graphical approach calls for the use of normal probability graph paper. In both cases, if the population is normal, a plot of the sample cumulative percentage frequencies should approximate a straight line. However, in both of these graphical approaches there will be difficulty in judging by eye how much departure from the ideal pattern we must expect. Modern statistical software generate these graphs and they also give the probabilities associated with the fit.

Transformations to Obtain Normality

Many physical situations produce data that are normally distributed, others produce data that follow some other known distribution, and still others produce data that can be transformed to normal data. For example, when the reaction rate in a chemistry experiment is proportional to the concentration of reacting substances, the distribution of rates is not likely to be normal, but the distribution of their *logarithms* may be theoretically normal. The most common transformations are $y = \log x$ (frequently used in sensitivity testing) and $y = x^{1/2}$. An easy way to decide whether one of these transformations is likely to produce normality is to make use of special graph papers that are commercially available. If a software package is used one may use advanced techniques such as Wilk's lambda (λ). In the first case, significance is shown by a straight line if normal and, in the second case, the determination is based on a preset probability. In the practice of Six Sigma transformation approaches, the utilization of Wilks lambda is very prevalent.

Confidence and Sensitivity

As we have said many times, the issue of confidence is very important in studying a sample and make projections about the population. Distributions are no different. The experimenter must be aware of the confidence and the sensitivity of that distribution.

For the confidence, the simplest approach is to use a graphical representation of the cumulative distribution. The graphical representation is on the basis of a random sample of size n, a $100(1 - a)\%$ confidence band for the cumulative distribution of the population. When the band is determined, draw two "staircase" lines parallel to the sample histogram, one with the + percentage units above the histogram, the other with the −percentage units below. In the long run, confidence bands constructed for random samples from populations will, in $100(1 - a)\%$ of the cases, completely contain the population cumulative distribution curve. (Obviously, the confidence may be defined as 90%, 95%, 99%, or any other confidence the experimenter feels that is applicable and appropriate for the study at hand.)

If we wish to test whether a distribution is of some general form, such as normal, but not to test specific values of constants (parameters), estimates of the latter are made from the sample. The test may still be made, but it lacks discrimination, in the sense that it will not reject the distribution unless its form is very different from that of the null hypothesis. It may be used correctly to reject such a distribution at a significance level no greater than alpha (α), but it will reject much too seldom.

On the other hand, "sensitivity testing" is the testing in which an increasing percentage of items fail, explode, or die as the severity of the test is increased. In such testing, we cannot measure the precise severity of test (i.e., the precise magnitude of the variable concerned) that would barely result in failure, but can only observe whether an applied severity results in failure or not.

For instance, in a test to discover the minimum range to target at which a particular drug dosage will function as intended, we might find a given range to be either too small or too large for proper functioning (but probably not the *minimum* dosage). The issue here is how we go about determining the lethal dosage. The lethal dosage is the dosage that 50% of the partakers of a specific drug will end in fatality.

Sensitivity testing is appropriate and applicable in cases where each of the items tested has its own critical level of severity so that in a whole population of items there is a probability distribution of critical levels.

Sensitivity tests for finding the mean, standard deviation, or percentiles in such cases may be analyzed by "staircase" methods or by "probit analysis." Both methods depend on an assumption of normality in the quantity being studied. This fitting provides a check of normality and may suggest a transformation of the independent variable that will give approximate normality if the data do not lie near a straight line. Probit analysis, on the other hand, does the fitting more precisely and gives a measure of the precision, but staircase methods are more efficient when they are applicable. The staircase methods are recommended when the effect on each item is known immediately, or soon after each application, and when the independent variable can be readily adjusted.

Probit analysis is recommended when it is not practical to measure one item at a time, for example, when the ultimate effects of given amounts of radiation are to be measured on a group of small animals, such as a pen of hamsters or mice. If it is inconvenient to adjust the independent variable, as in the case of tests of rockets at different temperatures, probit analysis is appropriate.

A valuable and frequently used method of sensitivity testing of the staircase type is the "up-and-down," or "Bruceton," method. It was developed by Dixon and Mood and is based on the normal distribution and ready adjustment of the independent variable to any prescribed level. The method is especially applicable to estimating the mean μ (50% point) of the distribution because it concentrates the observations in the neighborhood of the mean; it can also be used for estimating the standard deviation σ and other percentage points, though less accurately.

Other analysis methods are Neyer's d-optimal test and Dror and Steinberg sequential procedure. Bruceton analysis has an advantage over the modern techniques being very simple to implement and analyze—as it was designed to be performed without a computer. The modern techniques offer a great improvement in efficiency, needing a much smaller sample to obtain any desired significance level. Furthermore, these techniques enable the treatment of many other related experimental designs—such as when there is a need to learn the influence of more than one variable (say, testing the sensitivity of an explosive to both shock level and environment temperature), to models that are not only binary by nature (not only "detonate or not"), to experiments where you decide in advance (or "group") on more than one sample in each "run," and more. In fact, with the modern techniques, the experimenter is not even constrained to specify a single model and can reflect uncertainty as to the form of the true model.

Test for Independence in Contingency Tables

A contingency table is a table that represents data gathered from counting rather than measuring. The method of analyzing such data is an extension of differences among proportions. However, it applies to problems in which, unlike binomial problems, each trial permits more than two outcomes. In effect, the headings of the rows and columns need not be numerical groupings. The following combinations, for instance, are possible:

- Characteristic I, degree of damage: no damage, damaged slightly, completely demolished
- Characteristic II, age of structure: old, new
- Characteristic I, estimate of applicant's ability: excellent, good, fair, and poor
- Characteristic II, person making estimate: Jamey, Stacy, Caitlyn, and Dean

In the general case, a contingency table, $(r \times k)$, is used when there are k samples and each trial permits r alternatives. Tables of this sort are of two kinds: (a) tables in which the column totals are fixed (because they are the respective sample sizes) and the row totals vary as the result of chance and (b) tables in which both the column totals and the row totals depend on chance. These later tables, in which items are classified according to two characteristics and chance determines both the row and the column in which an item is entered, are called *contingency tables* or *cross-tab* or *cross-tabulation*.

The significance of the difference between the two proportions can be assessed with a variety of statistical tests, including Pearson's chi-square test, the *G*-test, Fisher's exact test, and Bernard's test, provided the entries in the table represent individuals randomly sampled from the population about whom we want to draw a conclusion. If the proportions of individuals in the different columns vary significantly between rows (or vice versa), we say that there is a *contingency* between the two variables. In other words, the two variables are *not* independent. If there is no contingency, we say that the two variables are *independent*.

Pearson's Chi-Square Test

Pearson's chi-square test is used to assess two types of comparison: tests of goodness of fit and tests of independence. A test of goodness of fit establishes whether or not an observed frequency distribution differs from a theoretical distribution. A test of independence assesses whether paired observations on two variables, expressed in a contingency table, are independent of each other, for example, whether people from different regions differ in the frequency with which they report that they support a political candidate.

The first step in the chi-square test is to calculate the chi-square statistic. In order to avoid ambiguity, the value of the test statistic is denoted by X^2 rather than χ^2 (i.e., uppercase chi instead of lowercase); this also serves as a reminder that the distribution of the test statistic is not exactly that of a chi-square random variable. However, some authors do use the χ^2 notation for the test statistic. An exact test that does not rely on using the approximate χ^2 distribution is Fisher's exact test: This is significantly more accurate in evaluating the significance level of the test, especially with small numbers of observation.

The chi-square statistic is calculated by finding the difference between each observed and theoretical frequency for each possible outcome, squaring them, dividing each by the theoretical frequency, and taking the sum of the results. A second important part of determining the test statistic is to define the degrees of freedom of the test: This is essentially the number of observed frequencies adjusted for the effect of using some of those observations to define the "theoretical frequencies."

Calculating the Test Statistic

The value of the test statistic is

$$X^2 = \sum_{i=1}^{n} \frac{(O_i - E_i)^2}{E_i}$$

where

X^2 = Pearson's cumulative test statistic, which asymptotically approaches a χ^2 distribution;
O_i = an observed frequency;
E_i = an expected (theoretical) frequency, asserted by the null hypothesis;
n = the number of cells in the table.

The chi-square statistic can then be used to calculate a *p*-value by comparing the value of the statistic to a chi-squared distribution. The number of degrees of freedom is equal to the number of cells, n, minus the reduction in degrees of freedom, p.

The result about the number of degrees of freedom is valid when the original data were multinomial, and hence, the estimated parameters are efficient for minimizing the chi-square statistic. More generally, however, when maximum likelihood estimation does not coincide with minimum chi-square estimation, the distribution will lie somewhere between a chi-square distribution with $n - 1 - p$ and $n - 1$ degrees of freedom (see, for instance, Chermoff and Lehmann, 1954).

G-Test

G-tests are likelihood ratio or maximum likelihood statistical significance tests that are increasingly being used in situations where chi-square tests were previously recommended.

The commonly used chi-squared tests for goodness of fit to a distribution and for independence in contingency tables are in fact approximations of the log-likelihood ratio on which the G-tests are based. This approximation was developed by K. Pearson because at the time it was unduly laborious to calculate log-likelihood ratios. With the advent of electronic calculators and personal computers, this is no longer a problem. G-tests are coming into increasing use, particularly since they were recommended in the Sokal and Rohlf (1994) edition of their popular statistics textbook. Dunning (1993) introduced the test to the computational linguistics community where it is now widely used.

The general formula for Pearson's chi-squared test statistic is

$$X^2 = \sum_{ij} \frac{(O_{ij} - E_{ij})^2}{E_{ij}}$$

where O_i is the frequency observed in a cell, E is the frequency expected on the null hypothesis, and the sum is taken across all cells. The corresponding general formula for G is

$$G = 2\sum_{ij} O_{ij}.ln(O_{ij}/E_{ij})$$

where ln denotes the natural logarithm (log to the base e) and the sum is again taken over all nonempty cells.

Fisher Exact Test

The Fisher exact test looks at a contingency table that displays how different treatments have produced different outcomes. Its null hypothesis is that treatments do not affect outcomes—that the two are independent. Reject the null hypothesis (i.e., conclude treatment affects outcome) if p is "small."

The usual approach to contingency tables is to apply the X^2 statistic to each cell of the table. You should probably use the chi-square approach, unless you have a special reason. The most common reason to avoid X^2 is because you have small expectation values.

Barnard's Test

Barnard's test is an exact test of the null hypothesis of rows and columns in a contingency table. It is an alternative to Fisher's exact test but is more time consuming to compute. The test was first published by G. A. Barnard (1945, 1947) who claimed that this test for 2×2 contingency tables is more powerful than Fisher's exact test. Mehta and Senchaudhuri (2003) have explained why Barnard's test can be more powerful than Fisher's under certain conditions.

When comparing Fisher's and Barnard's exact tests, the loss of power due to the greater discreteness of the Fisher statistic is somewhat offset by the requirement that Barnard's exact test must maximize over all possible p-values, by choice of the nuisance parameter π. For 2×2 tables, the loss of power due to the discreteness dominates over the loss of power due to the maximization, resulting in greater power for Barnard's exact test. But as the number of rows and columns of the observed table increase, the maximizing factor will tend to dominate, and Fisher's exact test will achieve greater power than Barnard's. Mato and Andres (1997) show how to compute the result of the test more quickly.

Yates' Continuity Correction

Yates' correction for continuity (or Yates' chi-square test) is used in certain situations when testing for independence in a contingency table. In some cases, Yates' correction may adjust too far, and so its current use is limited.

Using the chi-squared distribution to interpret Pearson's chi-squared statistic requires one to assume that the discrete probability of observed binomial frequencies in the table can be approximated by the continuous chi-squared distribution. This assumption is not quite correct and introduces some error.

To reduce the error in approximation, Frank Yates, an English statistician, suggested a correction for continuity that adjusts the formula for Pearson's chi-square test by subtracting 0.5 from the difference between each observed value and its expected value in a 2 × 2 contingency table (Yates, 1934). This reduces the chi-square value obtained and, thus, increases its p-value.

The effect of Yates' correction is to prevent overestimation of statistical significance for small data. This formula is chiefly used when at least one cell of the table has an expected count smaller than 5. Unfortunately, Yates' correction may tend to overcorrect. This can result in an overly conservative result that fails to reject the null hypothesis when it should. So it is suggested that Yates' correction is unnecessary even with quite small sample sizes (Sokal and Rohlf, 1981) such as:

$$\sum_{i=1}^{N} O_i = 20$$

The following is Yates' corrected version of Pearson's chi-squared statistic:

$$\chi^2_{Yates} = \sum_{i=1}^{n} \frac{(|O_i - E_i| - 0.5)^2}{E_i}$$

where

O_i = an observed frequency;
E_i = an expected (theoretical) frequency, asserted by the null hypothesis;
N = number of distinct events.

Summary

In this chapter, we discussed "how" and "why" we summarize data. We also addressed some issues of normality and gave some specific tests. In the next chapter, we will address the tests and confidence intervals for means.

References

Barnard, G. A (1945). A new test for 2 × 2 tables. *Nature* 156(3954):177.

Barnard, G. A. (1947). Significance tests for 2 × 2 tables. *Biometrika* 34(1/2):123–138.

Chermoff, H. and Lehmann, E. L. (1954). The use of maximum likelihood estimates in χ^2 tests for goodness-of-fit. *The Annals of Mathematical Statistics* 25:579–586.

Dunning, T. (1993). Accurate methods for the statistics of surprise and coincidence. *Computational Linguistics* 19(1):61–74.

Mato, S. A. and Andres, M. (1997). Simplifying the calculation of the P-value for Barnard's test and its derivatives. *Statistics and Computing* 7:134–143.

Mehta, C. R. and Senchaudhuri, P. (September 4, 2003). Conditional versus unconditional exact tests for comparing two binomials. Retrieved October 20, 2010, from http://www.cytel.com/Papers/twobinomials.pdf

Sokal, R. R. and Rohlf, F. J. (1981). *Biometry: The Principles and Practice of Statistics in Biological Research*. Oxford: W.H. Freeman.

Sokal, R. R. and Rohlf, F. J. (1994). *Biometry: The Principles and Practice of Statistics in Biological Research*, 3rd ed. New York: Freeman.

Yates, F. (1934). Contingency table involving small numbers and the χ^2 test. *Supplement to the Journal of the Royal Statistical Society* 1(2):217–235.

Selected Bibliography

____. (No date). *G-Test in Handbook of Biological Statistics*. Newark, DE: University of Delaware.

Anscombe, F. (1973). Graphs in statistical analysis. *The American Statistician* 195–199.

Anscombe, F. and Tukey, J. W. (1963). The examination and analysis of residuals. *Technometrics* 141–160.

Barnett, V. and Lewis, T. (1994). *Outliers in Statistical Data*, 3rd. New York: John Wiley and Sons.

Bartlett, M. S. (March 1947). The use of transformations. *Biometrics Bull* 3(1):39–51.

Bliss, C. I. (1935). The calculation of the dosage-mortality curve. *Annals Application of Biology* 22:134–167.

Bloomfield, P. (1976). *Fourier Analysis of Time Series*. New York: John Wiley and Sons.

Box, G. E. P. and Cox, D. R. (1964). An analysis of transformations. *Journal of the Royal Statistical Society* 211–243, discussion 244–252.

Box, G. E. P., Hunter, W. G. and Hunter, J. S. (1978). *Statistics for Experimenters: An Introduction to Design, Data Analysis, and Model Building*. New York: John Wiley and Sons.

Box, G. E. P. and Jenkins, G. (1976). *Time Series Analysis: Forecasting and Control*. San Francisco, CA: Holden-Day.

Bradley, J. V. (1968). *Distribution-Free Statistical Tests*. Upper Saddle River, NJ: Prentice Hall.

Bureau of Ordnance. (March 21, 1946). *"Staircase" Method of Sensitivity Testing.* T. W. Anderson, P. J. McCarthy, and J. W. Tukey (Eds.). Washington, DC: BuOrd. (NAVORD Report 65-46.)

Chakravarti, I. M., Laha, R. G. and Roy, J. (1967). *Handbook of Methods of Applied Statistics,* Vol. I. New York: John Wiley and Sons.

Chambers, J., Cleveland, W., Kleiner, B. and Tukey, P. (1983). *Graphical Methods for Data Analysis.* Belmont, CA: Wadsworth.

Chatfield, C. (1989). *The Analysis of Time Series: An Introduction,* 4th ed. New York, NY: Chapman & Hall.

Cleveland, W. (1985). *Elements of Graphing Data.* Belmont, CA: Wadsworth.

Cleveland, W. (1993). *Visualizing Data.* Lafayette, IN: Hobart Press.

Cleveland, W. and McGill, M. (Eds.). (1988). *Dynamic Graphics for Statistics.* Belmont, CA: Wadsworth.

Cramer, H. (1946). *Mathematical Methods of Statistics.* Princeton, NJ: Princeton University Press.

Dixon, W. and Massey, F. J., Jr. (1951). *An Introduction to Statistical Analysis.* New York: McGraw-Hill. pp. 184–191, tests for goodness of fit and independence; pp. 254–257, the run test and confidence bands for cumulative distribution curves; and pp. 279–287. up-and-down method of sensitivity testing.

Dixon, W. and Mood, A. M. (1948). A method for obtaining and analyzing sensitivity data. *Journal of the American Statistical Association.* 43:109–126.

Draper, N. and Smith, H. (1981). *Applied Regression Analysis,* 2nd ed. New York: John Wiley and Sons.

Dror, H. A. and Steinberg, D. M. (2008). Sequential experimental designs for generalized linear models. *Journal of the American Statistical Association.* 103(481):288–298.

Evans, M., Hastings, N. and Peacock, B. (2000). *Statistical Distributions,* 3rd. ed. New York: John Wiley and Sons.

Everitt, B. (1978). *Multivariate Techniques for Multivariate Data.* North-Holland.

Efron, B. and Gong, G. (1983). A leisurely look at the bootstrap, the Jackknife, and cross validation. *The American Statistician* 37(1):36–48.

Filliben, J. J. (1975). The probability plot correlation coefficient test for normality. *Technometrics* 111–117.

Finney, D. J. (1952). *Probit Analysis,* 2nd ed. Cambridge, England: Cambridge University Press.

Fisher, R. A. (1946). *Statistical Methods for Research Workers,* 10th ed. Edinburgh, Scotland: Oliver & Boyd, Ltd.

Granger, C. W. I. and Hatanaka, M. (1964). *Spectral Analysis of Economic Time Series.* Princeton, NJ: Princeton University Press.

Grubbs, F. (1969). Sample criteria for testing outlying observations. *Annals of Mathematical Statistics* 27–58.

Grubbs, F. (1969). Procedures for detecting outlying observations in samples. *Technometrics* 11(1):1–21.

Hawkins, D. M. (1980). *Identification of Outliers.* New York, NY: Chapman and Hall.

Iglewicz, B. and Hoaglin, D. (1993). Volume 16: How to detect and handle outliers. In *The ASQC Basic References in Quality Control: Statistical Techniques.* E. F. Mykytka (Ed.). Milwaukee, WI: Quality Press.

Jenkins, G. M. and Watts, D. G. (1968). *Spectral Analysis and Its Applications.* San Francisco, CA: Holden-Day.

Johnson, N. L., Kotz, S. and Balakrishnan, N. (1994). *Continuous Univariate Distributions,* Vols. I and II, 2nd ed. New York: John Wiley and Sons.

Johnson, N. L., Kotz, S. and Kemp, A. W. (1992). *Univariate Discrete Distributions,* 2nd ed. New York: John Wiley and Sons.

Latscha, R. (1953). Tests of significance in a 2-by-2 contingency table: Extension of Finney's table. *Biometrika* 40:74–86.

Levene, H. (1960). In *Contributions to Probability and Statistics: Essays in Honor of Harold Hotelling.* I. Olkin et al. (Eds.). Palo Alto, CA: Stanford University Press.

McNeil, D. (1977). *Interactive Data Analysis.* New York: John Wiley and Sons.

Mosteller, F. and Tukey, J. (1977). *Data Analysis and Regression.* Reading, MA: Addison-Wesley.

Nelson, W. (1982). *Applied Life Data Analysis.* Reading, MA: Addison-Wesley.

Neter, J., Wasserman, W. and Kunter, M. H. (1990). *Applied Linear Statistical Models,* 3rd ed. New York: Irwin.

Nelson, W. and Doganaksoy, N. (1992). A computer program POWNOR for fitting the power-normal and -lognormal models to life or strength data from specimens of various sizes, *NISTIR 4760.* U.S. Department of Commerce, Washington, DC: National Institute of Standards and Technology.

Neyer, B. T. (1994). A D-optimality-based sensitivity test. *Technometrics* 36:61–70.

Pearson, E. S. and Hartley, H. O. (1954). *Biometrika Tables for Statisticians.* Vol. 1. Cambridge, England: Cambridge University Press.

Plackett, R. L. (1983). Karl Pearson and the Chi squared test. *International Statistical Review* (International Statistical Institute (ISI)) 51(1):59–72.

Rosner, B. (1983). Percentage points for a generalized ESD many-outlier procedure. *Technometrics* 25(2):165–172.

Ryan, T. (1997). *Modern Regression Methods.* New York: John Wiley.

Scott, D. (1992). *Multivariate Density Estimation: Theory, Practice, and Visualization.* New York: John Wiley and Sons.

Snedecor, G. W. and Cochran, W. G. (1989). *Statistical Methods,* 8th ed. Ames, IA: Iowa State University Press.

Statistical Research Group, Columbia University. (1947). *Selected Techniques of Statistical Analysis.* C. Eisenhart, M. W. Hastay, and W. A. Wallis (Eds.). New York: McGraw·Hill, 1947. pp. 46–47, one-sided tolerance limits; pp. 95–110, symmetric tolerance limits; and pp. 249–253, contingency tables.

Stefansky, W. (1972). Rejecting outliers in factorial designs. *Technometrics* 14:469–479.

Stephens, M. A. (1974). EDF statistics for goodness of fit and some comparisons. *Journal of the American Statistical Association* 69:730–737.

Stephens, M. A. (1976). Asymptotic results for goodness-of-fit statistics with unknown parameters. *Annals of Statistics* 4:357–369.

Stephens, M. A. (1977). Goodness of fit for the extreme value distribution. *Biometrika* 64:583–588.

Stephens, M. A. (1977). *Goodness of Fit with Special Reference to Tests for Exponentiality,* Technical Report No. 262, Department of Statistics, Stanford University, Stanford, CA.

Stephens, M. A. (1979). Tests of fit for the logistic distribution based on the empirical distribution function. *Biometrika* 66:591–595.

Tietjen, G. and Moore, R. (1972). Some Grubbs-type statistics for the detection of outliers. *Technometrics* 14(3):583–597.

Tukey, J. (1977). *Exploratory Data Analysis*. New York: Addison-Wesley.

Tufte, E. (1983). *The Visual Display of Quantitative Information*. Ellicott City, MD: Graphics Press.

4

Tests and Confidence Intervals for Means

In the previous chapter, we discussed the summarizing of the data such as descriptive statistics and graphical presentations. We also provided a short introduction to normality and its application in the summarizing of the data. In this chapter, we address some of the tests and confidence intervals for means.

In the quality statistical world, quite often, we are interested in tests and confidence intervals for population means based on random samples from those populations. It is a common inquisition for most of us in the field. To make things easier for the experimenter let us begin by identifying the terms and their meaning. As used quite often, "test" means a test of a hypothesis, set up prior to the experiment (*a priori*), that the population mean has a specific numerical value. The test yields simply the conclusion that the data are or are not consistent with the hypothesis. The method of testing depends on whether the population standard deviation is known or unknown. In experimentation, the case of unknown standard deviation is the more important and usual; in repetitive work, such as production quality control and standard tests, the standard deviation may sometimes be calculated closely from the large amount of earlier data, including surrogate data.

Perhaps two of the most fundamental assumptions in all statistical tests and confidence intervals are based on (a) the samples are random and (b) the populations have normal distributions. We are rarely certain that these assumptions are perfectly satisfied. However, on the basis of historical empirical data, we are somewhat confident that the normality assumption is satisfied. If normality and randomness is in question, there are objective tests that may be performed.

It is imperative to recognize that no matter what the analysis and the results are, error is always present. This error depends on (a) the departure of assumptions; (b) measurement issues, that is, resolution, confidence of instrument; and (c) the operator who conducts the measurement.

Why do we worry about the population mean? The simple answer is that the mean enters into any question of measurement, both because of measurement errors and because of actual variation of the quantity measured. It is, then, highly desirable to have objective methods for testing whether an experimental mean value is consistent with some theoretical value and for stating a confidence interval for the population mean ("true" mean) on the basis of a sample. It is of interest to note here that it is practical to take observations successively without specifying beforehand the total sample size. The most economical method for testing a mean is provided by "sequential analysis."

Test for μ when σ Is Known

In the following discussion, we are focusing on the process of testing rather than the tests themselves. However, in this particular case where we test for μ when σ is known, we are presenting the process in a complete manner to serve as an example for the rest.

As we mentioned earlier, the focus is to understand the concepts and not the formulas or the calculations. In both cases, the computer can do those tasks much more efficiently and accurately. Therefore, we use formulas at a very minimum and basic level to show the reader how the process works. We begin by setting up the assumptions, determining the normal test, sample, and confidence.

Assumptions

1. The sample is a random sample.
2. The population is normally distributed with known standard deviation or else the sample size is large (>30), in which case s may be substituted for σ.

Normal Test

Given a sample of size n, to test at a given significance level a whether the population mean has the hypothetical value a, compute the statistic

$$z = \frac{\bar{x} - \alpha}{\sigma/\sqrt{n}}$$

The distribution of z is normal with unit standard deviation, for $\sigma/\sqrt{n}$ is the standard deviation of means $\bar{x}$. For an equal-tails test of the null hypothesis that $\mu = a$, reject the hypothesis if z falls in the rejection region $|z|>z_{\alpha/2}$, where $|z|$ is the absolute or numerical value of z. The value $z_{\alpha/2}$ is that exceeded with probability $\alpha/2$ by the normal variable z if the null hypothesis holds.

For a one-sided test of the null hypothesis that $\mu = a$, against the alternative that $\mu > a$, reject the hypothesis if $z > z_a$. For a one-sided test of the null hypothesis that $\mu = a$, against the alternative that $\mu < a$, reject the hypothesis if $z < -z_a$. If z falls in the rejection region for the test being made, reject the null hypothesis that the population mean equals a on the grounds that such a large or small value of z would occur with a probability of only 5% if the null hypothesis held.

If the population is finite of size N, rather than infinite, replace the aforementioned statistic z by

$$z = \frac{\bar{x} - \alpha}{\dfrac{\sigma}{\sqrt{n}} \sqrt{\dfrac{N-n}{N-1}}}$$

And continue the calculation as before. All the assumptions hold true as before. However, be very cognizant that rejection of the null hypothesis may not mean that the hypothesis itself was false; it may mean, instead, that some of the assumptions underlying the construction of the mathematical working model were not satisfied.

On the other hand, failure to reject the null hypothesis is not always strong evidence that it holds. If the sample size is small, only a large departure from the null hypothesis is likely to be adjudged significant. Hence, it is important to plan experiments of sufficient size to hold risks of error to acceptably low levels.

Determination of Sample Size

When an experiment is to be performed and the results are to be analyzed by a specific method, a decision must be made about the size of the sample to be used. Suppose we wish to guard against the possibility of not rejecting the null hypothesis $\mu = a$ if μ actually has a particular alternative value b. Let there be a specified risk β of making such an error, then using the classical formulas we have:

$$\alpha + \frac{k\sigma}{\sqrt{n}} = b \quad \text{or} \quad n = \left(\frac{k\sigma}{b-\alpha} \right)^2$$

where α is the probability of a Type I error, β is the probability of a Type II error, and b is an alternative value of μ. The value of the alternative mean generally is identified as

$$\mu = \alpha + k\sigma/\sqrt{n}$$

for a given risk β, k is the level of standard deviation (1, 2, 3, and so on), and σ is the standard deviation.

Confidence Intervals and Sample Size

As we mentioned earlier, every data and every analysis has error. The question is how much error? To address this issue, the experimenter should be aware of the confidence intervals for the population mean as well as the determination of sample size for such analysis. Here, we provide the classical formulas for each.

a. For confidence

$$\bar{x} \pm z_{\alpha/2} \, \sigma/\sqrt{n} \quad \text{For an infinite population}$$

$$\bar{x} \pm z_{\alpha/2} \, \sigma/\sqrt{n} \sqrt{\frac{N-n}{N-1}} \quad \text{For a finite N population}$$

b. For sample

$$n = \left(\frac{z_{\alpha/2} \, \sigma}{b} \right)^2$$

where $\bar{x}$ is the mean, $z_{\alpha/2}$ is the value in the standard normal distribution such that the probability of a random deviation is numerically greater than $z_{\alpha/2}$, N is the size of the population, n is the sample population, and b is the expected error.

Test for μ when σ Is Unknown

Two useful tests are applicable here: the τ_1 and the *t*-test. The first one is restricted to small samples $2 \le n \le 10$ and should not be used if the population is known, or suspected, to be nonnormal. It is often used for making a preliminary check on a portion of a larger sample. The *t*-test on the other hand involves considerably more computation, but it is the most powerful test available for testing the mean when the standard deviation is unknown. It is valid for any sample size, and it is a good approximation, especially for large samples, even though the population is nonnormal to a noticeable degree.

The formulas for each are:

$$\tau_1 = \frac{\bar{x} - \alpha}{w}$$

$$t = \frac{\bar{x} - \alpha}{s/\sqrt{n}}$$

where $\bar{x}$ is the average, w is the range of the sample, α is the probability, n is the sample, and s is the standard deviation of the sample.

In each case, we must calculate the sample size and determine the confidence intervals.

Tests for Proportion p

Generally speaking, when one deals with proportions one deals with the binomial distribution. Often the sample items are merely classified as having or not having a certain characteristic; for example, they function or fail to function, succeed or fail, are perfect or contain defects. We are interested in testing a hypothetical value, p_0, for the population proportion having the characteristic. The unknown parameter p is estimated from a sample as the ratio r/n of the number r of items having the characteristic to the sample size n.

Assumptions

While dealing with proportions, there are three assumptions that must be met. They are as follows:

(1) A proportion $p \le 0.5$ of the items of a population have a particular characteristic. The artificial restriction $p \le 0.5$, which is a convenience and conventional probability, is satisfied by considering p as the proportion *not* having the characteristic, if the proportion *having* the characteristic is greater than 0.5.

(2) The sample is a random sample, each item of which is tested for the characteristic.

(3) Ideally, the population is infinite. If the population is finite (as in a pilot lot), other methods are applicable. The most common ones are the situations where the sample size is $n < 150$ and $n > 5/p_0$.

Test for a Sample of Size $n < 150$

Here, we use the binomial expression, which may be used for any sample size especially when there is no replacement:

$$f(x) = \binom{n}{x} p^x (1-p)^{n-x}$$

where $\binom{n}{x} = \dfrac{n!}{x!(n-x)\,1}$, p is the constant probability of a success, n is independent trials, and x is the successes.

Test for a Sample of Size $n > 5/p_0$

Here, we focus on the x with an α significance level that hypothesizes $p = p_0$ against the hypothesis that $p > p_0$.

$$z = \frac{r - np_0 - 1/2}{\sqrt{np_0\,(1-p_0)}}$$

If we want to test the null hypothesis $p = p_0$ against the hypothesis that $p < p_0$ we use the same formula as above, but we change the $-1/2$ to $+1/2$ and reject the hypothesis if $x < -z_a$.

For an equal-tail test of the null hypothesis against alternatives greater or equal than p_0, we compute the following:

$$z = \frac{|r - np_0| - 1/2}{\sqrt{np_0\,(1-p_0)}}$$

where r is the number of defectives observed in a sample of n.

Sample and confidence should also be taken into consideration.

Test for $\mu_1 - \mu_2$ when σ_1 and σ_2 Are Known

Assumptions and Test Conditions

A random sample is drawn from each of two populations to determine whether the difference (d) between the two population means μ_1 and μ_2 is equal to d. The two samples may be independent, or the observations may be made in pairs, one from each population, extraneous variables being held constant for each pair. If we let y denote the difference $x_1 - x_2$ of the paired observations and if the standard deviation of y is known and is less than $\sqrt{\sigma_{x_1}^2 + \sigma_{x_2}^2}$, then it is desirable to pair during sampling. This situation will arise in general if the paired observations are positively correlated.

Of course, each population is assumed to be normally distributed with known standard deviation (s_i) or else the sample drawn from it is large (>30), in which case s_i may be substituted for σ_i.

Normal Test

There are three options here. They are as follows:

a. **Test for Paired Observations.** Subtract the second reading of each pair from the first and consider the resulting differences as sample values of a new variable y. Test the null hypothesis that $\mu_y = d$ by the same method that we discussed earlier, if σ_y is known or unknown. The reader will recall that we know from theory that $\mu_y = \mu_1 - \mu_2$. Confidence intervals for μ_y and determinations of sample size can also be obtained by the methods we discussed earlier.

b. **Test for Two Independent Samples.** In this situation, we test at the significance level a the hypothesis that $\mu_1 - \mu_2 = d$ with

$$z = \frac{\bar{x}_1 - \bar{x}_2 - d}{\left(\dfrac{\sigma_1^2}{n_1} + \dfrac{\sigma_1^2}{n_2} \right)^{1/2}}$$

Significance is tested via the already mentioned method. However, if the populations are finite of sizes N_1 and N_2, replace σ_1^2 by $\sigma_1^2 (N_1 - n_1)/(N_1 - 1)$ and σ_1^2 by $\sigma_1^2 (N_2 - n_2)/(N_2 - 1)$ in the z formula.

c. **Determination of Size of Independent Samples.** This third option is considered because it makes most effective use of a fixed number (N) of items. We use the following:

$$n_1 = \frac{\sigma_1}{\sigma_1 + \sigma_2} N \text{ and } n_2 = N - n_1 t$$

If $\sigma_1 = \sigma_2 = \sigma$, then choose $n_1 = n_2 = n$. Then, proceed with testing the significance.

Confidence Intervals and Determination of Size of Independent Samples

We reemphasize here the importance of confidence and size of sample. As such we show the formulas to make the reader understand why it is important to look at both issues.

For the confidence, we have:

$$\left(\bar{x}_1 - \bar{x}_2\right) \pm z_{\alpha/2}\left(\frac{\sigma_1^2}{n_1} + \frac{\sigma_2^2}{n_2}\right)^{1/2}$$

And for the independent sample size, we have:

$$N = n_1 + n_2 = \left[\frac{z_{\alpha/2}(\sigma_1 + \sigma_2)}{\text{error}}\right]^2$$

Where

$$n_1 = \left(\frac{z_{\alpha/2}}{\text{error}}\right)^2 \sigma_1(\sigma_1 + \sigma_2)$$

and

$$n_2 = \left(\frac{z_{\alpha/2}}{\text{error}}\right)^2 \sigma_2(\sigma_1 + \sigma_2)$$

Test for $\mu_1 - \mu_2$ when σ_1 and σ_2 Are Unknown

Assumptions and Design of Experiment

The assumptions vary with the different situations and tests given below. Random samples are necessary, as usual. Normal distributions are assumed except in the sign test. Using the common normality assumption, we must decide whether to sample the two populations independently or to pair observations so that conditions that usually fluctuate are the same for each pair, though changing from pair to pair. Thus, in a durability test of paints, panels of two or more paints would be exposed side by side and would thus give paired rather than independent observations.

If the conditions that are rendered the same by pairing are known to influence substantially the characteristic x under observation, then it is probably desirable to pair observations. The advantage of independent samples lies in the doubling of the number of degrees of freedom with no increase in the number of observations, but this advantage may or may not be overbalanced by the reduction in variance afforded by pairing; each problem must be examined carefully. Pairing has the additional advantage of eliminating any assumption about the equality of σ_1 and σ_2.

Test for Paired Observations

a. **Sign Test.** The sign test provides a quick and simple way of determining whether a set of paired readings shows a significant mean difference between pair members. Unlike the *t*-test described below, this test does not depend on an assumption of normality; however, when normality assumptions are valid, it is not as efficient as the *t*-test. For the sign test, it is unnecessary that the different pairs be observed under the same conditions, so long as the two members of each pair are alike. Quite often this approach is used with equal-tails test, but it may be also adapted for use as a one-sided test.

b. **The *t*-test on Differences.** The *t*-test, which depends on an assumption of normality and requires that all pairs be observed under the same conditions, gives more information than the sign test. The approach is to subtract the pair values and consider the resulting differences as the data to be analyzed by the earlier method discussed.

Test for Two Independent Samples when $\sigma_1 = \sigma_2$

The τ_d test is used with small samples ($n \leq 10$)

$$\tau_d = \frac{\bar{x}_1 - \bar{x}_2}{w_1 + w_2}$$

where w is the respective range of the sample. The significance is tested based on the traditional approaches that have been discussed.

The *t*-test is used with large samples ($n \geq 30$)

$$t = \frac{\bar{x}_1 - \bar{x}_2}{\sqrt{\dfrac{\sum (x_1 - \bar{x}_1)^2 + \sum (x_2 - \bar{x}_2)^2}{n_1 + n_2 - 2} \cdot \left(\dfrac{1}{n_1} + \dfrac{1}{n_2} \right)}}$$

The reader should notice that if we already have the sample variances this formula may be simplified by the appropriate substitution. Once we know what the *t*-value is, we can proceed to identify both the significance and the confidence.

Test of $\mu_1 - \mu_2 = 0$ for Two Independent Samples when $\sigma_1 \neq \sigma_2$

This approach to testing is prevalent when the two tests under consideration are of the type $n_1 \leq n_2 >$; when this situation exists, use the order of observation or randomize the readings (observations). It is preferred when doing this kind of analysis that the samples be of equal size, since the analysis makes use of extra items from either population. Specifically, the analysis takes the form of the following sequence:

First we define u_1, $\bar{u}$, and Q thus:

$$u_i = x_{1i} - x_{2i}(n_1/n_2)^{1/2} \text{ and } \bar{u} = \frac{1}{n_1}\sum_{i=1}^{n_1} u_i \text{ where } i = 1, 2, 3, \dots n_1$$

$$Q = n_1\sum_{i=1}^{n_1}(u_i - \bar{u})^2 = n_1\sum_{i=1}^{n_1} u_i^2 - \left(\sum_{i=1}^{n_1} u_i\right)^2$$

Then, we compute the *t*-value thus

$$t = \frac{\bar{x}_1 - \bar{x}_2}{\sqrt{\dfrac{Q}{n_1^2(n_1 - 1)}}}$$

where

$$\bar{x}_1 = \frac{1}{n_1}\sum_{i=1}^{n_1} x_{1i}$$

and

$$\bar{x}_2 = \frac{1}{n_2}\sum_{i=1}^{n_2} x_{2i}$$

Once we know what the value is, we proceed to calculate both the significance and confidence.

_____

Test for Differences between Proportions $p_1 - p_2$

One may also test whether or not the samples come from a binomial population having the same amount of defects. One may also use contingency tables or the chi-square (χ^2) statistic.

Test for Differences among Several Means

If there is more than two samples, one may use the *t*-test approach repetitive times until all samples have been tested, or the experimenter may use the more efficient way, analysis of variance (ANOVA).

Detection of a Trend in a Set of Means, with Standard Deviations Unknown but Assumed Equal

Quite often in the quality profession in many industries, we are interested in finding out whether or not there is a trend of the means in our samples. If that is the case, the assumption is that a random sample of size n is drawn from each of k normal populations with the same standard deviation. The common sample size n may be 1. The reader should remember that a trend in means is some relationship among the means that can be represented graphically by either a straight line or a curve. As such let us recall the variance among means in a k sample problem.

$$s^2 = \frac{1}{k-1}\left(\sum_{i=1}^{k}(\bar{x}_1 - \bar{\bar{x}}_2)\right)^2$$

where $\bar{\bar{x}}$ is the mean of k sample means (overall mean, grand mean) and $\bar{x}_i$ is the mean of the ith sample. Calculated as such this s^2 will include the effect of a trend in means. That is, if a trend exists. An estimate of variance that reduces the trend effect is $1/2\delta^2$. The ordinary simple method that one may use to calculate the δ^2 is the following:

$$\delta^2 = \frac{1}{k-1}\sum_{i=1}^{k-1}(\bar{x}_{i+1} - \bar{x}_i)^2 = \text{mean square successive difference}$$

The result of the calculations of both (s^2 and δ^2) will allow us to compare it with the critical value that we have set before the experiment was undertaken. The interpretation for this test—the value of δ^2/s^2—depends on three situations. They are as follows:

1. If the data show no trend, we expect δ^2/s^2 near 2.
2. If the data follow some curve, we expect a δ^2/s^2 less than 2.
3. If the data oscillate rapidly, we expect a δ^2/s^2 greater than 2.

Summary

In this chapter, we addressed the issue of testing and confidence interval for mean by giving both an explanation and a variety of common tests. In the next chapter, we will continue the discussion, but we will address the test and confidence interval for standard deviation

Selected Bibliography

Arsenal, F. (1951). *Statistical Manual: Methods of Making Experimental Inferences.* 2nd rev. ed. by C. W. Churchman. Philadelphia, Frankford Arsenal.

Dixon, W. J. and Massey, F. J., Jr. (1951). *An Introduction to Statistical Analysis.* New York, McGraw-Hill.

Freeman, H. (1942). *Industrial Statistics.* New York, Wiley.

National Bureau of Standards. (1949). *Tables of the Binomial Probability Distribution.* Washington, GPO. (Applied Mathematics Series 6.)

Romig, H. G. (1952). *50-100 Binomial Tables.* New York, Wiley.

Wilks, S. S. (1949). *Elementary Statistical Analysis.* Princeton, NJ, Princeton University Press.

5

Tests and Confidence Intervals
for Standard Deviations

In the previous chapters, we have given an overview of some statistical concepts with emphasis on the location of the mean and the significance of that mean for the experimenter. In fact, in the previous chapter, we addressed the issues of testing and confidence intervals for means. In this chapter, we extend the discussion to cover tests and confidence intervals for standard deviations. Specifically, we are focusing on the variation of the sample and its significance to the experimenter, by (a) introducing the testing method of the standard deviation from a random normal distribution whether or not it has a particular value that was specified on *a priori* basis, (b) calculating confidence limits for the estimate of the standard deviation (σ), and (c) comparing two or more standard deviations.

What is the importance of the population standard deviation? It is a measure of dispersion; in other words, a measure of lack of precision. As such, it includes both lack of reproducibility and whatever measurement error is present. A process cannot be used for tasks requiring precision if the standard deviation from a fixed target is large. On the other hand, if the standard deviation is small, but the average distance from the target line is large, it is possible that a simple adjustment (correction) will center the process on the target. If the correction can be made without increasing the dispersion, the process will be both accurate and precise.

In production, a sudden increase in variability may indicate the appearance of a production fault, such as change the operator, material, process itself, and so on. On the other hand, a drift in variability maybe because of wear or other inherent issues with the process itself.

The assumptions underlying these tests are that they are based on the notions of randomness and normality. However, these restrictions may be overlooked (somewhat), especially in cases involving large samples. Another point for the reader to notice here is that the significance tests mentioned (for simplicity) involve fixed sample sizes. However, if it is practical to experiment on one item at a time without knowing the total number that will be required, the most economical method is provided by "sequential analysis." For more advanced techniques one may use ANOVA, factorial analysis, or the Taguchi approach to experimentation.

Computation and Understanding s and σ

Standard deviation is variability. However, for samples it is noted as an s and for populations it is denoted by the Greek letter σ. The difference between the two is shown in the formulas. Specifically, one will notice that the s is calculated with a denominator of $(n-1)$, whereas the σ is calculated with a denominator of only n. The reason for this is that we lose 1 degree of freedom when we calculate the average. Since we are having small samples, we accommodate (think of it as weighing) the values to have a greater meaning for the variability. As the sample gets to be greater or equal to 30, there is no difference in the numerical value.

$$\text{Sample } s: s = \sqrt{\frac{\sum_{i=1}^{n}(x_i - \bar{x})^2}{(n-1)}} \qquad \text{population } \sigma: \sigma = \sqrt{\frac{\sum_{i=1}^{n}(x_i - \bar{x})^2}{n}}$$

If the means and standard deviations must be obtained from data that have been grouped together in intervals of length b, so that all observations lying in one interval are considered to be at the midpoint of the interval, we may employ "Sheppard's correction" to improve the approximation. This correction calls for subtraction of $b^2/12$ from the s^2 computed using midpoints. The correction is appropriate if the sample is taken from a distribution that tapers to zero at the ends.

Chi-Square Test for σ in a Normal Distribution

Assumptions

(1) The population has a normal distribution.

(2) The sample is a random sample.

Test

To test the null hypothesis that the normal population has standard deviation $\sigma = \sigma_0$ at the significance level α on the basis of a sample of size n, we compute the sum of the squared deviations from the sample mean, $(n-1)s^2$, and form the chi-square statistic.

$$\chi^2 = \frac{(n-1)^2}{\sigma_0^2}$$

If we are interested in confidence limits from a sample of size n, we have:

$$\left[\frac{(n-1)s^2}{\chi^2_{\alpha/2,n-1}}\right]^{1/2} \text{ and } \left[\frac{(n-1)s^2}{\chi^2_{1-\alpha/2,n-1}}\right]^{1/2} \text{ where } (n-1)s^2 = \sum(x_i - \bar{x})^2$$

The reader will remember that the confidence limits for the mean are symmetric about the mean estimate; however, these limits are not symmetric about the s.

F-Test for σ_1/σ_2 for Normal Distributions

Assumptions

1. The populations have normal distributions.
2. The samples are random samples drawn independently from the respective populations.

Test

To test the null hypothesis that the ratio of σ_1^2/σ_2^2 of the variances of two normal populations is 1 (i.e., σ_1/σ_2) at the significance level a, on the basis of a sample size n, from population 1 and an independent sample of size n_2 from population 2, compute the statistic

$$F = \frac{S_1^2}{S_2^2}$$

with the larger s^2 in the numerator so that the computed F is always greater than 1. The significance is measured with the traditional approach of comparing the calculated and the critical value.

If we are interested in a confidence, then we can compute the following:
Lower limit:

$$\frac{S_1}{S_2} \frac{1}{\sqrt{F_{\alpha/2}(n_1 - 1, n_2 - 1)}}$$

And upper limit is:

$$\frac{S_1}{S_2} \frac{1}{\sqrt{F_{1-\alpha/2}(n_1 - 1, n_2 - 1)}} = \frac{S_1}{S_2}\sqrt{F_{\alpha/2}(n_2 - 1, n_1 - 1)}$$

Special Note: Of interest is the notion that a *total* of slightly more than four times as many observations are needed to compare two unknown standard deviations with any given accuracy as are needed to compare one unknown standard deviation with a known value with the same accuracy.

M-Test for Homogeneity of Variances

An unusual but very worthy test for homogeneity of variances is the *M*-test. The test is analogous to the comparison test for determining the convergence of series of real or complex numbers.

Assumptions

1. The populations have normal distributions.
2. The samples are random samples drawn independently from the respective populations.

Test

To test the null hypothesis that k populations, from which k samples of sizes $n_1, n_2, \ldots, n_k$ have been drawn and have the same variances, we use

$$M = \varphi \ln \frac{\sum_{i=1}^{k} f_i s_i^2}{\varphi} - \sum_{i=1}^{k} f_i \ln s_i^2$$

where $\ln$ = natural log; $f_i = n_i - 1$, the number of degrees of freedom in computing s_i^2 for the ith sample; s_i^2 is the estimate of the variance of the ith population; and $\varphi = \sum_{i=1}^{k} f_i$.

Significance is determined by the same methods of critical value or a probability level. It should be noted here that if the variances are homogeneous, then the *M*-test can be applied for the homogeneity of the means as well.

Summary

In this chapter, we provided an overview of tests and confidence interval for the standard deviation with several specific tests. In the next chapter, we will address test for distributions and means.

Selected Bibliography

Dixon, W. J. and F. J. Massey, Jr. (1951). *An Introduction to Statistical Analysis.* New York, McGraw-Hill.

Eisenhart, C. (1947). Planning and interpreting experiments for comparing two standard deviations, in *Selected Techniques of Statistical Analysis,* edited by C. Eisenhart, M. W. Hastay, and W. A. Wallis. New York, McGraw-Hill.

Folland, G. (1999). *Real Analysis: Modern Techniques and Their Applications.* 2nd ed. New York, John Wiley & Sons, Inc.

Hoel, P. G. (1954). *Introduction to Mathematical Statistics.* 2nd ed. New York, Wiley.

Jahnke H. N. (2003). 6.7 The foundation of analysis in the 19th century: Weierstrass. *A history of analysis.* AMS Bookstore

Knopp, K. (1954). *Theory and Application of Infinite Series.* London, Blackie and Son. Reprinted by Dover Publications. (1990). New York.

Knopp, K. (1956). *Infinite Sequences and Series.* New York, Dover Publications, Inc.

Lakatos, I. (1976). *Proofs and Refutations.* Boston, MA, Cambridge University Press.

Rudin, W. (1976). *Principles of Mathematical Analysis.* 3rd ed. New York, McGraw-Hill.

Rudin, W. (1986). *Real and Complex Analysis.* New York, McGraw-Hill Series on Science/Engineering/Math.

Rudin, W. (1991). *Functional Analysis.* New York, McGraw-Hill Series on Science/Engineering/Math.

Thompson, C. M. and M. Maxine. (1943–1946). Tables for testing the homogeneity of a set of estimated variances. *Biometrica.* 33:296–304.

Whittaker, E. T. and G. N. Watson (1963). *A Course in Modern Analysis.* 4th ed. Boston, MA, Cambridge University Press.

6

Tests of Distributions and Means

In the previous two chapters, we focused on the tests and confidence intervals for means and standard deviations. In this chapter, we extend the discussion of testing of distributions and means.

Underlying Distribution Analysis

One of the fundamental bases in any statistical analysis of measurements is our ability to describe the data within the context of a model or probability distribution. These models are utilized primarily to describe the shape and area of a given process, so that probabilities may be associated with questions concerning the occurrence of scores of values in the distribution. Common probability distributions for discrete random variables include the binomial and Poisson distributions. Probability distributions employed to describe continuous random variables include the normal, exponential, Weibull, gamma, and lognormal.

What is not commonly understood, however, is that most techniques typically employed in statistical quality control and research are based on the assumption that the process(es) studied are approximated by a particular model. The selection of a specific formula or method of analysis may, in fact, be incorrect if this assumption is erroneous. If this erroneous assumption does occur, the decisions that are based on data studied may be incorrect, regardless of the quality of the calculations. Some examples of this situation are as follows:

- Many individuals have taken to describing the capability of a process as one where the process is in control, and ±3 (sigma) of the individual parts are within specifications. The use of ±3 (sigma) as a means of describing 99.73% of the area in the distribution, and therefore probability, is appropriate only if the process is normally distributed. If this is incorrect, the calculation of $\hat{\sigma}$ from $Rbar/d_2$ is misleading, and the results of the analysis spurious.

- The development of the X and Moving R control chart is heavily dependent on the assumption that the process is normally distributed in that the Central Limit Theorem does not apply to the X chart.

- Many statistical tests, such as the z, t, and ANOVA, depend on the assumption that the population(s) from which the data are drawn is normally distributed. Tests for comparing variances are particularly sensitive to this assumption. Tables referring to the values of t are found in many statistics books. In all software packages, the probability is actually calculated for significance.
- The assumption of a particular distribution is employed when computing tolerance (confidence) intervals and predicting product/part interference and performance.
- Life tests are often based on the assumption of an underlying exponential distribution.
- Time-to-repair system estimation methods often assume an underlying lognormal or exponential distribution.

Given that the validity of the statistical analysis selected is largely dependent on the correct assumption of a specific process distribution, it is desirable, if not essential, to determine whether the assumption we have made regarding an underlying distribution is reasonable.

While many statisticians would state this hypothesis as:

H_0: It is reasonable to assume that the sample data were drawn from a normal distribution.

Many others (Shapiro, 1980) believe that this is a misleading statement. R. C. Geary (1947, p. 241) once suggested that in the front page of all textbooks on statistics, the following statement should appear:

Normality is a myth. There never was and will never be, a normal distribution.

Therefore, Shapiro suggests, the hypothesis tested should actually be stated as:

H_0: It is reasonable to approximate our process/population data with a (e.g.) normal distribution model, and its associated analytical techniques.

Given that this is the approach with the most validity, these tests are often run at relatively higher levels of Type 1 error (0.10 is frequently suggested). This is due to the fact that, in this case, the consequences of committing a Type 1 error are relatively minor. Rejection of the null hypothesis will lead to one or more of the following actions:

1. Tests are run to find an alternative model and procedures that may be used to assess the data.
2. The data are transformed so that the assumed model is approximate. An example of this is the Box procedure for comparing the

logarithms of variances rather than the variances themselves when the assumption of normality may not be accepted.

3. Nonparametric, or supposedly "distribution-free," statistical analyses may be used in place of equivalent parametric methods. For example, the Mann–Whitney U rather than a t-test, or the Kruskal–Wallis test as a replacement for the ANOVA.

Methods of Testing for Underlying Distributions

There are many ways to test for underlying distributions and especially for normality. Here, we identify 47 such methods and 14 multivariate tests. However, we discuss only the most common ones. Some of these tests are not discussed because (a) they are not common and (b) they are considered to be too technical for average applications. However, the reader may want to pursue such reading in Cochran (1952), Duncan (1955, 1986), Geary (1947), Shapiro (1980), Stephens (1974), Wilk et al. (1962a, 1962b), Williams (1950), and other statistical books found in the bibliography.

Typical tests for underlying distributions are as follows:

- Pearson's method of moments
- Kolmogorov–Smirnov (K–S test)
- Anderson–Darling test
- Lilliefors test
- Kuiper V
- Pyke's C
- Brunk's B
- Cochran's Q
- Durbin's B
- Durbin's M
- Watson's U^2
- Fisher's method
- Hartley–Pfaffenberger S^2
- Cramer–von Mises W^2
- Geary's C
- Moran's I
- Shapiro–Wilks
- Bartlett's test
- Brown–Forsythe

- D'Agnostino's D
- Omnibus K^2 statistic
- Levene's test
- Dunnett test
- Freedman's test
- F-test
- Hodges–Ajne
- Jarque–Bera
- Kendall Tau
- Kruskal–Wallis
- Linear correlation test
- Linear regression test
- Multicorrelation test
- Nonparametric multiple-comparison test
- Nonparametric serial randomness test
- Spearman's rank correlation test
- BootStrap and Jackknife tests
- Scheffe's test
- Serial randomness test
- t-Test
- Tukey test
- Watson's U^2 test
- Watson–Williams test
- Wheeler Watson test
- Duncan's new multiple-range test (MRT)
- Wilcoxon rank test
- Weighted rank correlation test

Multivariate statistics

Hotelling's T^2
Discriminant analysis or canonical variate analysis
Linear discriminant analysis (LDA)
Artificial neural networks
Cluster systems
Recursive partitioning
Canonical correlation analysis

Redundancy analysis

Correspondence analysis (CA), or reciprocal averaging

Multidimensional scaling

Multivariate analysis of variance (MANOVA)

Principal components analysis (PCA)

Factor analysis

Probability Plotting

Although not an objective test, per se, this technique can be used to generally evaluate distributional assumptions based on sample data. Probability paper is widely available for normal, exponential, Weibull, lognormal, and extreme value distributions.

The basic concept underlying this method is that the scale has been structured so that (for each type of paper) the cumulative distribution of the data will plot as straight line if the assumed model applies. Deviations from absolute linearity will exist, but they diminish as sample sizes increase. If the distribution is not applicable, however, the plotted values will fluctuate in a nonlinear but systematic fashion. A judgment related to the approximate "degree of linearity" of the points is used, therefore, to assess the applicability of the assumed distribution.

In determining the degree of linearity for the plotted values, three factors should be considered:

1. The plotted, or observed, values represent continuous or discrete random variables and will, therefore, never fall on a straight line.

2. The values have been rank ordered and are, therefore, not statistically independent. The implication of this factor is that the observed values will not fall randomly around the line; one should anticipate runs of observed data around the line of best fit.

3. Based on the concept of statistical regression, the variances of the data at the extremes of the distribution will be less than those values at the center (or near the mean) for the normal distribution. For the distribution model, this will be reversed. When plotting the line of best fit appropriate weight should be given to the center (observed) points as compared to those at the extreme ends of the distribution.

The major disadvantage evident in the use of the probability plotting techniques for testing for normality (or the appropriateness of any other distribution) is the lack of objectivity due to the judgment of relative linearity. Two individuals might look at the same data, but may arrive at different conclusions regarding the appropriateness of the model. Indeed, two individuals rarely develop lines of best fit over the same data with exactly uniform slopes, and therefore, will arrive at different conclusions based on the analysis of the plot.

As a result, there are numerous other objective methods for the determination of the best fit for an underlying distribution/model—see Table 6.1.

Selected Tests

The remainder of this chapter is devoted to a review on some of the common tests used for either hypothesis or means testing. The reader, if interested in the others, is encouraged to see the selected bibliography.

- **Pearson's method of moments**: It is the most popular measure of correlation for measuring the linear relationship between two numerically valued random variables. There are four requirements for its use: (1) relationship is linear, (2) scores of the population form a normal distribution curve, (3) scattergram is homoscedastic, and (4) scores are at interval level of measurement.

- **Kolmogorov–Smirnov (K–S test)**: It is a form of minimum distance estimation used as a nonparametric test of equality of one dimension (one-sample K–S test) or to compare two samples (two-sample K–S test). In essence, this test quantifies a distance between the empirical distribution function of the sample and the cumulative distribution function of the reference distribution, or between the empirical distribution functions of two samples. The null distribution of this statistic is calculated under the null hypothesis that the samples are drawn from the same distribution (in the two-sample case). In each case, the distributions considered under the null hypothesis are continuous distributions but are otherwise unrestricted. The two-sample K–S test is one of the most useful and general nonparametric method for comparing two samples, as it is sensitive to differences in both location and shape of the empirical cumulative distribution functions of the two samples.

  The Kolmogorov–Smirnov test can also be modified to serve as a goodness-of-fit test. In the special case of testing for normality of the distribution, samples are standardized and compared with a standard normal distribution. This is equivalent to setting the mean and variance of the reference distribution equal to the sample estimates,

TABLE 6.1

Summary of Statistics in Determining Goodness of Fit

Descriptive statistics	Continuous data	Location: Mean (arithmetic, geometric, harmonic), median, mode Dispersion: Range, standard deviation, coefficient of variation, percentile Shape: Variance, skewness, kurtosis, moments, L-moments
	Categorical data	Frequency, contingency table
Statistical graphs		Bar chart, Biplot, histogram, Pareto, box plot, control chart, correlogram, Q-Q plot, forest plot, run chart, scatter plot, stem and leaf plot, radar chart
Statistical inference	Inference	Confidence interval (frequentist inference), credible inference (Bayesian inference), hypothesis testing, meta-analysis
	Design of experiment	Control experiments, natural experiment, observational study, replication, blocking, sensitivity and specificity, optimal design
	Sample size determination	Statistical power, effect size, standard error
	General estimation	Maximum spacing, density estimation. Method of moments, minimum distance, maximum likelihood, Bayesian estimator
	Specific tests	Z-test (normal), t-test, F-test, chi-square test, Pearson's chi-square test, Wald test, Mann–Whitney U, Shapiro Wilk test, Wilcoxon signed rank test
Correlation and regression analysis	Correlation	Pearson product–moment correlation Rank correlation (Spearman's rho, Kendall's tau)
	Linear regression	Partial correlation, confounding variable
	Nonstandard predictors	Simple linear regression, ordinary least squares general linear model, analysis of variance, analysis of covariance
	No normal conditions	Nonlinear regression, nonparametric, semiparametric robust, isotonic Generalized linear model, binomial, Poisson, logistic
Survival analysis		Survival function, Kaplan–Meier, Logrank test, failure rate, proportional hazard models, accelerated failure time model
Multivariate analysis		Multivariate analysis, cluster analysis, factor analysis, principle component
Time series analysis		Trend estimation, decomposition, Box–Jenkins, ARMA models, spectral density estimation
Social statistics		Census, surveys, sampling, stratifying sampling, opinion poll, psychometrics, official statistics, questionnaire

and it is known that using the sample to modify the null hypothesis reduces the power of a test. Correcting for this bias leads to the Lilliefors test. However, even Lilliefors' modification is less powerful for testing normality than the Shapiro–Wilk test or the Anderson–Darling test.

- **Anderson–Darling test**: It is a test of whether there is evidence that a given sample of data did not arise from a given probability distribution. In its basic form, the test assumes that there are no parameters to be estimated in the distribution being tested, in which case the test and its set of critical values are distribution free. However, the test is most often used in contexts where a family of distributions is being tested, in which case the parameters of that family need to be estimated and account must be taken of this in adjusting either the test statistic or its critical values. When applied to testing whether a normal distribution adequately describes a set of data, it is one of the most powerful statistical tools for detecting most departures from normality. In addition to its use of fit for distributions, it can be used in parameter estimation as the basis for a form of minimum distance estimation procedure.

- **Lilliefors test**: It is used to test the null hypothesis that data come from a normally distributed population, when the null hypothesis does not specify which normal distribution. In other words, the application of this test is useful when the expected value and its variance are not specified. If the maximum discrepancy is large enough to be statistically significant, then the null hypothesis is rejected.

- **Kuiper V**: This test is closely related to the well-known Kolmogorov–Smirnov (K–S) test. As with the K–S test, the discrepancy statistics D^+ and D^- represent the maximum deviation above and below the cumulative distribution functions being compared. The trick with Kuiper's test is to use the quantity D^+ and D^- as the test statistic. This small change makes the Kuiper's test as sensitive in the tails as at the median and also makes it invariant under cyclic transformations of the independent variable. The Anderson–Darling test is another test that provides equal sensitivity at the tails and median, but it does not provide the cyclic invariance. This invariance under cyclic transformation makes Kuiper's test invaluable when testing for cyclic variations by time of year or day of the week or time of month.

- **Cochran's Q**: It performs a test on a randomized block or repeated-measures dichotomous data. The operation computes Cochran's statistic and compares it to a critical value from a chi-squared distribution that, interestingly enough, depends only on the significance level and the number of groups (columns). The chi-square distribution is appropriate when there are at least four columns and at least 24 categories of total data.

- **Durbin's test**: This is a nonparametric test for balanced incomplete designs that reduces to the Friedman test in the case of a complete block design.

- **Durbin–Watson test**: It is a test statistic used to detect the presence of autocorrelation in the residuals from a regression analysis. Durbin and Watson (1950, 1951) applied this statistic to the residuals from least squares regressions and developed bounds tests for the null hypothesis that the errors are serially independent against the alternative that they follow a first-order autoregresive process. Later, J. D. Sargan and A. Bhargava developed several von Neumann–Durbin–Watson-type test statistics for the null hypothesis that the errors on a regression model follow a process with a unit root against the alternative hypothesis that the errors follow a stationary first-order autoregression (Sargan and Bhargava, 1983).

If e_t is the residual associated with the observation at time t, then the test statistic is:

$$d = \frac{\sum_{t=2}^{T}(e_t - e_{t-1})^2}{\sum_{t=1}^{T} e_t^2}$$

Since d is approximately equal to $2(1-r)$, where r is the sample autocorrelation of the residuals, $d = 2$ indicates no autocorrelation. The value of d always lies between 0 and 4. If the Durbin–Watson statistic is substantially less than 2, there is evidence of positive serial correlation. As a rough rule of thumb, if Durbin–Watson is less than 1.0, there may be cause for alarm. Small values of d indicate successive error terms are, on average, close in value to one another or positively correlated. If $d > 2$ successive error terms are, on average, much different in value to one another, that is, negatively correlated. In regressions, this can imply an underestimation of the level of statistical significance.

To test for positive autocorrelation at significance α, the test statistic d is compared to lower and upper critical values ($d_{L,\alpha}$ and $d_{U,\alpha}$):

- If $d < d_{L,\alpha}$, there is statistical evidence that the error terms are positively autocorrelated.

- If $d > d_{U,\alpha}$, there is statistical evidence that the error terms are *not* positively autocorrelated.

- If $d_{L,\alpha} < d < d_{U,\alpha}$, the test is inconclusive.

To test for negative autocorrelation at significance α, the test statistic $(4 - d)$ is compared to lower and upper critical values ($d_{L,\alpha}$ and $d_{U,\alpha}$):

- If $(4 - d) < d_{L,\alpha}$, there is statistical evidence that the error terms are negatively autocorrelated.
- If $(4 - d) > d_{U,\alpha}$, there is statistical evidence that the error terms are *not* negatively autocorrelated.
- If $d_{L,\alpha} < (4 - d) < d_{U,\alpha}$, the test is inconclusive.

The critical values, $d_{L,\alpha}$ and $d_{U,\alpha}$, vary by level of significance (α), the number of observations, and the number of predictors in the regression equation. Their derivation is complex—statisticians typically obtain them from the appendices of statistical texts.

An important note is that the Durbin–Watson statistic, while displayed by many regression analysis programs, is not relevant in many situations. For instance, if the error distribution is not normal, if there is higher-order autocorrelation, or if the dependent variable is in a lagged form as an independent variable, this is not an appropriate test for autocorrelation.

- **Watson's U^2**: This test is particularly suited to test for goodness of fit when the observations are taken on the circumference of a circle.
- **Fisher's method**: Also known as Fisher's combined probability test, it is a technique for data fusion or meta-analysis (analysis of analyses). It was developed by and named after R. Fisher. In its basic form, it is used to combine the results from several independent tests bearing upon the same overall hypothesis (H_0). Fisher's method combines extreme value probabilities from each test, commonly known as p-values, into one test statistic (X^2) using the formula

$$X^2 = -2\sum_{i=1}^{k} \log_e(p_i)$$

where p_i is the p-value for the ith hypothesis test. When the p-values tend to be small, the test statistic X^2 will be large, which suggests that the null hypotheses are not true for every test. When all the null hypotheses are true, and the p_i (or their corresponding test statistics) are independent, X^2 has a chi-square distribution with $2k$ degrees of freedom, where k is the number of tests being combined. This fact can be used to determine the p-value for X^2.

The null distribution of X^2 is a chi-squared distribution for the following reason. Under the null hypothesis for test i, the p-value p_i follows a uniform distribution on the interval $[0,1]$. The negative natural logarithm of a uniformly distributed value follows an exponential distribution. Scaling a value that follows an exponential distribution by a factor of 2 yields a quantity that follows a chi-squared distribution with 2 degrees of freedom. Finally, the sum of k independent chi-square values, each with 2 degrees of freedom, follows a chi-square distribution with $2k$ degrees of freedom.

- **Cramer–von Mises W²**: It is an alternative to the Kolmogorov–Smirnov test for testing the hypothesis that a set of data come from a specified continuous distribution. The test has been adapted for use with discrete random variables, especially for cases where parameters have to be estimated from the data, and for comparing two samples. A modification leads to the Anderson–Darling test.

- **Geary's C**: It is a special autocorrelation. Autocorrelation means that adjacent observations of the same phenomenon are correlated. In other words, it is about proximity in time. On the other hand, special autocorrelation is about proximity in two-dimensional space. Geary's C statistic measures this proximity from a value of 0 and 2: 1 means no spatial autocorrelation, 2 means autocorrelation, and less than 1 indicates negative spatial autocorrelation. If the value of any one zone is spatially unrelated to another zone, the expected value of C will be 1.

- **Moran's I**: It is a statistic that is related to Geary's C, but it is not identical. Geary's statistic is more sensitive to local spatial autocorrelation, whereas Moran's I statistic is a measure of global spatial autocorrelation. It compares the sum of the cross-products of values at different locations, two at a time weighted by the inverse of the distance between the locations. The statistic value varies from -1.0 to $+1.0$. Specifically, when autocorrelation is high, the coefficient is also high. This means that when the I value is high (+1.0) it indicated positive autocorrelation. The significance is tested with the Z statistic.

- **Shapiro–Wilks**: It is a statistic that tests the null hypothesis that a sample $x_1...x_n$ came from a normally distributed population. The user may reject the null hypothesis if W is too small. It is generally used with small- to medium-size samples.

- **Bartlett's test**: It is a test that is used if k samples are from populations with equal variances. Equal variances between samples are called *homoscedasticity* or *homogeneity of variances*. Some statistical tests such as ANOVA assume that variances are equal across groups or samples. The Bartlett test can be used to verify that assumption. The reason for its use is because it is sensitive to departures from normality. That is, if the samples come from nonnormal distributions, then Bartlett's test may simply be testing for nonnormality. It must be mentioned here that the Levene and Brown–Forsythe tests are alternatives; however, they are less sensitive to departures from normality. Bartlett's test is used to test the null hypothesis (H_0) that all k population variances are equal against the alternative that at least two are different.

- **Brown–Forsythe**: It is a statistical test for the equality of group variances based on performing an ANOVA on a transformation of the response variable. Brown and Forsyth (1974) test statistic is the F-statistic resulting from an ordinary one-way ANOVA on the absolute deviations from the median. Keep in mind that the Levene's test uses the mean instead of the median. Although the optimal choice depends on the underlying distribution, the definition based on the median is recommended as the choice that provides good robustness against many types of nonnormal data while retaining good statistical power. If one has knowledge of the underlying distribution of the data, this may indicate using one of the other choices. Brown and Forsythe performed Monte Carlo studies that indicated that using the trimmed mean performed best when the underlying data followed a Cauchy distribution (heavy-tailed distribution) and the median performed best when the underlying data followed a chi-square distribution with 4 degrees of freedom (a heavily skewed distribution). Using the mean provided the best power for symmetric, moderate-tailed, distributions.

- **D'Agnostino's D**: It is a goodness-of-fit measure of departure from normality. That is, the test aims to establish whether the given sample comes from a normally distributed population. The test is based on transformations of the sample kurtosis and skewness and has power only against the alternatives that the distribution is skewed and/or kurtic. The reader should recall that the skewness and kurtosis are both asymptotically normal. However, the rate of their convergence to the distribution limit is frustratingly slow, especially for kurtosis. When that happens, D'Agostino (1970) has suggested that both kurtosis and skewness be transformed in such a way that makes their distribution as close to standard normal as possible. Similarly Anscombe and Glynn (1983) have suggested a transformation for kurtosis that works quite well for sample sizes of 20 or greater.

- **Omnibus K^2 statistic**: Sometimes when it is necessary to detect deviations from normality due to either skewness or kurtosis, D'Agostino, Belanger, and D'Agostino (1990) have suggested the omnibus test, which is a combination of the standardized skewness and kurtosis. If the null hypothesis of normality is true, then K^2 is approximately chi-square distributed with 2 degrees of freedom. A warning is applicable here: Skewness and kurtosis are not independent, only uncorrelated. Therefore, their transformations to a standardized statistic will also be dependent on each other, rendering the validity of chi-square approximation questionable (Shenton and Bowman, 1977).

- **Levene's test**: It is a test used to evaluate the equality (homogeneity) of variances. This is called homoscedasticity. The variance of data in groups should be the same. In fact, most model-based approaches usually assume that the variance is constant. The constant variance property also appears in the randomization (design-based) analysis of randomized experiments, where it is a necessary consequence of the randomized design and the assumption of unit treatment additivity (Hinkelmann and Kempthorne, 2008): If the response of a randomized balanced experiment fail to have constant variance, then the assumption of unit treatment additivity is necessarily violated. In essence, the Lavene's test is typically used to examine the plausibility of homoscedasticity.

- **Dunnett test**: Performs the Dunnett test of comparing multiple groups to a control group.

- **Freedman's test**: Performs Friedman's test on a randomized block of data. The test is a nonparametric analysis of data contained in either individual 1D waves or in a single 2D wave.

- **F-test**: A test on two distributions contained in wave 1 and wave 2. The waves can be of any real numeric type. They can have arbitrary number of dimensions, but they must contain at least two data points each. Traditionally, the *F*-test is a test that compares the explained variance with the unexplained variance given a specific probability. Typical *F*-tests include the following:

  - The hypothesis that the means of several normally distributed populations, all having the same standard deviation, are equal. This is perhaps the best-known *F*-test and plays an important role in the ANOVA.

  - The hypothesis that a proposed regression model fits the data well.

  - The hypothesis that a data set in a regression analysis follows the simpler of two proposed linear models that are nested within each other.

  - Scheffe's method for multiple comparisons adjustment in linear models.

- **Hodges-Ajne**: A nonparametric test for uniform distribution around a circle.

- **Jarque-Bera**: A test for normality on numeric data in a single wave.

- **Kendall Tau**: Performs the nonparametric Mann–Kendall test, which computes a correlation coefficient τ (somewhat similar to

Spearman's correlation) from the relative order of the ranks of the data.

- **Kruskal–Wallis**: Performs the nonparametric Kruskal–Wallis test, which examines variances using the ranks of the data.

- **Linear correlation test**: Performs correlation tests on the input waves. Results include the linear correlation coefficient with its standard error, the statistics t and F, Fisher's Z and its critical value.

- **Linear regression test**: Performs regression analysis on the input wave(s). Options include Dunnett's multicomparison test for the elevations and Tukey-type tests on multiple regressions.

- **Multicorrelation test**: Performs various tests on multiple correlation coefficients. These include multiple comparisons with a control, multiple-contrasts test, and a Tukey-type multicomparison testing among the correlation coefficients

- **Nonparametric multiple-comparison test**: Performs a number of nonparametric multiple-comparison tests. These include Dunn–Holland–Wolfe test, Student Newman–Keuls test, and the Tukey-type (Nemenyi) multiple-comparison test. The results are saved in the current data folder in the wave(s) corresponding to the optional flags. You can perform one or more of the supported tests depending on your choice of flags. Note that some tests are only appropriate when you have the same number of samples in all groups.

- **Nonparametric serial randomness test**: Performs a nonparametric serial randomness test for nominal data consisting of two types. The null hypothesis of the test is that the data are randomly distributed.

- **Spearman's rank correlation test**: The operation ranks the two inputs and then computes the sum of the squared differences of ranks for all rows. Ties are handled by assigning an average rank and computing the corrected Spearman rank correlation coefficient with ties.

- **BootStrap and Jackknife tests**: Resample the input wave by drawing (with replacement) values from the input and storing them in the wave W resampled. Flag options allow you to iterate the process and to compute various statistics on the drawn samples.

- **Scheffe's test**: It is a test for the equality of the means. The operation supports two basic modes. The default consists of testing all possible combinations of pairs of waves. The second mode tests a single combination where the precise form of H_0 is determined by the coefficients of a contrast wave.

- **Serial randomness test**: Performs a parametric or nonparametric serial randomness tests. The null hypothesis of the tests is that the data are randomly distributed. The parametric test for serial randomness is due to Young and the critical value is obtained from mean square successive difference distribution. The nonparametric test consists of counting the number of runs that are successive positive or successive negative differences between sequential data. If two sequential data are the same, the operation computes two numbers of runs by considering the two possibilities where the equality is replaced with either a positive or a negative difference. The results of the operation include the number of runs up and down, the number of unchanged values, the size of the longest run and its associated probability, the number of converted equalities, and the probability that the number of runs is less than or equal to the reported number. A separate option in this operation is to run Marsaglia's GCD test on the input.

- **T-test**: There are two kinds. The first kind compares the mean of a distribution with a specified mean value and the second compares the means of the two distributions contained in wave 1 and wave 2.

- **Tukey test**: Performs multiple-comparison Tukey (HSD) test, and optionally, the Newman–Keuls test.

- **Watson's U^2 test**: Watson's nonparametric two-sample U^2 test for samples of circular data. The Watson U^2 H_0 postulates that the two samples came from the same population against the different populations' alternative. The operation ranks the two inputs while accounting for possible ties. It then computes the test statistic U^2 and compares it with the critical value.

- **Watson–Williams test**: It is conducted for two or more sample means. The Watson–Williams H_0 postulates the equality of the means from all samples against the simple inequality alternative. The test involves the computation of the sums of the sine and cosine of all data from which a weighted r value (rw) is computed. According to Mardia et al. (1980) you should use different statistics depending on the size of rw: if $rw > 0.95$ it is safe to use the simple F-statistic, while for $0.95 > rw > 0.7$ you should use the F-statistic with the K correction factor. Otherwise you should use the t-statistic. The operation computes both the (corrected) F-statistic and the t-statistic as well as their corresponding critical values.

- **Wheeler–Watson test**: Performs the nonparametric Wheeler–Watson test for two or more samples, which postulates that the samples came from the same population. The extension of the test to more than two samples is due to Mardia. The test is not valid for data with ties.

- **Duncan's new multiple-range test (MRT):** The test is a multiple-comparison procedure to compare sets of means. It is a variation of the Student Newman–Keuls method that uses increasing alpha levels to calculate the critical values in each step of the Newman–Keuls procedure. Specifically, the MRT attempts to control family-wise error rate (FWE) at $\alpha_{ew} = 1 - (1 - \alpha_{pc})^{k-1}$ when comparing k, where k is the number of groups. This results in higher FWE than unmodified Newman–Keuls procedure, which has FWE of $\alpha_{ew} = 1 - (1 - \alpha_{pc})^{k/2}$. The idea of the MRT is to have a test that has greater power than the Newman–Keuls approach. It is especially protective against false-negative (Type II) error at the expense of having a greater risk of making false-positive (Type I) error.

- **Wilcoxon rank test**: Performs the nonparametric Wilcoxon–Mann–Whitney two-sample rank test or the Wilcoxon signed rank test on data contained in wave A and wave B.

- **Weighted rank correlation test**: Performs a weighted rank correlation test. The input waves contain the ranks of sequential factors. The test computes a top-down correlation coefficient using Savage sums.

- **Monte Carlo methods**: Monte Carlo methods are often used in simulating physical and mathematical systems. Because of their reliance on repeated computation of random or pseudorandom numbers, these methods are most suited to calculation by a computer and tend to be used when it is unfeasible or impossible to compute an exact result with a deterministic algorithm (Hubbard, 2007). Monte Carlo simulation methods are especially useful in studying systems with a large number of coupled degrees of freedom, such as fluids, disordered materials, strongly coupled solids, and cellular structures. More broadly, Monte Carlo methods are useful for modeling phenomena with significant uncertainty in inputs, such as the calculation of risk in business.

Multivariate Statistics

It is a form of statistics encompassing the simultaneous observation and analysis of more than one statistical variable. The application of multivariate statistics is multivariate analysis. Methods of bivariate statistics, for example, ANOVA and correlation, are special cases of multivariate statistics in which two variables are involved. There are many different models, each with its own type of analysis:

Hotelling's T^2: It is a generalization of the Student's t statistic and is used in multivariate hypothesis testing.

Discriminant analysis or canonical variate analysis: It attempts to establish whether a set of variables can be used to distinguish between two or more groups.

Linear discriminant analysis (LDA): It computes a linear predictor from two sets of normally distributed data to allow for classification of new observations.

Artificial neural networks: They are extended regression methods to nonlinear multivariate models.

Cluster systems: They assign objects into groups (called clusters) so that objects from the same cluster are more similar to each other than objects from different clusters.

Recursive partitioning: It creates a decision tree that strives to correctly classify members of the population based on a dichotomous dependent variable.

Canonical correlation analysis: It finds linear relationships among two sets of variables; it is the generalized (i.e., canonical) version of correlation.

Redundancy analysis: It is similar to canonical correlation analysis but derives a minimal set of synthetic variables from one set of (independent) variables that explain as much variance as possible in the other (dependent) set. It is a multivariate analogue of regression.

Correspondence analysis (CA), or reciprocal averaging: It is a methodology that finds (like PCA) a set of synthetic variables that summarize the original set. The underlying model assumes chi-squared dissimilarities among records (cases). There is also canonical correspondence analysis (CCA) for summarizing the joint variation in two sets of variables.

Multidimensional scaling: It covers various algorithms to determine a set of synthetic variables that best represent the pair-wise distances between records. The original method is principal coordinate analysis (based on PCA).

Multivariate analysis of variance (MANOVA): It is a method to extend analysis of variance methods to cover cases where there is more than one dependent variable to be analyzed simultaneously.

Principal components analysis (PCA): It is an analytical approach to find a set of synthetic variables that summarize the original set. It rotates the axes of variation to give a new set of ordered orthogonal axes that summarize decreasing proportions of the variation.

Factor analysis: It is similar to PCA but attempts to determine a smaller set of synthetic variables that could explain the original set.

Summary

As in the previous two chapters, this chapter provided an anthology of several tests and explanations for testing distributions and means. In the next chapter, we will address the requirements of planning an experiment and we will give a short introduction to ANOVA.

References

Anscombe, F.J. and Glynn, W.J. (1983). Distribution of the kurtosis statistic b_2 for normal statistics. *Biometrika*. 70(1):227–234.

Brown, M.B. and Forsythe, A.B. (1974). Robust tests for equality of variances. *Journal of the American Statistical Association*. 69:364–367.

Cochran, W.G. (1952). The χ^2 test of goodness of fit. *Annals of Mathematical Statistics*. 23:315–345.

D'Agostino, R.B. (1970). Transformation to normality of the null distribution of $g1$. *Biometrika*. 57(3):679–681.

D'Agostino, R.B., Albert, B., and D'Agostino, R.B. Jr (1990). A suggestion for using powerful and informative tests of normality. *The American Statistician*. 44(4):316–321.

Duncan, A.J. (1986). *Quality Control and Industrial Statistics*. 5th ed. Homewood, IL: Irwin.

Duncan, D.B. (1955). Multiple range and multiple F tests. *Biometrics*. 11:1–42.

Durbin, J. and Watson, G.S. (1950) Testing for serial correlation in least squares regression, I. *Biometrika*. 37:409–428.

Durbin, J. and Watson, G.S. (1951) Testing for serial correlation in least squares regression, II. *Biometrika*. 38:159–179.

Geary, R.C. (1947). Testing for normality. *Biometrika*. 34:209–241.

Hinkelmann, K. and Kempthorne, O. (2008). *Design and Analysis of Experiments*. Vol. I and II 2nd ed. New York: J. Wiley.

Hubbard, D. (2007). *How to Measure Anything: Finding the Value of Intangibles in Business*. New York: J. Wiley & Sons.

Mardia, V., Kent, J.T. and Bibby, J.M. (1980). *Multivariate Analysis*. New York: Academic Press.

Sargan, J.D. and Bhargava, A. (1983). Testing residuals from least squares regression for being generated by the Gaussian random walk. *Econometrica*. 51:153–174.

Shenton, L.R. and Bowman, K.O. (1977). A bivariate model for the distribution of $\sqrt{b_1}$ and b_2. *Journal of the American Statistical Association*. 72(357):206–211.

Shapiro, S.S. (1980). *How to Test Normality and Other Distributional Assumptions*. Vol. 3. Milwaukee, WI: American Society for Quality Control.

Stephens, M.A. (1974). EDF Statistics for Goodness of Fit and Some Comparisons. *Journal of the American Statistical Association*. 69:730–737.

Wilk, M.B., Gnanadesikan, G. and Huyett, M.J. (1962a). Separate maximum likelihood estimation of scale or shape parameters of the gamma distribution using order statistics. *Biometrika*. 49:525–545.

Wilk, M.B., Gnanadesikan G. and Huyett, M.J. (1962b). Probability plots for the gamma distribution. *Technometrics*. 4:1–20.

Williams, C.A. (1950). On the choice of the number and width of classes for the chi-square test of goodness of fit. *Journal of American Statistical Association*. 45:77–86.

Selected Bibliography

Agresti, A. (2002). *Categorical Data Analysis*. New York: Wiley.

Anderson, T.W. and Darling, D.A. (1952). Asymptotic theory of certain "goodness-of-fit" criteria based on stochastic processes. *Annals of Mathematical Statistics*. 23:193–212.

Bartlett, M.S. (1937). Properties of sufficiency and statistical tests. *Proceedings of the Royal Statistical Society Series A*. 160:268–282.

Blair, R.C. and Higgins, J.J. (1985). A comparison of the power of the paired samples rank transform statistic to that of Wilcoxon's signed ranks statistic. *Journal of Educational and Behavioral Statistics*. 10(4):368–383.

Blair, R.C., Sawilowsky, S.S. and Higgins, J.J. (1987). Limitations of the rank transform in factorial ANOVA. *Communications in Statistics: Computations and Simulations*. B16:1133–1145.

Box, H. and Hunter, J. (2005). *Statistics for Experimenters*. 2nd ed. New York: Wiley.

Caliński, T. and Kageyama, S. (2000). *Block designs: A Randomization approach, Volume I: Analysis*. Lecture Notes in Statistics. 150. New York: Springer-Verlag.

Cameron, A.C. and Trivedi, P.K. (1998). *Regression Analysis of Count Data*. Boston, MA: Cambridge University Press.

Christensen, R. (2002). *Plane Answers to Complex Questions: The Theory of Linear Models*. 3rd ed. New York: Springer.

Cohen, J. (1988). *Statistical Power Analysis for the Behavior Sciences*. 2nd ed. New York: Routledge Academic.

Cohen, J. (1992). Statistics a power primer. *Psychology Bulletin*. 112:155–159.

Conover, W.J. (1974). Some reasons for not using the Yates continuity correction on 2×2 contingency tables. *Journal of the American Statistical Association*. 69:374–376.

Conover, W.J. and Iman, R.L. (1976). On some alternative procedures using ranks for the analysis of experimental designs. *Communications in Statistics*. A5:1349–1368.

Conover, W.J. and Iman, R.L. (1981). Rank transformations as a bridge between parametric and nonparametric statistics. *American Statistician*. 35:124–129.

Corder, G.W. and Foreman, D.I. (2009). *Nonparametric Statistics for Non-Statisticians: A Step-by-Step Approach*. New York: Wiley.

Cox, D.R. (1958). *Planning of Experiments*. New York: Wiley.

Cox, D.R. and Reid, N.M. (2000). *The Theory of Design of Experiments*. Boca Raton, FL: Chapman & Hall.

Fergusonm, S.S. and Tkane, Y. (2005). *Statistical Analysis in Psychology and Education*. 6th ed. Montreal, Quebec: McGraw-Hill Ryerson Limited.

Fisher, R.A. (1934). *Statistical Methods for Research Workers*. 5th ed. Edinburgh: Oliver and Boyd.

Freedman, D. (2005). *Statistical Models: Theory and Practice*. Boston, MA: Cambridge University Press.

Freedman, D., Pisani, R. and Purves, R. (2007). *Statistics*. 4th ed. New York: W.W. Norton & Company.

Gates, C. E. (Nov. 1995). What really is experimental error in block designs? *The American Statistician*. 49(4):362–363.

Grizzle, J.E. (1967). Continuity correction in the $\times 2$ test for 2×2 tables. *The American Statistician*. 21:28–32.

Haber, M. (1980). A comparison of some continuity corrections for the chi-squared test on 2 × 2 tables. *Journal of the American Statistical Association.* 75:510–515.

Haber, M. (1982). The continuity correction and statistical testing. *International Statistical Review.* 50:135–144.

Headrick, T.C. (1997). *Type I error and power of the rank transform analysis of covariance (ANCOVA) in a 3 × 4 factorial layout.* Unpublished doctoral dissertation. University of South Florida.

Hettmansperger, T.P. and McKean, J.W. (1998). *Robust Nonparametric Statistical Methods.* 5th ed. London: Kendall's Library of Statistics.

Hilbe, J.M. (2007). *Negative Binomial Regression.* Boston, MA: Cambridge University Press.

Hinkelmann, K. and Kempthorne, O. (2005). *Design and Analysis of Experiments, Volume 2: Advanced Experimental Design.* New York: Wiley.

Hinkelmann, K. and Kempthorne, O. (2008). *Design and Analysis of Experiments, Volume I: Introduction to Experimental Design.* 2nd ed. New York: Wiley.

Hotelling, H. (1931). The generalization of student's ratio. *Annals of Mathematical Statistics.* 2(3):360–378.

Hubbard, D. (2009). *The Failure of Risk Management: Why It's Broken and How to Fix It.* New York: J. Wiley & Sons.

Iman, R.L. (1974). A power study of a rank transform for the two-way classification model when interactions may be present. *Canadian Journal of Statistics.* 2:227–239.

Iman, R.L. and Conover, W.J. (1976). *A Comparison of Several Rank Tests for the Two-Way Layout* (SAND76-0631). Albuquerque, NM: Sandia Laboratories.

Kempthorne, O. (1979). *The Design and Analysis of Experiments.* (Corrected reprint of (1952) NY. Wiley. Republished by Robert E. Krieger, R. Malabar, FL.

Kendall, M.G. and Stuart, A. (1967). *The Advanced Theory of Statistics.* Vol. 2, 2nd ed. London: Griffin.

Keppel, G. and Wickens, T.D. (2004). *Design and analysis: A researcher's handbook.* 4th ed. Upper Saddle River, NJ: Pearson Prentice-Hall.

King, B. M. and Minium, E.W. (2003). *Statistical Reasoning in Psychology and Education.* 4th ed. Hoboken, NJ: John Wiley & Sons, Inc.

Kittler, J.E., Menard, W. and Phillips, K.A. (2007). Weight concerns in individuals with body dysmorphic disorder. *Eating Behaviors.* 8:115–120.

Lentner, M. and Thomas, B. (1993). *Experimental Design and Analysis.* 2nd ed. Blacksburg, VA: Valley Book Company.

Lindman, H.R. (1974). *Analysis of Variance in Complex Experimental Designs.* San Francisco: W. H. Freeman & Co. Hillsdale.

Marsaglia, G., Tsang, W.W. and Wang, J. (2003). Evaluating Kolmogorov's distribution. *Journal of Statistical Software.* 8(18):1–4.

Maxwell, E.A. (1976). Analysis of contingency tables and further reasons for not using Yates correction in 2 × 2 tables. *Canadian Journal of Statistics.* 4:277–290.

Nanna, M.J. (2002). Hoteling's T^2 vs. the rank transformation with real Likert data. *Journal of Modern Applied Statistical Methods.* 1:83–99.

Pearson, E.S. (1947). The choice of statistical tests illustrated on the interpretation of data classed in a 2 × 2 table. *Biometrika.* 34:139–167.

Pearson, E.S. and Hartley, H. O. (Eds.). (1972). *Biometrica Tables for statisticians.* Volume II. Published by Biometrica Trustees at the CUP. Cambridge, England Cambridge University Press.

Pearson, E.S. (1931). Note on tests for normality. *Biometrika.* 22(3/4):423–424.

Pierce, C.A., Block, R.A. and Aguinis, H. (2004). Cautionary note on reporting eta-squared values from multifactor ANOVA designs. *Educational and Psychological Measurement.* 64(6):916–924.

Plackett, R.L. (1964). The continuity correction in 2×2 tables. *Biometrika.* 51:327–337.

Richardson, J.T.E. (1990). Variants of chi-square for 2×2 contingency tables. *British Journal of Mathematical and Statistical Psychology.* 43:309–326.

Richardson, J.T.E. (1994). The analysis of 2×1 and 2×2 contingency tables: an historical review. *Statistical Methods in Medical Research.* 3:107–133.

Salsburg, D. (2002). *The Lady Tasting Tea: How Statistics Revolutionized Science in the Twentieth Century.* London: Owl Books.

SAS Institute. (1985). *SAS/stat Guide forPersonal Computers.* 5th ed. Cary, NC: Author.

SAS Institute. (1987). *SAS/stat Guide for Personal Computers.* 6th ed. Cary, NC: Author.

SAS Institute. (2008). *SAS/STAT 9.2 User's guide: Introduction to Nonparametric Analysis.* Cary, NC: Author.

Sawilowsky, S. (1985a). *Robust and power analysis of the $2 \times 2 \times 2$ ANOVA, rank transformation, random normal scores, and expected normal scores transformation tests.* Unpublished doctoral dissertation, University of South Florida.

Sawilowsky, S. (1985b). A comparison of random normal scores test under the F and Chi-square distributions to the 2x2x2 ANOVA test. *Florida Journal of Educational Research.* 27:83–97.

Sawilowsky, S. (1990). Nonparametric tests of interaction in experimental design. *Review of Educational Research.* 60(1):91–126.

Sawilowsky, S. (2000). Review of the rank transform in designed experiments. *Perceptual and Motor Skills.* 90:489–497.

Sawilowsky, S., Blair, R.C. and Higgins, J.J. (1989). An investigation of the type I error and power properties of the rank transform procedure in factorial ANOVA. *Journal of Educational Statistics.* 14:255–267.

Scholz, F.W. and Stephens, M.A. (1987). *K-sample Anderson-Darling Tests. Journal of the American Statistical Association.* 82:918–924.

Shapiro, S.S. and Wilk, M.B. (1965). An analysis of variance test for normality (complete samples). *Biometrica.* 52(3–4):591–611.

Shorak, G.R. and Wellner, J.A. (1986). *Empirical Processes with Applications to Statistics.* New York: Wiley.

Siegel, S. and Castellan, N.J. (1988). *Nonparametric Statistics for the Behavioural Sciences.* 22nd ed. New York: McGraw-Hill.

Snedecor, G. W. and Cochran, W.G. (1989). *Statistical Methods.* 5th ed. Iowa: Iowa State University Press.

Stephens, M.A. (1979). Test of fit for the logistic distribution based on the empirical distribution function. *Biometrik.* 66(3):591–595.

Strang, K.D. (2009). Using recursive regression to explore nonlinear relationships and interactions: A tutorial applied to a multicultural education study. *Practical Assessment, Research & Evaluation.* 14(3):1–13.

Thompson, G.L. (1991). A note on the rank transform for interactions. *Biometrika.* 78(3):697–701.

Thompson, G.L. and Ammann, L.P. (1989). Efficiencies of the rank-transform in two-way models with no interaction. *Journal of the American Statistical Association.* 4(405):325–330.

Wilk, M.B. (1955). The randomization analysis of a generalized randomized block design. *Biometrika.* 42(1–2):70–79.

Yates, F. (1934). Contingency tables involving small numbers and the $\times 2$ test. *Journal of the Royal Statistical Society.* (Suppl. 1):217–235.

Yates, F. (1984). Tests of significance for 2×2 contingency tables. *Journal of the Royal Statistical Society,* Series A. 147:426–463.

Zyskind, G. (1963). Some consequences of randomization in a generalization of the balanced incomplete block design. *The Annals of Mathematical Statistics.* 34(4):1569–1581.

7

Understanding and Planning the Experiment and Analysis of Variance

In the previous six chapters, we have been addressing fundamental issues of data, statistics, summarizing the data for better understanding, tests and confidence for means and standard deviation as well as distributions and means. In that effort, we presented several tests and explanation. In this chapter, we are focusing on planning the experiment and analysis of variance (ANOVA) and some of the specific items that should be addressed before experimentation.

Understanding the Fundamentals of Experimentation

Usually an experiment is undertaken to test something or to investigate an unknown situation. Crow, Davis, and Maxfield (1960, pp. 109–111) have crystallized these two ideas by saying that an experiment is generally conducted (1) to determine whether some conjectured effect exists and (2) if the effect does exist, to determine its size. If the experimenter is interested only in the effect of one *factor* (or variable; sometimes is called parameter) on one other variable, the design of his experiment may be simple. But if the experimenter is concerned with more than one factor, as is frequently the case, his or her experiment or experiments will be more complex and may be conducted in various ways. The classical method is to restrict attention to one variable at a time, the other factors being held fixed; however, this method also restricts conclusions. When several factors are of interest, more reliable and more general conclusions can be reached by introducing further *levels* of these factors in the initial design. For example, if pressure is a concern for study, it is identified as a factor or parameter. However, if we are interested in finding more information about this temperature we may say that we need to study pressures between 20 psi and 35 psi and between 40 psi and 60 psi. This separation will allow the experimenter to see how the outcome of the experiment is affected by the different groupings of the selected pressures. The groupings in this case are the levels, whereas the factor remains the same (pressure). Of course, the experimenter is not bound by two levels only. More may be selected as needed. In this example, we may decide to have three levels (low, medium, and high pressure), or even more. The selection of the levels depends on the objective of the study. What is important here in the selection process is to remember that the levels must be sufficiently apart so that differences may be observed. If the levels are too close or too far, important information will be lost. The levels are also called *treatment*.

On the other hand, if we have two or more factors we also have to be concerned about their own effect on each other. This effect is called *interaction*, and sometimes it is more important than the factors themselves. For example, if our factors are temperature and pressure we may want to know what is the effect of one on the other and what happens as both temperature and pressure increase.

It is also important before the experiment begins to take into consideration as much background information as possible, which may affect the results of the experiment. Certainly, one must be aware of the theoretical outcomes but also the knowledge gained from experience. Both must be considered and taken into account in the design of the experiment. A caution here is applicable. As one increases the factors and levels, the experiment becomes more complex and costly. Therefore, the factors and levels, should be chosen with the outmost attention so that they may provide fruitful results.

Part of this background information should be the considerations of the following: (a) Measurement. Can we measure what we intent to measure? Is our instrument capable of measuring enough discrimination? Can we repeat the measurement with consistency other than random error? Is the measurement cost-effective? Quite often we are trying to measure things beyond the limits of our instruments and beyond the requirements of our customers. In both cases we incur unnecessary cost. (b) Replication versus repetition. Replication is a complete repetition of the experiment—from scratch. That means set up the experiment and run it. Repetition, on the other hand, means once the set up has been set, you run the experiment several times. Obviously repetition is easier to do, but replication provides more information.

Planning the Process

Any experiment aiming to find the significance of any difference must be planned in advance. The typical approach includes the following steps:

- Define and understand the problem at hand. A good way to do that is to use either "real" common words that everyone understands and or statistical terms.

- Determine the kind of data that is required for analysis and how that data are going to be collected.
- Determine what tools or statistics are going to be used in evaluating the decision. Make sure you are daring here to ask the question: Why? The type of data selected must be analyzed with the tool or methodology selected.
- Understand (make sure) the difference between data analysis and statistical analysis and their implications. How one uses the tools and methodologies to reach a statistical solution is important and should never be underestimated.
- Transfer the knowledge of the analysis to the process. Make sure you review the experimentation process; review the things that were learned (both positive and negative) and review and evaluate what is still needed for further improvement.

Data Analysis versus Statistical Analysis

Data analysis is a way of partitioning, slicing, and configuring the data into different logical subgroups to allow you to determine sources of variation. The easiest and most effective way to do this is through the use of graphs. Graphs quite often are tools used to identify and help visualize sources of variation without any great effort. This is true especially for evaluating continuous data. Plotting the data allows the investigator to evaluate predictive values of the data, especially when correlation and regression are used. Furthermore, graphs allow rapid analysis and visualization of differences in means and/or variation. By utilizing graphs and identifying differences through grouping or summarizing the data in different ways, it helps to identify the variation source.

On the other hand, statistical analysis is a way of analyzing a statistical test of some sort. Usually, this test is based on:

$\bar{X}$ estimate of population mean μ

s estimate of population standard deviation σ

In either case you are analyzing statistics, not the data. Of course the statistics calculated are from the data, but regardless of the data, once the statistics are calculated, the statistical test assumes that the distribution around the calculated estimates follow a defined distribution.

In essence, recognize that in the process of analyzing a statistical test, you are comparing an *Observed* value to an *Expected* value (O-E). The ability to detect a difference between what is observed and what is expected depends on the preciseness of the parameter estimate for the observed data $(s/\sqrt{n})$. Preciseness depends on the spread of the data and the amount of data used to calculate the point estimate for the sample mean. However, remember that the observed mean and standard deviation are estimates of parameters and they could be wrong! A typical example may prove the point. In a *t*-test, the model that is used is the *t*-distribution based on the distribution of the sample means. Both t_{calc} and t_{crit} are based on the distribution. Therefore, keep in mind that when the actual distribution of data is not matching the model distribution, results may not be accurate.

Typical statistical tools used in most processes and most applications are shown in Table 7.1.

TABLE 7.1

Typical Statistical Tools

Statistical Testing Methods	Items	Application
Graphs	Pareto, histogram, box plot, stem and leaf, and many others	Good exploratory tools for choosing key input variables
Correlation	Contingency coefficient (nominal data), Spearman rank (ordinal data), Pearson product-moment (interval data)	Good exploratory tools for choosing key input variables. The first two are used in nonparametric situations and the last one is used with parametric statistics
Regression	Simple multiple step	Used to optimize a process set-up, regression analysis estimates process parameters and predicts outputs. Sometimes as part of the regression analysis subanalyses are performed such as correlation, scatterplot analysis, residual analysis, and many more
Comparative experiments	*t*-tests (1 or 2 factor)	Test for mean shift
ANOVA	One way 1 factor, 2 or more levels; balanced (>1 factor, balanced (=) design); GLM (used for unbalanced design) Two way Nested designs	Test for mean shift
Variance tests	Barlett's (normal data) Levene's (nonnormal)	Test for variation

Analysis of Variance

The data obtained from an experiment involving several levels of one or more factors are analyzed and tested with a null hypothesis that several population means are equal. This technique is called analysis of variance because it examines the variability in the sample and, based on the variability, it determines whether there's reason to believe the population means are unequal. Therefore, we draw conclusions about means by looking at variability. The strength of ANOVA is that it allows the experimenter to make efficient use of limited data and to draw the strongest possible inferences. This is explained by the model of ANOVA which is:

Variation in Y – around the mean (dependent variable) = Explained variation (due to factors, including interaction) + Unexplained variation (due to random error)

The reader will notice that this model (technique) enables us to break down the variance of the measured variable into the portions caused by the several factors, varied singly or in combination, and a portion caused by experimental error. More precisely, Crow et al. (1960) have suggested that the ANOVA consists of (1) a partitioning of the *total* sum of squares of deviations from the mean into two or more *component* sums of squares, each of which is associated with a particular factor or with experimental error and (2) a parallel partitioning of the total number of degrees of freedom.

All software statistical packages contain several different procedures that can perform ANOVA. Here, we'll begin with the simplest format of ANOVA, which is the ONE-WAY procedure. It's called one-way ANOVA because cases fall into different groups based on their values for one variable.

Necessary Assumptions

The data must meet two conditions for you to use ANOVA:

1. Each of the groups must be a random sample from a normal population.
2. In the population, the variances in all groups must be equal.

You can visually check these conditions by making a histogram of the data for each group and seeing whether the data are approximately normal. To check whether the groups have the same variance in the population, you can examine the histograms as well as compute the variance for each of the groups and compare them. In practice, ANOVA gives good results even if the normality assumption doesn't quite hold. If the number of observations in each of the groups is fairly similar, the equal-variance assumption is also not too important. The assumption of random samples, however, is always important and cannot be relaxed.

In ANOVA, the observed variability in the sample is divided, or *partitioned,* into two parts: variability of the observations *within* a group (around the group mean) and variability *between* the group means themselves. Why are we talking about variability? Aren't we testing hypotheses about means? Previously, we mentioned that there is a relationship between variability of observations (in the population) and variability of sample means. If you know the standard deviation of the observations, you can estimate how much the sample means should vary. If in your study, you have several different groups—for example, in a typical horse ride with several groups, some individuals will find it comfortable, normal, and rough. If the null hypothesis is true (i.e., if all groups have the same mean in the population), you can estimate how much observed means should vary due to sampling variation alone. If means you actually observe vary more than you'd expect from sampling variation, you have reason to believe that this extra variability is due to the fact that some of the groups don't have the same mean in the population.

Within-Groups Variability

Let's look a little more closely now at the two types of variability we need to consider. *Within-groups* variability is a measure of how much the observations within a group vary. It's simply the variance of the observations within a group in your sample and is used to estimate the variance within a group in the population. (Remember, ANOVA requires the assumption that all of the groups have the same variance in the population.) Since you don't know whether all of the groups have the same mean, you can't just calculate the variance for all of the cases together. You must calculate the variance for each of the group individually and then combine these into an "average" variance.

For example, suppose you have 3 groups of 20 cases each. All 20 cases in the first group have a value of 100, all 20 cases in the second group have a value of 50, and all 20 cases in the third group have a value of 0. Your best guess for the population variance within a group is 0. It appears from your sample that the values of the cases in any particular group don't vary at all. But if you'd computed the variance for all of the cases together, it wouldn't even be close to zero. You'd calculate the overall mean as 50, and cases in the first and third groups would all vary from this overall mean by 50. There would be plenty of variation.

Between-Groups Variability

Remember the earlier discussion that there is a relationship between the variability of the observations in a population and the variability of sample means from that population. If you divide the standard deviation of the observations by the square root of the number of observations, you have an estimate of the standard deviation of the sample means, also known as the *standard error*. So if you know what the standard error of the mean is, you can estimate what the standard deviation of the original observations must be. You just multiply the standard error by the square root of the number of cases to get an estimate of the standard deviation of the observations. You square this to get an estimate of the variance.

Let's see how this insight can give you an estimate of the variance based on *between-groups* variability. You have a sample mean for each of the groups, and you can compute how much these means vary. If the population mean is the same in all three groups, you can use the variability between the sample means (and the sizes of the sample groups) to estimate what the variability of the original observations is. Of course, this estimate depends on whether the population means really are the same in all three groups, which is the null hypothesis. If the null hypothesis is true, the between-groups estimate is correct. However, if the groups have different means in the population, then the between-groups estimate, the estimate of variability based on the group means, will be too large.

Hypothesis Testing

Frequently in practice decisions about populations based on sample information are required. In an earlier section, the utility of sample data was primarily aimed at estimating parameters either by point estimates or interval estimates (confidence intervals). In this section, the focus will be on determining whether processes or populations differ significantly or whether a certain change in a process significantly affected a system. Appendix D shows some typical examples using statistical hypothesis testing.

Statistical Hypothesis/Null Hypothesis

In an attempt to reach decisions, it is often useful to make assumptions or guesses about a process. Such an assumption, which may or may not be true, is called a *statistical hypothesis* and is really just a statement about the population distribution or a process.

Many times, a statistical hypothesis is formulated for the sole purpose of trying to reject or nullify it. For example, if one wishes to decide whether one procedure is better than another, a hypothesis might be formulated that "no difference exists between the two procedures." Such a hypothesis is called a *null hypothesis* and is most frequently denoted by H_0. On the other hand, any hypothesis that is different from a stated hypothesis is termed an *alternative hypothesis*. A hypothesis that is alternative to a null hypothesis is usually denoted by H_1 or H_a.

Tests of Hypothesis

Once a particular hypothesis is stated, it is assumed to be true until proven otherwise. If the results observed in a randomly drawn sample differ markedly from those expected under the hypothesis, the hypothesis would be rejected. The question as to what constitutes a marked difference is the key. It would be great if we could always make correct decisions, but with any decision the potential of being incorrect exists. The best for which one can hope is to control the risk at a reasonably low probability that an error will occur. Two types of errors may exist: Type I and Type II and they are shown in Table 7.2.

In plain terms, a Type I error is incorrectly rejecting a null hypothesis, often termed the manufacturer's (or producers) risk because it may reject a good product. A Type II error is frequently called the consumer's risk, which is incorrectly accepting the null hypothesis—see Figure 7.1.

In testing a hypothesis, the maximum probability with which one would be willing to risk a Type I error is the *level of significance*, designated by α. If a hypothesis was rejected at a stated level of significance of 0.05 ($\alpha = 0.05$), a 0.05 probability or 5% chance exists that the decision could be wrong.

For any tests of hypothesis to be good, they must be designed to minimize errors of decision. This is not a simple matter, since for a given sample size, an attempt to decrease one type of error is most often accompanied by an increase of the other type of error. In general, the only way to reduce both types of error is to increase the sample size, which may not be practical or possible. In practice, one type error may be more "serious" than the other, which might lead to a compromise in favor of a limitation of the more serious type error—see Figure 7.2.

To reduce these thoughts to the real world, control charts might be considered. The process is always assumed to be stable until proven

TABLE 7.2

Error Types

Decision	If the Null Hypothesis Is:	
	True	**False**
Accept null hypothesis	Correct decision $(1-\alpha)^*$	Type II error (β)
Reject null hypothesis	Type I error (α)	Correct decision $(1-\beta)$

* Associated probability.

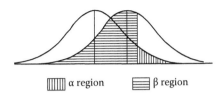

FIGURE 7.1
α and β critical regions.

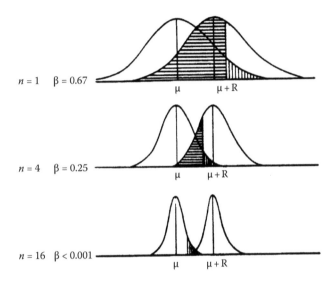

FIGURE 7.2
An example of error compromise given different n- and β-values. Reduction of β risk as a function of sample size; α risk constant at 0.05.

otherwise. Hence, the null hypothesis is essentially that the parameters of the processes are given values (whatever has been established as centerlines, based on 25 or more samples). Each sample statistic that is plotted simply tests that hypothesis. A rejection is called a signal of an out-of-control condition.

Since the control chart activity is confirmatory in nature (i.e., we wish to maintain the stability of nonchanging process variables and are looking for something that we, in general, do not wish to find), strong evidence that the process had changed (out-of-control) should be required before action is taken. In other words, the control chart is set up to be fairly conservative in that very few false alarms of out-of-control conditions should emerge. Such false alarms are in the language of statistics called Type I errors. Confirmatory studies typically are set up with a small risk of a Type I error (e.g., $\alpha \leq .01$) and are characterized as seeking to confirm ideas that have been developed through prior history or research.

TABLE 7.3

Investigative Purpose Considerations regarding Risk

Investigation Type	Typical Example	Knowledge Desired	Assumption	Wish to Avoid	Typical α
Confirmatory	Control chart	Verification of no change in process variables (Stability)	Variables are in correct setting	Type I error False signals of causes	0.01
Exploratory	Designed experiment	Effects of all variables purposely changed	Analysis will identify both strong and weak real effects	Type II error real effect not identified	0.05 or 0.10

With exploratory studies, which describe that setting for many designed experiments, the situation is often the reverse. Variables are purposely changed in the process and some corresponding changes in the measured responses are hopefully the result. The challenge is not to determine whether there has been a change, but rather to determine the nature of the effect(s) of the changes that were purposely staged.

Exploratory studies differ in their approach with confirmatory studies. Instead of attempting to avoid false alarms, or Type I errors, exploratory experimentation is aimed at avoiding real effects not being identified or Type II errors. The exploratory studies purport to find all real effects, including weak ones. Therefore, reconciliation of a larger Type I error (typically $\alpha = 0.05$ or $\alpha = 0.10$) is common—see Table 7.3.

Hypothesis Testing Steps

Several general steps are common when testing a hypothesis. The procedure can be stated as follows:

Step 1: State assumptions

Step 2: State null and alternative hypothesis

Step 3: Choose a significance level

Step 4: Select an appropriate test statistic

Step 5: Determine critical region or value(s)

Step 6: Calculate statistic

Step 7: Determine decision

On the other hand, carrying out the experiment is a process that follows seven steps. They are as follows:

1. *State the objective of the experiment.*
   a. Be aware of the theoretical constraints as well as the practical ones. The first one depends on educational knowledge, and the second one depends on experience, including previous experimental results. Use both for best definition of the problem and optimum results.
   b. Make sure the experimentation is representative of the conditions that the results will apply. Make it as real as possible. Statistically, this can be through randomization or some kind of systematic choice of the experimenter.
   c. Make sure that the hypothesis reflects what you are about to test.
2. *Draw up a preliminary design.*
   a. Make sure that as an experimenter you account for the experimental error, *df*, for estimating the error and if you are dealing with measurement issues (as in a gauge R&R study) make sure that you have more than five distinct categories for proper evaluation. For an appropriate design format see Chapter 10.
   b. Do not forger or underestimate the cost of the experimentation. You should consider both the actual cost and the opportunity cost of the experimentation. Both are important and they may have a direct influence of what kind of experimentation will take place—if any at all.
3. *Review the design with all colleagues and stakeholders.*
   a. Keep notes throughout the experimentation process. They will come very handy when you analyze the results. They will definitely resolve procedural conflicts based on memory. Notes will eliminate the guessing of what happened during the experiment.
   b. Make sure the null hypothesis is correct and understand the consequences of the alternative hypotheses. Depending on the understanding of both, the decisions that are made may differ considerably. Therefore, before you begin the analysis, review both the assumption and the process of the experiment and make sure that all stakeholders are in agreement.
   c. Encourage all stakeholders to contribute in the discussion of identifying factors and levels for the experiment. In fact, time permitting, a good idea is to have a discussion about the anticipated results, the appropriateness of the levels, and any bias that may be inserted in the experiment. If problems are identified in this discussion in any form remove them or plan for a contingency. No excuse not to.
   d. Successful experiments depend heavily on ingenuity in spotting possible sources of bias and eliminating them by a well-organized

plan. *Before the experiment,* we should decide upon and record the procedure that will be followed for every contingency that may arise during the experiment and the conclusion that will be drawn from every possible outcome. One way to minimize, if not completely eliminate, bias is through randomization. Complete randomization is recommended for small experiments on material known to be homogeneous and, because of ease in analysis compared with other designs, for any experiment likely to contain failures that must be omitted from the analysis.

4. *Draw up the final design.*

   a. Discuss and review the design in clear terms one more time to assure that it is doable without confusion or misunderstandings. Review Chapter 10.

   b. Of major concern in this final stage is to make sure that validity and reliability of the experiment have indeed been accounted for. If not make sure you include the conditions that will make the experiment valid and reliable.

5. *Carry out the experiment.*

   a. During the course of the experiment, communication is the key. Make sure all stakeholders are informed of what is going on.

   b. In addition, make sure management is kept informed, so that questions arising from unforeseen experimental conditions or results may be answered. No need to have experimental surprises.

6. *Analyze the data.*

   a. Stick with the plan that you have developed in step 4. Do not change the significance and the *a priori* probability now that you have seen the results. It is very tempting to do that. However, avoid it. If you do not, the results will not be applicable to what you have attempted to study.

7. *Write a report.*
   a. Be clear and concise.

   b. Avoid lengthy verbiage. As much as possible present the data and results in appropriate tables and graphs.

   c. Analyze the results based on the original stated objective of step 1.

   d. Do not be intimidated with negative or insufficient or totally unexpected results. It is not unusual. In fact most experiments need maybe 3 to 4 iterations for good results—in some cases even more. In either case, when you review the outcome of the experimentation make sure the appropriate follow-up is recognized and planned.

e. In case of the ANOVA results, the traditional form of presenting the data is shown in Table 7.4, where k = number of groups; n = samples in the within groups; SS(Tr) = sum of squares for the treatment; SSE = sum of squares error; SST = sum of squares total; MS(Tr) = mean square for the treatment; MSE = mean square error; SSB = sum of squares within treatment (blocks); MSB = mean square within treatment (blocks); T.. = the overall sum of squares. For more information on the formulas and their applications see Appendix F.

$$SST = \sum_{i=1}^{k} \sum_{j=1}^{n} x_{ij}^2 - \frac{1}{kn}.T_{..}^2 \; ; \; SS(Tr) = \frac{1}{n} \sum_{i=1}^{k} T_{i.}^2 - \frac{1}{kn}.T_{..}^2 \; ; SSE = SST -$$

SS(Tr). Table 7.5 shows a typical two-way ANOVA table, where

$$SSE = SST - SS(Tr) - SSB; \; SSB = \frac{1}{k}.\sum_{j=1}^{n} T_{i.j}^2 - \frac{1}{kn}.T_{..}^2 .$$

TABLE 7.4

Analysis of Variance for One-Way Classification

Source of Variation	Sum of Squares	Degrees of Freedom	Mean Square	F-test
Between treatments (treatment)	SS(Tr)	$k-1$	$MS(Tr) = \frac{SS(Tr)}{k-1}$	$\frac{MS(Tr)}{MSE}$
Within treatment (error)	SSE	$K(n-1)$	$MSE = \frac{SSE}{k(n-1)}$	
Total	SST	$nk-1$		

TABLE 7.5

Analysis of Variance for Two-Way Classification

Source of Variation	Sum of Squares	Degrees of Freedom	Mean Square	F-test
Between treatments (treatment)	SS(Tr)	$k-1$	$MS(Tr) = \frac{SS(Tr)}{k-1}$	$\frac{MS(Tr)}{MSE}$
Within treatment (blocks)	SSB	$(n-1)$	$MSB = \frac{SSB}{n-1}$	$\frac{MSB}{MSE}$
Error	SSE	$(n-1)(k-1)$	$MSE = \frac{SSE}{(k-1)(n-1)}$	
Total	SST	$nk-1$		

Calculating the *F*-Ratio

You now have two estimates of the variability in the population: the within-groups mean square and the between-groups mean square. The within-groups mean square is based on how much the observations within each of the groups vary. The between-groups mean square is based on how much the group means vary among themselves. If the null hypothesis is true, the two numbers should be close to each other. If we divide one by the other the ratio should be close to 1.

The statistical test for the null hypothesis that all of the groups have the same mean in the population is based on computing such a ratio. It is called an *F*-statistic. You take the between-groups mean square and divide it by the within-groups mean square as shown in the following formula:

$$F = \text{between-groups mean square/within-groups mean square}$$

Multiple-Comparison Procedures

A significant *F*-value only tells you that the population means are probably not all equal. It doesn't tell you *which* pairs of groups appear to have different means. You can reject the null hypothesis that all means are equal in several different situations. For example, people who find the bike ride comfortable differ in, say, age from people who find the ride rough, but not from people who find the ride normal. Or people who find the bike ride comfortable may differ from both other groups. In most situations, you want to pinpoint exactly where the differences are. To do this, you must use *multiple-comparison procedures.*

Why do you need yet another statistical technique? Why not just calculate *t*-tests for all possible pairs of means? The reason for not using many *t*-tests is that when you make a lot of comparisons involving the same means, the probability that one out of the bunch will turn out to be statistically significant increases. For example, if you have five groups and compare all pairs of means, you're making 10 comparisons. When the null hypothesis is true (i.e., all of the means are equal in the population), the probability that at least one of the 10 observed significance levels will be less than 0.05 is about 0.29. *If you keep looking, even unlikely events will happen.* Therefore, when you're testing statistical significance, the more comparisons you make the more likely it is that you'll find one or more pairs to be statistically different, even if all means are equal in the population.

Multiple-comparison procedures protect you from calling too many differences significant. They adjust for the number of comparisons you are making. The more comparisons you're making, the larger the difference between pairs of means must be for a multiple-comparison procedure to report a significant difference. So, you can get different results from multiple *t*-tests and from multiple-comparison procedures. Differences that the *t*-tests find significant may not be significant based on multiple-comparison procedures. When you use a multiple-comparison procedure, you can be more confident that you're finding true differences.

There are several different procedures that can be used when making multiple comparisons. The procedures differ in how they adjust the observed significance level for the fact that many comparisons are being made. Some require larger differences between pairs of means than others. For further discussion of multiple comparisons, see R. Kirk's (1968).

Interactions

ANOVA allows you to test not only for the individual variables but also for their combinations. This is an important concern. As you've noticed in previous discussions, combinations of variables sometimes have a different effect than you'd expect from variables alone. That effect is important and as experimenters we want to know. In statistical terms, we say that there is an *interaction* effect between variable X and variable Y.

How Can You Test the Null Hypothesis That Several Population Means Are Equal?

ANOVA, analysis of covariance, and regression analysis are specializations of the more general theory of testing linear hypotheses. As such there are many ways one can test the null hypothesis as well as set up the experiments for maximum efficiency and powerful results. In Chapter 10, we address some of these approaches. Here, however, we focus specifically on ANOVA.

- ANOVA can be used to test the null hypothesis that several population means are equal.
- To use ANOVA, your groups must be random samples from normal populations with the same variance.
- In ANOVA, the observed variability in the samples is subdivided into two parts: variability of the observations within a group about the group mean (within-groups variation) and variability of the group means (between-groups variation).
- The *F*-statistic is calculated as the ratio of the between-groups estimate of variance to the within-groups estimate of variance.

- The ANOVA *F*-test does not pinpoint which means are significantly different from each other.
- Multiple-comparison procedures, which protect you against calling too many differences significant, are used to identify pairs of means that appear to be different from each other.

Special Considerations

As we have said many times thus far, there are many ways to conduct an experiment and to chose a particular technique whether it is for sampling or analysis. Within the range of ANOVA, there are many ways that one may set up the experiment. Some examples are shown in Chapter 10. Here, we will attempt to discuss with simplicity and brevity at the same time some of the most common approaches to advanced ANOVA analysis. Obviously, our discussion is focused on familiarizing the reader to use them with confidence with any statistical software program.

Complete Factorial Design for Treatments

The complete factorial design is used for arranging the treatments to be tested call for the use of every combination of the different levels of the factors. For example, in an experiment dealing with plastic extrusion, we have to deal with extrusion rate and die temperature, pressure, material, machine, operator(s), and so on. All these have to be investigated simultaneously, whereas in the classical method all variables but one remain fixed. The advantages of complete multiple factorial design are primarily two: (1) the opportunity to test all possible combinations of factors and levels and (2) the opportunity to test for *interactions* (i.e., the effects produced by crossing two or more factors together. In some cases, the interactions may be more important than the individual factor effect as in the chemical industry and many others).

Complete Randomization for Background Conditions

Complete randomization is recommended for small experiments on material known to be homogeneous and, because of ease in analysis compared with other designs, for any experiment likely to contain failures that must be omitted from the analysis. In essence, a randomized design allows the experimenter to assign the experimental units at random to the treatments since any lack of homogeneity among the level units increase the error mean square and decrease the chances of detecting differences among treatments.

Randomized Block Design for Background Conditions

A randomized block design allocates treatments to experimental units that have first been sorted into homogeneous groups called blocks. In cases where the experiment is very large, we may not have homogeneity, and thereby, we separate our experiment into blocks (which we can randomize) and then compare each of the blocks. This approach helps us avoid an unusual amount of error. The reason why we may use a randomized block in our design is to remove from the experiment a source of variation that is not of interest in order to improve our measurement of the effects that are of interest. In the final analysis, blocking reduces the probability of not detecting differences in effects when differences do exist. The actual mechanics of this approach follows the two-way ANOVA analysis because both the treatment effects and the block effects can be estimated and tested.

Latin Square Design for Background Conditions

Although a Latin square is a simple object to a mathematician, it is multi-faceted to an experimental designer. The same Latin square can be used in many different situations and in many industries. The proliferation of statistical software packages has made the Latin square methodology popular and accessible to many. Remember that an experimental design consists in the allocation of treatments to experimental units; both the set of experimental units and the set of treatments may themselves be structured in some way. Therefore, in a situation when we want to test k treatments under k choices for one background condition and k choices for a second, independent background condition, we can use a Latin square design; for example, a plastic manufacturer wants to compare the tensile strength of plastic sheet made by four different processing methods. The current knowledge lets the manufacturer to assume that there are two main sources of extraneous variation: (a) the technician who mixes the formula and (b) the equipment used in the manufacturing process.

Advantages of the Latin square design are primarily three: (1) The ANOVA is simple, (2) elimination of extraneous variation more often than not leads to a smaller error mean square than a completely randomized design, and (3) if complications arise, there are simple ways to handle them.

The main disadvantage for the Latin square methodology is complexity, if it is done by hand, and its inflexibility. The inflexibility is apparent when the rows, columns, and treatments must be equal and the same number of choices and degrees of freedom is required and determined for each variable. Also, the requirement that the extraneous information be known and the factor to be tested be independent is sometimes a hard requirement to meet.

Other Designs

In addition to the aforementioned designs one has a choice of many others. Typical are the nested and the Greco-Latin designs. They are used with very complicated situations and with uneven samples. To accommodate the unevenness, the general linear model of analysis is used. However, because their applications are somewhat difficult it is beyond the scope of the book. The reader is encouraged to see Peng (1967), Hicks (1982), and Green and Tukey (1960) for more information. Both approaches have become somewhat easier to apply since the advent of high-speed electronic computing machines.

Types of Analysis of Variance

There are two mathematical models to which the ANOVA applies: a model for Type I problems and a model for Type II problems (Crow et al., 1960). In this book, we do not address the more complicated cases, such as mixed or split designs. In Chapter 10, we only provide an overview perspective of the setup for such designs without a complete and thorough discussion.

Type I Problems

These are problems that arise when the ANOVA is mentioned; it is referred to the actual finite number of treatments and the variation is called systematic. Fundamentally, these kinds of problems are selected because they offer the most promise for a fast and practical solution and the analysis is basically that of a comparison of the mean effect of the selected levels. The F-tests are made to see whether the observed differences in mean effects are real or random. If the differences are real, the population constants or parameters (main effects and interactions) may be estimated easily as averages Crow et al. (1960).

Type II Problems

Crow et al. (1960) suggest that these tests for Type II problems are of the same form as those used for Type I problems; however, the objective in Type II problems is the estimation of the component of variance contributed by each factor to the total variance of the variable under study, rather than the estimation of the main effects and interactions at particular levels of each factor, as in Type I problems.

Taguchi Approach to Design of Experiments

The Taguchi approach for quality has become popular over the past 20 years in the United States. Kirkland (1988) reported that Japanese companies (e.g., Sony and Toyota) have practiced elements of the Taguchi approach since the early 1960s. These companies have led the successful invasion into the US electronics and automotive markets, and their success is primarily due to aggressive use of the Taguchi approach in their overall quality.

The methodology used by Taguchi may be used in any product design or manufacturing operation, regardless of volume or type of production or market served. Port (1987) identified the basic tenant of Taguchi's contribution to quality as a design of a product robust enough to achieve high quality despite fluctuations on the production line. This sounds simple; however, the mathematical formulas associated with robust quality are too intensive to be discussed here. For a detailed analysis of the topic see Roy (1990), Ross (1988), and Taguchi (1986, 1987). The goal of this section is to address the basic philosophy of Taguchi and explain some of his ideas as they relate to the quality issue.

On the question of how Taguchi is able to do what he claims to do (i.e., a robust design), the answer is very simple: He does it by consistency throughout. He feels so strongly about consistency of performance that in fact he uses it as the sole definition of quality. Please note that his definition of quality is quite different from that of many people in the field of quality. By the same token, one must admit that the aspect of quality that engineers may affect can be generally measured in terms of performance of the product or process under study.

When Taguchi speaks of a product or process that consistently performs the intended function (at target), it is considered to be of quality. When the performance is consistent, its deviation from the target is minimum. As a consequence, the lower the deviation of performance from the target, the better the quality. The secret of Taguchi's philosophy and approach is to select an appropriate quality characteristics for performance evaluation. For example, if we want to hire an engineer with an excellent command of the language, then we look for one who has a better track record in English. On the other hand, if we are looking for an athlete, we put more emphasis on the candidate's ability and/or performance in sports activity. In selecting the best combination of product or process parameters that are consistent with his definition of quality, Taguchi looks for a combination that produces the least average deviation from the target. In doing this, he neglects the interaction of the variables and this has become one of the strongest arguments against the method (Box, Hunter, and Hunter, 1978).

Taguchi (1987) points out that in all cases interaction is not considered in his book. This is not because there is no interaction, but rather since there can be interaction we perform experiments only on the main effects, having canceled interactions. If the interactions are large, then no assignment will work except experiments on a certain specific combination. Good results are obtained using an orthogonal array if interactions have been omitted. This is because minimizing interaction is not a matter of assignment, but should be handled by techniques of the specific technology and appropriate analysis.

Taguchi uses a statistical quantity that measures deviation from the target, and he calls it *mean square deviation* (MSD). For a robust design of product or process, the combination with the least MSD should be selected. For convenience of analysis and to accommodate a wider range of observations, MSD is transformed into a signal-to-noise (S/N) ratio before analysis.

So far we looked at Taguchi from an overall point of view. Now let us address some specific philosophical issues.

1. An important dimension of the quality of a manufactured product is the total loss generated by the product to society. Barker and Clausing (1984) interpret Taguchi to mean that any product's quality can be defined as the monetary loss that a product imparts on society once it is shipped. The idea of quality as a loss to society is unique in that quality is typically defined in positive terms. Taguchi's definition implies that a product's desirability is inversely related to the amount of societal loss. Wood (1988) believes that in addition to the above, loss to society should include raw materials, energy, and labor consumed during the production of the product. However, Taguchi specifically means losses such as failure to meet ideal performance, and harmful side effects caused by the product. Taguchi has shown this idea in a graphical form by plotting a parabola, which he calls the *loss function*. This concept was used by Kackar (1986) to redefine the aim of quality control as a concept to reduce total societal cost and its function to be discovering and implementing innovative techniques that produce a new savings to society.

2. In a competitive economy, continuous quality improvement and cost reduction are necessary for staying in business. In today's and future competitive market environment, a business must earn a reasonable profit to survive. Profit is a function of reducing manufacturing costs and increasing revenue. Market share may be increased by providing high-quality products at competitive costs. Customers want high quality and low cost. Taguchi believes that the customer is willing to pay a little more for a higher-quality product. He hints

(but does not say explicitly) that quality must have a reference frame of price to be meaningful. It would not be fair to compare the quality of a Ford to a Rolls Royce. The Rolls is perceived to be a much higher-quality car and is sold at a substantially higher price. Yet, because of their price differential, Ford sells a significantly higher volume of vehicles. Taguchi insists that companies determined to stay in business must use high quality and low cost in their business strategy. The quest for increasing quality at decreasing cost must be a never-ending proposition.

3. A continuous quality improvement program includes incessant reduction in the variation of product performance characteristics about their target values. A product's quality cannot be improved unless the associated quality characteristics are quantifiable. Kackar (1986) points out that continuous quality improvement depends on knowledge of the ideal values of these quality characteristics. However, because most products have numerous quality characteristics the most economical procedure to improve a product's quality is to concentrate on its primary or performance characteristics. Performance characteristics are values such as the life span of a tire or the braking distance of an automobile. These are measurable quantities. Taguchi believes that the smaller the performance variation about the target value, the better the resulting quality. In contrast to this, the classical approach has been to have target values, which for all practical purposes are ranges. This has led to the erroneous idea that the quality within specification intervals is equal. This concept has been known as the "goal post philosophy." Taguchi suggests that the target value be defined as the ideal state of the performance characteristic. All performance characteristics may not be measurable on a continuous scale and subjective evaluation may be necessary. Taguchi recommends the use of categorical scale such as poor, fair, good, or whatever fits the product and/or process under study.

4. The customer's loss due to a product's performance variation is often approximately proportional to the square of the deviation of the performance characteristic from its target value. Taguchi proposes that customer's economic loss due to the performance variation can be estimated using a quadratic approximation. The derivation of the loss function is fully developed and discussed by Kackar (1986), Ross (1988), and Taguchi (1986, 1987). For our discussion here, it is only important to recognize that as a performance characteristic moves away from the target value (in either direction), the customer's (or societal) monetary loss increases quadratically.

5. The final quality and cost of a manufactured product is determined to a large extent by the engineering designs of the product and its

manufacturing process. Earley (1989) pointed out that Taguchi's method of experimental design is becoming increasingly popular because they are speedy, dependable, and cost-effective. Taguchi believes that a product's field performance is affected by environmental variables, human variables in operating the product, product deterioration, and manufacturing imperfections. He believes that countermeasures caused by environmental variables and product deterioration can only be built into the product at the design stage. The implication that Taguchi makes here is that the manufacturing cost and manufacturing imperfections in a product are a function of the design of the manufacturing process.

Process control built into the process design will significantly reduce manufacturing imperfections. Process controls are expensive, but are justifiable as long as the loss due to manufacturing imperfections is more than the cost of the controls. Optimally, both manufacturing imperfections and the need for process controls should be reduced.

6. The performance variation of a product (or process) can be reduced by exploiting the nonlinear effects of the product (or process) parameters on the performance characteristics. Taguchi's basic premise of product or process robustness is best acquired by implementing quality control in all the steps of the product development cycle. To do this, a two-pronged approach is utilized. First, off-line quality-control methods are used. These are technical aids for quality and cost control in product and process design. They may include prototype testing, accelerated life tests, sensitivity tests, and reliability tests. These tests are used to improve product quality and manufacturability through evaluation, and specification of parameters should be given in terms of ideal values and corresponding tolerances. Second, online quality control methods are used. These are technical aids for quality and cost control in the actual manufacturing process and/or customer service stage. Taguchi indicates that by stressing off-line quality control in the development process of a product, the potential of ongoing loss associated with a product is dramatically reduced.

   To assure off-line quality, Taguchi introduced a three-step approach:

   a. *System design.* This is the process of developing a prototype design using engineering knowledge. In this stage, the definition of the product or process parameters is established.

   b. *Parameter design.* This is the stage where product and/or process parameters are established for an optimized sensitivity.

   c. *Tolerance design.* This is the scientific determination of the acceptable tolerance around the ideal setting.

7. Statistically planned experiments can be used to identify the settings of product (and process) parameters that reduce performance variation. Taguchi established that statistically planned experiments are essential to successful parameter design. Statistically designed experiments have previously been used in industry; however, Taguchi's contribution brings this experimentation to a new height by providing a quick yet accurate way of determining optimization. Taguchi believes that his form of experimentation is the only method for identifying settings of design parameters while paying attention to costs.

Summary

In this chapter, we addressed some of the preliminary items of concern in conducting an ANOVA. We also discussed some specific items that one should be aware in the process of conducting and reporting the results of an ANOVA analysis. A brief introduction of the Taguchi methodology was also given. In the next chapter, we will expand the discussion of ANOVA to include fitting functions, for example, regression methods and other multivariate techniques.

References

Barker, T. B. and D. P. Clausing. (March 1984). *Quality engineering by design—the Taguchi approach.* Presented at the 40th annual RSQC conference. Detroit, MI.
Box, G. E. P., W. G. Hunter and J. S. Hunter. (1978). *Statistics for experimenters.* John Wiley, New York.
Crow, E. L., F. A. Davis and M. W. Maxfield. (1960). *Statistics manual.* Dover Publications, Inc., New York.
Earley, L. (February 1989). Kanban car design. *Automotive industries.* pp. 8–11.
Green, B. F. and Tuey, J. W. (1960). Complex analyses of variance: general problems. *Psychometrika.* 25. pp. 127–152.
Hicks, C. R. (1982). *Fundamental concepts in the design of experiments.* 3rd ed. Holt, Reinhart and Winston, New York.
Kackar, R. N. (December 1986). Taguchi's philosophy: Analysis and commentary. *Quality progress.* pp. 45–48.
Kirk, R. E. (1968). *Experimental design: procedures for the behavioral sciences.* Brooks/Cole, Belmont, CA.
Kirkland, C. (February 1988). Taguchi methods increase quality and cut costs. *Plastics world.* pp. 42–47.

Peng, K. C. (1967). *The design and analysis of scientific experiments*. Addison-Wesley, Reading, MA.

Port, O. (June 1987). How to make it right the first time. *Business week*. P. 32.

Ross, P. J. (1988). *Taguchi techniques for quality engineering*. McGraw Hill, New York.

Roy, R. (1990). *A primer on the Taguchi method*. Van Nostrand Reinhold, New York.

Taguchi, G. (1986). *Introduction to quality engineering*. Asian Productivity Organization, Available in North America through Kraus International Publications, White Plains, New York.

Taguchi, G. (1987). *System of experimental design*. Vol. 1 and 2. Kraus International Publications, White Plains, NY.

Wood, R. C. (Fall/Winter 1988). The prophets of quality. *Quality Review*. pp. 12–17.

Selected Bibliography

The following sources are old but are considered classic

Anderson, R. L. and T. A. Bancroft. (1952). *Statistical Theory in Research*. New York, McGraw-Hill.

Bartlett, M. S. (1947). The use of transformations. *Biometrics*. 3: 39–52.

Cochran, W. G. (1947). Some Consequences When the Assumptions for the Analysis of Variance Are Not Satisfied. *Biometrics*. 3: 22–38.

Cochran, W. G. and G. M. Cox. (1950). *Experimental designs*. New York, Wiley.

Dixon, W. J. and F. J. Massey, Jr. (1951). *Introduction to Statistical Analysis*. New York, McGraw-Hill.

Eisenhart, C. (1947). The assumptions underlying the analysis of variance. *Biometrics*. 3: 1–21.

Hald, A. (1952). *Statistical Theory with Engineering Applications*. New York, Wiley.

Kempthorne, O. (1952). *The Design and Analysis of Experiments*. New York, Wiley.

Mann, H. B. (1949). *Analysis and Design of Experiments*. New York, Dover.

Mood, A. M. (1950). *Introduction to the Theory of Statistics*. New York, McGraw-Hill.

Snedecor, G. W. (1946). *Statistical Methods*. 4th ed. Ames, Iowa, Iowa State College Press.

Villars, D. S. (1951). *Statistical Design and Analysis of Experiments*. Dubuque, Iowa, William C. Brown Co.

Youden, W. J. (1951). *Statistical Methods for Chemists*. New York, Wiley.

Newer Sources

Bhote, K. R. (1991). *World class quality: Using design of experiments to make it happen*. American Management Association, New York.

Lawson, J. (2010). *Design and analysis of experiments with SAS*. CRC Press, Boca Raton, FL.

Lesic, S. A. (2009). *Applied statistical inference with MINITAB*. CRC Press. Boca Raton, FL.

Montgomery, D. C. (1991). *Design and analysis of experiments*. 3rd ed. Quality Press, Milwaukee, WI.

Morris, M. (2010). *Design of experiments: An introduction based on linear models*. CRC Press, Boca Raton, FL.

Onyiah, L. C. (2009). *Design and analysis of experiments: Classical and regression approaches with SAS*. CRC Press, Boca Raton, FL.

Stamatis, D. H. (2003). *Six sigma and beyond: Design of experiments*. St. Lucie Press, Boca Raton, FL.

8

Fitting Functions

In the previous chapter, we discussed the issue of planning an experiment and introduced the analysis of variance and some of the critical components for understanding it. This chapter introduces an advanced statistical technique to the quality professional. Specifically, the topic of regression and some of its derivative techniques are discussed. In addition, a short discussion is given on the issues and concerns of all statistical techniques. The intent of the discussion is awareness rather than competency in implementing the techniques themselves.

Consequently, readers are encouraged to pursue the individual topics on their own. For more details and understanding, see Stamatis (1997, 2002, 2003); Duncan (1986); Judge et al. (1985); Studenmund and Casidy (1987); Wittink (1988); Gibbons (1985); Cohen and Cohen (1983); Box and Draper (1987); Hanson (1988); Makridakis, Wheelwright and McGee (1983); Kreitzer (1990); Mirer (1988); Thomas (1990); Mullet (1988); and Pedhazur (1982).

Selecting Dependent Variables

The selection of the dependent variable is the variable of interest—the response. In other words, it is the item that we want to find something about. Once we decide as to what the variable of interest is, then we move on to identifying the independent variables that will affect this variable. Generally, this response is denoted by a Y.

Selecting Independent Variables

Often you do not know which independent variables together are good predictors of the dependent variable. You want to eliminate variables that are of little use from your equation so you will have a simple, easy-to-interpret model. You can do this with the assistance of what are called *variable selection methods.* Based on statistical considerations, such as the percent of variance explained by one variable that is not explained by any other variables, some software packages (e.g., SPSS) select a set of variables for inclusion in a regression model. Although such procedures often result in a useful model, the selected model is not necessarily best in any absolute sense. It is up to the experimenter to use both theoretical and experiential knowledge to optimize the identification process for these predictors. Independent variables are usually denoted with X.

Correlation Analysis

Correlation analysis is the simplest of the regression methodologies and by far the most often used statistical analysis in the quality profession; its focus is to define quantifiable relationships between variables. The relationship is always identified with the letter r and is called the correlation coefficient. Whereas the scatter plot shows pictorially the relationship, the correlation analysis defines the relationship with a numerical value. The numerical value varies from $r = -1$ (perfect negative relationship: as one variable increases the other one decreases), to $r = 0$ (no relationship), to $r = +1$ (perfect positive relationship: as one variable increases the other one increases). In using correlation analysis, care must be exercised not to assume that just because there is a relationship between the variables, there is also causation. That is a major error and should be avoided. A reality check may help in avoiding this assumption. Instead of the correlation coefficient, the experimenter should use the coefficient of determination, which, of course, is r^2. The r^2 identifies the true explanation of the relationship. For example, if $r = 0.8$, then the $r^2 = 0.64$, which is much less.

Regression

Regression analysis is the grandfather of all multivariate analytical techniques. It is a methodology that provides many individual techniques to identify variation and relationships. Specifically, what regression analysis usually does is to find an equation that relates a variable of interest (the dependent or criterion variable), to one or more other variables (the independent or predictor variables), of the following form:

$$y = a + b_1 x_1 + b_2 X_2 + b_3 X_3 \ldots b_n X_n + e$$

where

y	= the dependent variable
a	= the y intercept
b_1, b_2, b_3, b_n	= the coefficient of the variables
$X_1, X_2, X_3, ..., X_n$	= the independent variables

The important thing to recognize in regression analysis is that the dependent variable is supposed to be a quantity such as how much, how many, how often, how far, and so on. (If you are using a computer package, be warned! The computer will not tell you if you have defined the variable of interest wrong, it is up to you—the experimenter.)

Most regression models will leave you with an equation that shows only the predictor variables that are statistically significant. One misconception that many people have is that the statistically significant variables are also those that are substantive from a quality perspective. This is not necessarily true. It is up to the experimenter to decide which are which.

Regression analysis is one of the most underutilized statistical tools in the field of quality. It provides a tool offering all of the analysis potential of the analysis of variance (ANOVA), with the added ability of answering important questions that ANOVA is ill-equipped to address.

It is true that there are many statistical techniques for determining the relevance of one measure to another. The strongest of these techniques are analysis of variance, goodness of fit, and regression analysis. Each is capable of answering the same basic question of whether or not variation in one measure can be statistically related to variation in one or more other measures. However, only regression analysis can specify just how the two measures are related. That is, only regression can provide quantitative as well as qualitative information about the relationship.

For example, suppose that you need to address the question of whether or not a particular nonconformance is related to factor X. Suppose further that you use ANOVA and find a statistically significant relationship does exist. That is all! You have gone as far as ANOVA can take you. Regression analysis offers additional information that neither ANOVA nor goodness of fit can provide, namely, how much of factor X will change the specific nonconformity. This information provides an objective basis for the infamous "what if" problems, so central to Lotus-type simulations.

A second advantage of regression analysis is its ready application to graphic imagery. Regression is sometimes referred to as curve or line fitting. The regression output, indeed, yields an equation for a line that can be plotted. The old adage about "a picture is worth a thousand words" holds especially true for regression lines. The image of actual data plotted against predicted values is instantly accessible to even the staunchest statistical cynic. A regression that has been well specified provides its own pictorial justification.

So why would anyone not use regression analysis? There are at least two reasons why regression analysis might not be the first choice of technique for a quality professional and/or researcher. Regression analysis assumes (indeed requires) the error terms to be normally distributed. The normality requirement is sometimes more than a researcher is willing to take on. Nonparametric techniques, which do not make such a restrictive assumption, are better suited to the temperament of these individuals. In general, however, the normality assumption is not outrageous if a sufficiently large sample can be obtained.

The second reason regression might be avoided is paradoxically related to regression's strong suit, quantification of relationships. In some cases one might wish to determine whether or not two measures are significantly related, yet it does not make immediate sense to quantify the relationship between qualitative measures.

To understand regression, there are a few terms that must be understood before undertaking the actual analysis; once understood, they facilitate the understanding of multivariate analysis. The terms are as follows:

- *Multicollinearity.* The degree to which the predictor variables are correlated or redundant. In essence, it is a measure of the extent that two or more variables are telling you the same thing.

- R^2. A measure of the proportion of variance in, say, the amount consumed that is accounted for by the variability in the other measures that are in your final equation. You should not ignore it, but it is probably overemphasized. There are a variety of ways to get to the final equation for your data, but the thing to recognize for now is that if you want to build a relationship between a quantitative variable and one or more (either quantitative or qualitative), regression analysis will probably get you started.

- Adj R^2. The ADJUSTED R^2 is most useful when you have a model with several independent variables. This statistic adjusts the value of R^2 to take into account the fact that a regression model always fits the particular data from which it was developed better than it will fit the population. When only one independent variable is present and the number of cases is reasonably large, the adjusted R^2 will be very close to the unadjusted value. (This statistic is very useful when we check for measurement error.)

Step Regression

Step regression is a methodology that may be used for exploratory analysis because it can give insight into the relationships between the independent variables and the dependent or response variable. There are five ways that the experimenter may want to use step regression. They are as follows:

1. *Forward.* The forward technique begins with no variables in the model. For each of the independent variables, the forward approach calculates F-statistics, reflecting the variable's contribution to the model if it is included. These F-statistics are compared with the preset significant value that is specified in the model. If no F-statistic has a significant level greater than the preset value, the inclusion of new variables stops. If that is not the case, the variable with the largest F-statistic is added and the evaluation process is repeated. Thus, variables are added one by one to the model until no remaining variable produces a significant F-statistic. Once the variable is in the model, it stays.

2. *Stepwise.* This method is a modification of the forward technique and differs in that variables already present in the model do not necessarily stay there. As in the forward selection method, variables are added one by one to the model, and the F-statistic for a variable to be added must be significant at the preset level. After a variable is added, however, the stepwise method looks at all the variables already included in the model and deletes any variable that does not produce an F-statistic significant at the preset level. Only after this check is made and the necessary deletions accomplished can another variable be added to the model. The stepwise process ends when none of the variables outside the model has an F-statistic significant at the preset level, and every variable in the model is significant at the preset level equal to the level or when the variable to be added to the model is one just deleted from it.

3. *Backward.* The backward elimination technique begins by calculating statistics for a model, including all of the independent variables. Then, the variables are deleted from the model one by one until all the variables remaining in the model produce F-statistics significant at the preset level. At each step, the variable showing the smallest contribution to the model is deleted.

4. *Maximum R^2 (MAXR).* As a general rule, this approach of step regression is considered to be superior to the stepwise technique and almost as good as all possible regressions. Unlike the previous three techniques, this method does not settle on a single model; instead, it tries to find the best one-variable model, the best two-variable model, and so on. One of the greatest shortcomings is the fact that the use of MAXR does not guarantee to find the model with the largest R^2 for each size.

5. *Minimum R^2 improvement (MINR).* This method closely resembles the MAXR, but the switch chosen is the one that produces the smallest increase in R^2. Of note is the consideration that for a given

number of variables in the model, MAXR and MINR usually produce the same "best" model, but MINR considers more models of
each size.

In any software program, the experimenter can specify several METHODS
subcommands after a single DEPENDENT subcommand. The methods are
applied one after the other. Once that is completed, now the experimenter is
interested in testing the hypothesis. How can you test hypotheses about the
population regression line, based on the values you obtain in a sample? Here
are the important principles:

- To draw conclusions about the population regression line, you must
  assume that for each value of the independent variable, the distribution of values of the dependent variable is normal, with the same
  variance. The means of these distributions must all fall on a straight
  line.
- The test of the null hypothesis that the slope is zero is a test of
  whether a linear relationship exists between the two variables.
- The confidence interval for the population slope provides you with
  a range of values that, with a designated likelihood, includes the
  population value.
- When a single independent variable is present, the analysis of variance table for the regression is equivalent to the test that the slope
  is zero.

General Linear Model

The general linear model (GLM) procedure uses the method of least squares
to fit general linear models. Among the many options within the GLM family of statistical methods, one may use the following, especially if unbalanced data are present:

- Analysis of variance (especially for unbalanced data)
- Analysis of covariance
- Response surface models
- Weighted regression
- Polynomial regression
- Partial correlation
- Multivariate analysis of variance (MANOVA)
- Repeated measures analysis of variance

Even though the GLM approach and application to data analysis is very complicated, with the use of computer software the applicability of GLM in quality applications is within reach. There are many reasons why one may want to use the GLM model; the following nine are the most important as identified by the SAS Institute in explaining the features of their specific software (SAS, 1985):

1. When more than one dependent variable is specified, GLM can automatically group those variables that have the same pattern of missing values within the data set or within a group. This ensures that the analysis for each dependent variable brings into use all possible observations.

2. GLM allows the specification of any degree of interaction (i.e., crossed effects) and nested effects. It also provides for continuous-by-continuous, continuous-by-class, and continuous-nesting effects.

3. Through the concept of estimability, GLM can provide tests of hypothesis for the effects of a linear model regardless of the number of missing cells or the extent of confounding data. GLM can produce the general form of all estimable functions.

4. GLM can create an output data set containing predicted and residual values from the analysis and all of the original variables.

5. The MANOVA statement allows you to specify both the hypothesis effects and the error effect to use for a multivariate analysis of variance.

6. The REPEAT statement lets you specify effects in the model that represent repeated measurements on the same experimental unit and provides both univariate and multivariate tests of hypotheses.

7. The RANDOM statement allows you to specify random effects in the model; expected mean squares are printed for each Type I, Type II, Type III, Type IV, and contrast mean square used in the analysis.

8. You can use the ESTIMATE statement to specify an **L** vector for estimating a linear function of the parameter **Lb**.

9. You can use the CONTRAST statement to specify a contrast vector or matrix for testing the hypothesis that **Lb** = 0.

Discriminant Analysis

Discriminant analysis is very similar to regression analysis except that here the dependent variable will be a category such as good/bad, or heavy, medium, or light product usage, and so on. The output from a discriminant analysis will be one or more equations that can be used to put an item with a given profile into the appropriate slot.

As with regression, the predictor variables can be a mixed bag of both qualitative and quantitative. Again, the available computer packages will not be of any help in telling you when to use regression analysis and when to use discriminant analysis. A discriminant analysis is a perfect tool for an organization to use when there is a question of which supplier to audit or survey as opposed to those that they need not bother.

As with regression, you need to be concerned with statistical versus substantive significance, multicollinearity, and R^2 (or its equivalent). If used correctly, it is a powerful tool since much quality data are categorical in nature.

Log-Linear Models

Using a cross-classification table and the chi-square statistic, you were able to test whether two variables that have a small number of distinct values are independent. However, what if you wanted to know the effect of additional variables on the relationships that you are examining? You could always make a cross-tabulation table of all of the variables, but this would be very difficult to interpret. You would have hundreds or thousands of cells and most of them would contain few cases, if any.

One way to study the relationships among a set of categorical variables is with log-linear models. With a log-linear model, you try to predict the number of cases in a cell of a cross-tabulation, based on the values of the individual variables and on their combinations. You see whether certain combinations of values are more likely or less likely to occur than others. This tells you about the relationships among the variables.

Logistic Regression

Logistic regression does the same things as regression analysis as far as sorting out the significant predictor variables from the chaff, but the dependent variable is usually a 0–1 type, similar to discriminant analysis. However, rather than the usual regression-type equation as output, a logistic regression gives the user an equation with all of the predicted values constrained to be between 0 and 1.

Most users of logistic regression use it to develop such things as probability of success from concept tests (a very good tool for product development).

If a given respondent gives positive success intent, they are coded as "1" in the input data set; a negative intent yields a "0" for the input. The logistic regression can also be used instead of a discriminant analysis when there are only two categories of interest.

Factor Analysis

There are several different methodologies that wear the guise of factor analysis. Generally, they are all attempting to do the same thing. That is, to find groups, chunks, clumps, or segments of variables that are corrected within the chunk and uncorrelated with those in the other chunks. The chunks are called factors.

Most factor analyses depend on the correlation matrix of all pairs of variables across all of the respondents in the sample. An excellent application of factor analysis is in the areas of customer satisfaction, value engineering, and overall quality definition for an organization; after all, quality is indeed a bundle of things. The question is, however, what makes that bundle, and this is where factor analysis can help. Also, as it is commonly used, factor analysis refers to grouping the variables or items in the current study or the questionnaire together. However, Q-factor refers to putting the respondents together, again by similarity of their answers to a given set of questions. Two of the most troubling spots in any factor analysis are as follows:

1. *Eigenvalue.* Eigenvalues are addressed here from a practical approach and in no way represents a mathematician's and/or a pure statistician's explanation. All that is necessary for the user to know about eigenvalues in a factor analysis is that he or she add up the number of variables that he or she started with and each one is proportional to the amount of variance explained by a given factor. Analysts use eigenvalues to help decide when a factor analysis is a good one and also how many factors they will use in a given analysis. Usually that comes from a visual inspection of the graph output called "spree."

2. *Rotation.* Rotating an initial set of factors can give a result that is much easier to interpret. It is a result of rotation that labels, such as "value sensitive," "quality," "basic characteristics," and so on, are applied to the factor.

Factor analysis should be done with quantitative variables, although it is possible to conduct an experiment with categorical variables. As with most multivariate procedures, that seems to be the bottom line for factor analysis: Does it make sense? If yes, it is a good one; otherwise, it is probably not, irrespective of what the eigenvalues say.

Cluster Analysis

Now the points of interest are respondents, instead of variables. As with factor analysis, there are a number of algorithms around to do cluster analysis. Also, clusters are usually not formed on the basis of correlation coefficients. Cluster analysis usually looks at squared differences between respondents on the actual variables you are using to cluster. If two respondents have a large squared difference (relative to other pairs of respondents), they end up in different clusters. If the squared differences are small, they go into the same cluster. Cluster analysis may be used in conjunction with a quality function deployment in defining customer requirements. For example, if you question sick patients about their symptoms you will undoubtedly have a very long list of complaints.

You may think that you will find as many combinations of symptoms as you have patients. However, if you study the types of symptoms that frequently occur together, you will probably be able to put the patients into groups—those who have respiratory disturbances, those who have gastric problems, and those who have cardiac difficulties. Classifying the patients into groups of similar individuals may be helpful both for determining treatment strategies and for understanding how the body malfunctions.

In statistics, the search for similar groups of objects or people is called *cluster analysis.* By forming clusters of objects and then studying the characteristics the objects share, as well as those in which they differ, you can gain useful insights. For example, cluster analysis has been used to cluster skulls from various archeological digs into the civilizations from which they originated. Cluster analysis is also frequently used in market research to identify groups of people for whom various marketing pitches may be particularly attractive.

Testing Hypothesis about Many Means

Previously, we tested the hypothesis about the equality of population means. We wanted to know whether people find the ride comfortable, normal, or rough. We used the analysis of variance procedure to test hypotheses that more than two population means are equal. We tested whether there was a difference in education among the three excitement groups.

What if we have several interrelated dependent variables, such as education and income, about which we wish to test hypotheses? How can we test hypotheses that *both* education and income do not differ among the three excitement groups in the population?

MANOVA is used to test such hypotheses. Using MANOVA, you can compare four instructional methods based on student achievement levels, satisfaction, anxiety, and long-term retention of the material. Or you compare five new ice-cream flavors based on the amount consumed, a preference rating, and the price people say they would pay.

If the same variable is measured on several different occasions, special "repeated measures" analysis of variance techniques can be used to test hypotheses. These can be thought of as extensions of the simple paired *t* test.

The previous chapters explained some of the more widely used statistical techniques, and this chapter has attempted to give an idea of the more sophisticated methods available. Still others exist. You can use nonparametric procedures that do not require such stringent assumptions about the distributions of variables. You can use procedures for analyzing specialized types of data such as test scores or survival times. There are often many different ways to look at the same problem. No one way is best for every problem; each view tells you something new.

Conjoint Analysis

Marketing professionals have used conjoint analysis for more than 20 years to focus on the relationship between price differentiation and customer preferences. Quality professionals can and should use the conjoint analysis especially in the areas of supplier selection and customer satisfaction, as part of the QFD approach and product development. Whereas it is beyond the scope of this book to address the step-by-step mechanics and methodology of conjoint analysis, in this section we are going to give some potential applications in the field of quality. The reader is encouraged to see Sheth (1977) and any advanced marketing research book for a detailed discussion.

Conjoint analysis, a sophisticated form of trade-off analysis, provides useful results that are easy for managers to embrace and understand. It aims for greater realism, grounds attributes in concrete descriptions, and results in greater discrimination between attribute importance. In other words, it creates a more realistic context. Some of the potential applications are as follows:

1. Design or modifying products and services. One needs to develop suitable mathematical formulations that translate subjective evaluations of new product or service concepts into objectives (e.g., short-term and long-term profits, or rate of return) of the firm. Such formulations will then enable the determination of optimum new product subject to several technological, economic, social, or other constraints. Specification of the objectives as well as constraints can be quite complex, particularly for the case of public goods or services. For initial formulations of this problem see Rao and Soutar (1975).

2. Comparing the importance and value of technical product features. Engineers tend to fall in love with innovative ideas. Conjoint analysis, on the other hand, places the importance and value of specific features in perspective. A very strong application in this area may be to study the balance between precision performance of a product and durability and/or longevity.

Multiple Regression

A special class of statistical techniques, called *multivariate methods*, is used for studying the relationships among several interrelated variables. The goals of such multivariate analyses may be quite different from that of a univariate analysis, but they share many common features. Here, we will take a look at some of the most popular ones.

You can use *multiple linear regression* analysis to study the relationship between a single dependent variable and several independent variables of the form.

$$Y = a + B_1X_1 + B_2X_2 + B_3X_3 \text{ ... } B_nX_n + \text{error}$$

The model looks like the regression model we have seen earlier. The difference is that you now have several variables on the independent variable side of the model. The independent variables are indicated by $X_1, X_2, X_3, ..., X_n$ and the coefficients by $B_1, B_2,$ and B_3. As before, the method of least squares can be used to estimate all of the coefficients.

Perhaps the most important issue with multiple linear regression is that because the variables are measured in different units, you cannot just compare the magnitudes of the coefficients to one another. Because of this characteristic, the experimenter must standardize the variables in some fashion. That standardization takes place with a statistic called BETA (β) and contains the regression coefficients when all variables are standardized to a mean of 0 and a standard deviation of 1. Of course, just like before we still use the significance test for the test of the null hypothesis that the value of a coefficient is zero in the population. You can see that the null hypothesis can be rejected for all of the variables.

When you build a model with several interrelated independent variables, it is not easy to determine how much each variable contributes to the model. You cannot just look at the coefficients and say *this* is an important variable for predicting the dependent variable and *this* one is not. The contributions of the variables are "shared." The goodness-of-fit statistics we considered for a regression model with one independent variable can easily be extended to a model with multiple independent variables.

Partial Regression

Partial regression is a recent technique that generalizes and combines features from principal component analysis and multiple regression. It is particularly useful when we need to predict a set of dependent variables from a (very) large set of independent variables (i.e., predictors).

The goal of partial regression is to predict **Y** from **X** and to describe their common structure. When **Y** is a vector and **X** is full rank, this goal could be accomplished using ordinary multiple regression. When the number of predictors is large compared to the number of observations, **X** is likely to be singular and the regression approach is no longer feasible (i.e., because of multicollinearity). Several approaches have been developed to cope with this problem. One approach is to eliminate some predictors (e.g., using stepwise methods); another one, called principal component regression, is to perform a principal component analysis (PCA) of the **X** matrix and then use the principal components of **X** as regressors on **Y**. The orthogonality of the principal components eliminates the multicolinearity problem. But the problem of choosing an optimum subset of predictors remains. A possible strategy is to keep only a few of the first components. But they are chosen to explain **X** rather than Y, and so, nothing guarantees that the principal components, which "explain" **X**, are relevant for **Y**.

By contrast, partial regression finds components from **X** that are also relevant for **Y**. Specifically, partial regression searches for a set of components (called latent vectors) that performs a simultaneous decomposition of **X** and **Y** with the constraint that these components explain as much as possible of the covariance between **X** and **Y**. This step generalizes PCA. It is followed by a regression step where the decomposition of **X** is used to predict **Y**.

After Regression

When the experiment or study is complete, the experimenter should evaluate or at least review the following:

Residuals

When you begin studying the relationship between two variables you usually do not know whether the assumptions needed for regression analysis are satisfied. You do not know whether a linear relationship exists between the two variables, much less whether the distribution of the dependent variable is normal and has the same variance for all values of the independent variable. One of the goals of regression analysis is to check whether the required assumptions of linearity, normality, and constant variance are met. To do this, we do an analysis of residuals.

A quantity called the *residual* plays a very important role when you are fitting models to data. You can think of a residual as what is left over after a model is fit. In a linear regression, the residual is the difference between the observed and predicted values of the dependent variable. If a person has 12 years of education and your model predicts 9, the residual for the case is 12 − 9 = 3. You have 3 years of education left over (not explained by the model).

By looking at the residual for each case you can see how well a model fits. If a model fits the data perfectly, all of the residuals are zero. Cases for which the model does not fit well have large residuals. Obviously, you can use the REGRESSION procedure to calculate the residuals for all of the cases.

Judging the Size of the Residuals

How can you tell whether a residual is big or small? If I tell you that a case has a residual of 500, can you say whether the model gives a reasonably good prediction for the case? On first thought, 500 seems to be a pretty large number—an indication that a model does not fit that case. However, if you are predicting income in dollars, a residual of 500 may not be all that large. Predicting a person's income to the nearest 500 dollars is pretty good. On the other hand, if you're predicting years of education, a residual of 500 should send you searching for a new and improved model. One way to modify the residuals so that they would be easier to interpret is to standardize them. That is, divide each residual by an estimate of its standard deviation.

How come you are only dividing the residual by its standard deviation? Why are you not first subtracting off the mean, as you did before, when computing standardized values? You do not need to subtract the mean of the residuals before dividing by the standard deviation because the mean of the residuals is zero. If you add up all of the residuals, you will find that their sum, and therefore their mean, is zero. That is always true for a regression model that includes a constant.

For most cases, the standardized residuals range in value from −2 to +2 and they are identified as *ZRESID. (Remember that in a normal distribution with a mean of 0 and a standard deviation of 1, about 95% of the cases fall between +2 and −2.) Whenever you see a standardized residual larger than +2 or smaller than −2, you should examine the case to see whether you can find some explanation for why the model does not fit.

Checking Assumptions with Residuals

Residuals are the primary tools for checking whether the assumptions necessary for linear regression appear to be violated. We can draw histograms of the residuals, plot them against the observed and predicted values, recompute them excluding certain cases, and manipulate them in other ways. By examining the resulting plots and statistics, we can learn much about how appropriate the regression model is for a particular data set.

Looking for Outliers

If you have a large number of cases, you may not want to look at the values of the residuals for all of them. Instead you may want to look only at the cases with "large" residuals. Such cases are called *outliers*. This is easy to do with the REGRESSION procedure. Just leave off the keyword ALL on the CASEWISE subcommand. In most software applications, you will find that if you do not tell the program to print all cases, it prints only those whose standardized residuals are greater than 3 or less than –3. By looking at the characteristics of the outliers, you can see situations where the model does not work well.

Normality

If the relationship is linear and the dependent variable is normally distributed for each value of the independent variable (in the population), then the distribution of the residuals should also be approximately normal. A simple histogram can demonstrate this.

On the other hand, when the distribution of residuals does not appear to be normal, you can sometimes transform the data to make them appear more normal. When you "transform" a variable, you change its values by taking square roots, or logarithms, or some other mathematical function of the data. If the distribution of residuals is not symmetric but has a tail in the positive direction, it is sometimes helpful to take logs of the dependent variable. If the tail is in the negative direction and all data values are positive, taking the square root of the data may be helpful.

The distribution of your residuals may not appear to be normal for several reasons, besides a population in which the distributions are not normal. If you have a variance that is not constant for different values of the independent variable, or if you simply have a small number of residuals, your histogram may also appear not to be normal. So it is possible that after you have remedied some of these problems, the distribution of residuals may look more normal. To check whether the variance appears to be constant, you can plot the residuals against the predicted values and also against the values of the independent variable.

You may find some common transformations useful when the variance does not appear to be constant. If the variance increases linearly with the values of the independent variable and all values of the dependent variable are positive, take the square root of the dependent variable. If the standard deviation increases linearly with values of the independent variable, try taking logs of the data.

Prediction and Coefficient Standardization

Generally, b is used as a symbol for the statistic and β as a symbol for the parameter. There is, however, another way in which these symbols are frequently used: b is the unstandardized regression coefficient and β is the standardized regression coefficient. Unfortunately, there is no consistency in the use of symbols. For example, some authors use b^* as the symbol for the standardized regression coefficient, others use $\hat{\beta}$ as the symbol for the estimator of β (the unstandardized coefficient) and $\hat{\beta}$ *as the symbol for the standardized coefficient. To add to the confusion, it will be recalled that β is also used as a symbol for Type II error and, in the Taguchi methodology for calculating the signal-to-noise ratio, it will be identified as the slope of the ideal function (or the value for sensitivity). While the use of the different symbols is meant to avoid confusion, it is believed that they are unnecessarily cumbersome and may, therefore, result in even greater confusion. Consequently, in subsequent discussions in this work, b is used as the symbol for the sample's unstandardized regression coefficient and β as the symbol for the sample's standardized coefficient. Another way to think of the difference is that b is for explaining samples and β is to predict populations.

When raw scores are used, bs are calculated and applied to the Xs (raw scores) in the regression equation. If, however, one were to first standardize the scores for the Y and the Xs (i.e., convert them to z scores), βs would be calculated and applied to zs in the regression equation. For simple regression, the equation in which standard scores are used is

$$z'_y = \beta z_x$$

where z'_y = predicted standard score of Y, β = standardized regression coefficient, and z_x = standard score of X. As in the case of b, β is interpreted as the expected change in Y associated with a unit change in X. But, because the standard deviation of z scores is equal to 1.00, a unit change in X, when it has been standardized, refers to a change of 1 standard deviation in X. With one dependent variable, the formula for the calculation of β is

$$\beta = \frac{\sum z_x z_y}{\sum z_x^2}$$

The calculation for a, the *intercept*, and b, the *coefficient*, is

$$a = \frac{\sum(y)(\sum x^2) - (\sum(y)(\sum xy))}{n\sum(x^2) - \sum(x)^2}$$

$$b = \frac{n\sum xy - (\sum x)(\sum y)}{n(\sum x^2) - (\sum x)^2}$$

The reader should note that there is a similarity between the β and b. Whereas sum of cross products and sum of squares of standard scores are used in the former, the latter requires the deviation sum of cross products and sum of squares. It is, however, not necessary to carry out the β calculations because b and β are related as follows:

$$\beta = b\frac{S_x}{S_y} \text{ and } b = \frac{S_y}{S_x}$$

where β = standardized regression coefficient, b = unstandardized regression coefficient, and S_x, S_y = standard deviations of X and Y, respectively. Substituting the b formula and the above formulas for the standard deviation of X and Y, we obtain:

$$\beta = b\frac{S_x}{S_y} = \frac{\Sigma xy\sqrt{\Sigma x^2}\sqrt{N-1}}{\Sigma x^2\sqrt{N-1}\sqrt{\Sigma y^2}} = \frac{\Sigma xy}{\sqrt{\Sigma x^2}\sqrt{\Sigma y^2}} = r_{xy}$$

Note that, with one independent variable, $\beta = r_{xy}$. Note also that, when using standard scores, the intercept, a, is 0. The reason for this is readily seen when it is recalled that the mean of z scores is 0. Therefore, $a = \bar{Y} - \beta \bar{X} = 0$.

For two independent variables, X_1, and X_2, the regression equation with standard scores is:

$$z'_y = \beta_1 z_1 + \beta_2 z_2$$

where β_1 and β_2 are the standardized regression coefficients, and z_1 and z_2 are standard scores on X_1 and X_2, respectively. The formulas for calculating the βs when two independent variables are used are:

$$\beta_1 = \frac{r_{y1} - r_{y2}r_{12}}{1 - r_{12}^2} \text{ and } \beta_2 = \frac{r_{y2} - r_{y1}r_{12}}{1 - r_{12}^2}$$

Note that when the independent variables are not correlated, then $r_{12} = 0$, $\beta_1 = r_{y1}$, and $\beta_2 = r_{y2}$, as in the case in simple linear regression. This holds true for any number of independent variables. When there is no correlation among the independent variables, β for a given independent variable is equal to the product–moment correlation coefficient (r) of that variable with the dependent variable.

Special note on weights: The magnitude of the b is affected, in part, by the scale of measurement that is being used to measure the variable with which the b is associated. Assume, for example, a simple linear regression in which X is the length of objects measured in feet. Suppose that, instead of using feet, one were to express X in inches. The nature of the regression of Y on X will, of course, not change, nor will the test of significance of the b. The magnitude of the b, however, will change drastically. In the present case, the b associated with X when measured in inches will be 1/12 of the b obtained when X is measured in feet. This should alert you to two things:

1. A relatively large b may be neither substantively meaningful nor statistically significant, whereas a relatively small b may be both meaningful and statistically significant.
2. In multiple regression analysis, one should not compare the magnitude of the bs associated with different Xs when attempting to determine the relative importance of variables.

Let us assume that, in a numerical example both bs were found to be statistically significant. It may not be appropriate to compare the magnitude of the bs because they could be based on different scales of measurement. Incidentally, because the b is affected by the scale being used, it is necessary to calculate bs to several decimal places. For a given scale, the b may be 0.0003 and yet be substantively meaningful and statistically significant. Had one solved to only two decimal places, the b would have been declared to be 0. In general, it is recommended that you carry out the calculations of regression analysis to as many decimal places as feasible. Further rounding can be done at the end of the calculations.

Because of the incomparability of bs, experimenters who wish to speak of relative importance of variables resort to comparisons among βs, as they are based on standard scores.

Confidence Intervals for Regression Coefficients

The sample values for the slope and intercept are our best guesses for the population values. However, we know it is unlikely that they are exactly on target. As we have discussed before, it is possible to calculate a confidence interval for the population value. A confidence interval is a range of values that, with a designated likelihood, contains the unknown population value. To obtain 95% confidence intervals for the slope and intercept using the SPSS/PC software, we identify the command REGRESSION, but we must add an additional specification to the command. It is called the STATISTICS subcommand, and it tells the system what values we want to see printed. That is all. The output is going to give us not only the line of the regression but also the confidence intervals.

Remember what 95% confidence means: If we draw repeated samples from a population, under the same conditions, and compute 95% confidence intervals for the slope and intercept, 95% of these intervals should include the unknown population values for the slope and intercept. Of course, since the true population values are not known, it is not possible to tell whether any particular interval contains the population values. Quite often, neither the confidence interval for the slope nor the one for the intercept contains the value zero. An interval will only include zero if you cannot reject the null hypothesis that the slope or intercept is zero, at an observed significance level of 0.05 or less.

Linearity

To see whether it is appropriate to assume a linear relationship, you should always plot the dependent variable against the independent variable. If the points do not seem to cluster around a straight line, you should not fit a linear regression model.

Another way to see whether a relationship is linear is to look at the plots of the residuals against the predicted values and the residuals against the values of the independent variable. If you see any type of pattern to the residuals, that is, if they do not fall in a horizontal band, you have reason to suspect that the relationship is not linear.

Sometimes when the relationship between two variables does not appear to be linear, it is possible to transform the variables and make it linear. Then you can study the relationship between the transformed variables using linear regression.

It may seem that when you transform the data, you are cheating or at least distorting the picture. But this is not the case. All that transforming a variable does is change the scale on which it is measured. Instead of saying that a linear relationship exists between work experience and salary, you say that it exists between work experience and the log of salary. It is much easier to build models for relationships that are linear than for those that are not. That is why transforming variables is often a convenient tactic.

How do you decide what transformation to use? Sometimes you might know what the mathematical formula is that relates two variables. In that case, you can use mathematics to figure out what transformation you need. This situation happens more often in engineering or the physical or biological sciences than in the social sciences. If the true model is not known, you choose a transformation by looking at the plot of the data. Often, a relationship appears to be nearly linear for part of the data but is curved for the rest. The log transformation is useful for "straightening out" such a relationship. Sometimes taking the square root of the dependent variable may also straighten a curved relationship. These are two of the most common transformations, but others can be used.

When you try to make a relationship linear, you can transform the independent variable, the dependent variable, or both. If you transform only the independent variable you are not changing the distribution of the dependent variable. If it was normally distributed with a constant variance for each value of the independent variable, it remains unchanged. However, if you transform the dependent variable, you change its distribution. For example, if you take logs of the dependent variable, then the log of the dependent variable—not the original dependent variable—must be normally distributed with a constant variance. In other words, the regression assumptions must hold for the variables you actually use in the regression equation. Another assumption that we made was that all observations are independent (The same person is not included in the data twice on separate occasions. One person's values do not influence the others'.) When data are collected in sequence, it is possible to check this assumption. You should plot the residuals against the sequence variable. If you see any kind of pattern, you should be concerned.

Finally, it is important to examine the data for violation of the assumptions since significance levels, confidence intervals, and other regression tests are sensitive to certain types of violations and cannot be interpreted in the usual fashion if serious departures exist. If you carefully examine the residuals, you will have an idea of what sorts of problems might exist in your data. Transformations provide you with an opportunity to try to remedy some of the problems. You can then be more confident that the regression model is appropriate for your data.

How can you tell whether the assumptions necessary for a regression analysis appear to be violated? Here are the key points to remember:

- A residual is the difference between the observed value of the dependent variable and the value predicted by the regression model.

- To check the assumption of normality, make a histogram of the residuals. It should look approximately normal.

- To check the assumption of constant variance, plot the residuals against the predicted values and against the values of the independent variable. There should be no relationship between the residuals and either of these two variables. If you note a pattern in the plots, you have reason to suspect that the assumption of constant variance is violated.

- To check whether the relationship between the two variables is linear, plot the two variables. If the points do not cluster about a straight line, you have reason to believe that the relationship is not linear.

- If any of the assumptions appear to be violated, transforming the data may help. The choice of the transformation depends on which assumption is violated and in what way.

Summary

In this chapter, we have discussed some advanced techniques of fitting functions such as regression and multivariate analysis. We also addressed some of the common steps that any experimenter should follow after such an analysis. The next chapter focuses on some basic yet important sampling techniques.

━━━━━━━━━

References

Box, G. E. P. and Draper, N. R. (1987). *Empirical model-building and response surfaces.* John Wiley and Sons. New York.

Cohen, J. and Cohen, P. (1983). *Applied regression/correlation analysis for the behavior sciences.* 2nd ed. Lawrence Erlbaum Associates. Hillsdale, NJ.

Cox, D. R. and Snell, E. J. (1981). *Applied statistics: Principles and examples.* Chapman. London, England

Duncan, A. J. (1986). *Quality control and industrial statistics.* 5th ed. Irwin. Homewood, IL.

Fuller, W. A. (1987). *Measurement error models.* John Wiley. New York.

Gibbons, J. D. (1985). *Nonparametric statistical inference.* 2nd ed., rev. and expanded. Marcel Dekker. New York.

Hanson, R. (1988). Factor analysis: A useful tool but not a panacea. *Quirk's Marketing Research Review.* May: 20–22, 33.

Hays, W. L. (1981). *Statistics.* 3rd ed. Holt, Rinehart Winston. New York.

Judge, G. G., Griffiths, W. E., Hill, R. C., Lutkepohl, H., and Lee, T. (1985). *The theory and practice of econometrics.* 2nd ed. John Wiley and Sons. New York.

Kreitzer, J. L. (1990). To progress you must first regress. *Quirk's Marketing Research Review.* Jan.: 8–9, 22–26.

Lehmann, E. L. (1986). *Testing statistical hypotheses.* 2nd ed. John Wiley and Sons. New York.

Makridakis, S., Wheelwright, S. C., and McGee, V. E. (1983). *Forecasting: Methods and applications.* 2nd ed. John Wiley and Sons. New York.

Milliken, G. A. and Johnson, D. E. (1984). *Analysis of messy data.* Van Nostrand Reinhold Co. New York.

Mirer, T. W. (1988). *Econometric statistics and econometrics.* 2nd ed. Macmillian Publishing Co. New York.

Mullet, G. (1988). Multivariate analysis—some vocabulary. *Quirk's Marketing Research Review.* Mar.: 18, 21, 33.

Pedhazur, E. J. (1982). *Multiple regression in behavioral research.* 2nd ed. Holt, Rinehart and Winston. New York.

Rao, V. R. and Soutat, G. N. (1975). Subjective evaluations for product design decisions. *Decision Sciences.* 6: 120–134.

Rousseeuw, P. J. and Leroy, A. M. (1987). *Robust regression and outlier detection.* John Wiley and Sons. New York.

SAS Institute. (1985). The GLM procedure. In: *SAS User's Guide: Statistics.* Version 5 Edition. Cary. North Carolina.

Sheth, J. N., Ed. (1977). *Multivariate methods for market and survey research.* American Marketing Association. Chicago.

Snell, E. J. (1987). *Applied statistics: A handbook of BMDP analyses.* Chapman and Hall. New York.

Stamatis, D. H. (2003). *Six sigma and beyond: Statistics and probability*. St. Lucie. Boca Raton. FL.

Stamatis, D. H. (2002). *Six sigma and beyond: Design of experiments*. St. Lucie. Boca Raton. FL.

Stamatis, D. H. (1997). *TQM engineering handbook*. Marcel Dekker. New York.

Studenmund, A. H. and Casidy, H. J. (1987). *Using econometrics: A practical guide*. Little Brown and Company. Boston.

Thomas, T. (1990). New findings with old data using cluster analysis. *Quirk's Marketing Research Review*. Feb.: 9–18, 36.

Wittink, D. R. (1988). *The application of regression analysis*. Allyn and Bacon. Boston.

Selected Bibliography

Dobson, A. J. and A. C. Barnett. (2008). *An introduction to generalized linear models*. 3rd ed. CRC Press. Boca Raton, FL.

Hilbe, J. M. (2009). *Logistic regression models*. CRC Press. Boca Raton, FL.

Khuri, A. I. (2010). *Linear model methodology*. CRC Press. Boca Raton, FL.

Madsen, H. and P. Thyregod. (2010). *Introduction to general and generalized linear models*. CRC Press. Boca Raton, FL.

Monahan, J. F. (2008). *A primer on linear models*. CRC Press. Boca Raton, FL.

Panik, M. (2009). *Regresion modeling: Methods, theory and computation with SAS*. CRC Press. Boca Raton, FL.

Raghavarao, D., J. B. Wiley and P. Chitturi. (2010). *Choice-based conjoint analysis: Models and designs*. CRC Press. Boca Raton, FL.

9

Typical Sampling Techniques

In the previous chapter, we discussed how data can fit into mathematical presentations such as regression analysis to make decisions easier and quantifiable. In this chapter, we summarize some key items of understanding sampling techniques. The focus is to make the reader understand that it all starts with sampling and if the sampling is not appropriate and applicable, the results will not be accurate or reflective of what is going on. There are many ways that the sample may be selected; however, here we will discuss the lot acceptance sampling plan (LASP), which is a sampling scheme and a set of rules for making decisions. The decision, based on counting the number of defects in a sample, can be to accept the lot, reject the lot, or even for multiple or sequential sampling schemes, to take another sample and then repeat the decision process.

In any sampling technique we also must be aware of the issue of whether there are any replacement items in the lot after inspection or not. Depending on the situation two distinct experiments must be followed: (a) the hypergeometric experiment and (b) the binomial experiment.

A *hypergeometric experiment* is a statistical experiment that has the following properties:

- A sample of size n is randomly selected without replacement from a population of N items.
- In the population, k items can be classified as successes, and $N - k$ items can be classified as failures.

Consider the following statistical experiment. You have an urn of 10 marbles, 5 red and 5 green. You randomly select two marbles without replacement and count the number of red marbles you have selected. This would be a hypergeometric experiment.

A *hypergeometric random variable* then is the number of successes that result from a hypergeometric experiment. The probability distribution of a hypergeometric random variable is called a *hypergeometric distribution*. With the hypergeometric distribution the assumptions are as follows:

- The mean of the distribution is equal to $n * k/N$.
- The variance is $n * k * (N - k) * (N - n)/[N^2 * (N - 1)]$.

In contrast to this, there is also a binomial experiment. A binomial experiment requires that the probability of success be constant on every trial. In other words, the selection of the sample is with replacement. With the above marble experiment, the probability of a success changes on every trial. In the beginning, the probability of selecting a red marble is 5/10. If you select a red marble on the first trial, the probability of selecting a red marble on the second trial is 4/9. And if you select a green marble on the first trial, the probability of selecting a red marble on the second trial is 5/9. Note further that if you selected the marbles with replacement, the probability of success would not change. It would be 5/10 on every trial.

Notation and Formula for the Hypergeometric Distribution

The following notation is helpful, when we talk about hypergeometric distributions and hypergeometric probability:

- N: The number of items in the population.
- k: The number of items in the population that are classified as successes.
- n: The number of items in the sample.
- x: The number of items in the sample that are classified as successes.
- $_kC_x$: The number of combinations of k things, taken x at a time.
- $h(x; N, n, k)$: *hypergeometric probability* — the probability that an n-trial hypergeometric experiment results in exactly x successes, when the population consists of N items, k of which are classified as successes.

Given then x, n, N, and k, we can compute the hypergeometric probability based on the following formula. Suppose a population consists of N items, k of which are successes, and a random sample drawn from that population consists of n items, x of which are successes. Then the formula is

$$h(x; N, n, k) = [_kC_x]\,[_{N-k}C_{n-x}]/[_NC_n]$$

Example 1
Suppose we randomly select five cards without replacement from an ordinary deck of playing cards. What is the probability of getting exactly two red cards (i.e., hearts or diamonds)?

Solution
This is a hypergeometric experiment in which we know the following:

- $N = 52$, since there are 52 cards in a deck.
- $k = 26$, since there are 26 red cards in a deck.

- $n = 5$, since we randomly select five cards from the deck.
- $x = 2$, since two of the cards we select are red.

We substitute these values into the hypergeometric formula as follows:

$$h(x; N, n, k) = [{}_kC_x] [{}_{N-k}C_{n-x}]/[{}_NC_n]$$

$$h(2; 52, 5, 26) = [{}_{26}C_2] [{}_{26}C_3]/[{}_{52}C_5]$$

$$h(2; 52, 5, 26) = [325] [2{,}600]/[2{,}598{,}960] = 0.32513$$

Thus, the probability of randomly selecting two red cards is 0.32513.

A *cumulative hypergeometric probability*, on the other hand, refers to the probability that the hypergeometric random variable is greater than or equal to some specified lower limit and less than or equal to some specified upper limit.

For example, suppose we randomly select five cards from an ordinary deck of playing cards. We might be interested in the cumulative hypergeometric probability of obtaining two or fewer hearts. This would be the probability of obtaining zero hearts plus the probability of obtaining one heart plus the probability of obtaining two hearts, as shown in the example below.

Example 2

Suppose we select five cards from an ordinary deck of playing cards. What is the probability of obtaining two or fewer hearts?

Solution

This is a hypergeometric experiment in which we know the following:

- $N = 52$, since there are 52 cards in a deck.
- $k = 13$, since there are 13 hearts in a deck.
- $n = 5$, since we randomly select five cards from the deck.
- $x = 0$ to 2, since our selection includes zero, one, or two hearts.

We substitute these values into the hypergeometric formula as follows:

$$h(x \le x; N, n, k) = h(x \le 2; 52, 5, 13)$$

$$h(x \le 2; 52, 5, 13) = h(x = 0; 52, 5, 13) + h(x = 1; 52, 5, 13) + h(x = 2; 52, 5, 13)$$

$$h(x \le 2; 52, 5, 13) = [({}_{13}C_0) ({}_{39}C_5)/({}_{52}C_5)] + [({}_{13}C_1) ({}_{39}C_4)/({}_{52}C_5)] + [({}_{13}C_2) ({}_{39}C_3)/({}_{52}C_5)]$$

$$h(x \le 2; 52, 5, 13) = [(1)(575{,}757)/(2{,}598{,}960)] + [(13)(82{,}251)/(270{,}725)] + [(78)(9{,}139)/(22{,}100)]$$

$$h(x \le 2; 52, 5, 13) = [0.2215] + [0.4114] + [0.2743]$$

$$h(x \le 2; 52, 5, 13) = 0.9072$$

Thus, the probability of randomly selecting at most two hearts is 0.9072.

Binomial Experiment

A binomial experiment (also known as a *Bernoulli trial*) is a statistical experiment that has the following properties:

- The experiment consists of n repeated trials.
- Each trial can result in just two possible outcomes. We call one of these outcomes a success and the other, a failure.
- The probability of success, denoted by p, is the same on every trial.
- The trials are independent; that is, the outcome on one trial does not affect the outcome on other trials.

Here is an example of a binomial experiment. You flip a coin two times and count the number of times the coin lands on heads. This is a binomial experiment because:

- The experiment consists of repeated trials. We flip a coin two times.
- Each trial can result in just two possible outcomes—heads or tails.
- The probability of success is constant—0.5 on every trial.
- The trials are independent; that is, getting heads on one trial does not affect whether we get heads on other trials.

Notation and Formula

The following notation is helpful when we talk about binomial probability.

- x: The number of successes that result from the binomial experiment.
- n: The number of trials in the binomial experiment.
- P: The probability of success on an individual trial.
- Q: The probability of failure on an individual trial. (This is equal to $1 - P$.)
- $b(x; n, P)$: Binomial probability – the probability that an n-trial binomial experiment results in *exactly* x successes, when the probability of success on an individual trial is P.
- $_nC_r$: The number of combinations of n things, taken r at a time.

Therefore, given x, n, and P, we can compute the binomial probability based on the following formula: Suppose a binomial experiment (without replacement) consists of n trials and results in x successes. If the probability of success on an individual trial is P, then the binomial probability formula is

$$b(x; n, P) = {}_nC_x * P^x * (1 - P)^{n-x}$$

Binomial Distribution

A *binomial random variable* is the number of successes x in n repeated trials of a binomial experiment. The probability distribution of a binomial random variable is called a *binomial distribution* (also known as a Bernoulli distribution).

Suppose we flip a coin two times and count the number of heads (successes). The binomial random variable is the number of heads, which can take on values of 0, 1, or 2. The binomial distribution is presented below.

Number of heads	Probability
0	0.25
1	0.50
2	0.25

The binomial distribution has the following properties:

- The mean of the distribution (μ_x) is equal to $n * P$.
- The variance (σ_x^2) is $n * P * (1 - P)$.
- The standard deviation (σ_x) is sqrt[$n * P * (1 - P)$].

The *binomial probability* refers to the probability that a binomial experiment results in *exactly* x successes. For example, in the above table, we see that the binomial probability of getting exactly one head in two coin flips is 0.50.

Example 1

Suppose a die is tossed five times. What is the probability of getting exactly two 4s?

Solution

This is a binomial experiment in which the number of trials is equal to 5, the number of successes is equal to 2, and the probability of success on a single trial is 1/6 or about 0.167. Therefore, the binomial probability is

$$b(2; 5, 0.167) = {}_5C_2 * (0.167)^2 * (0.833)^3 \quad b(2; 5, 0.167) = 0.161$$

Cumulative Binomial Probability

A *cumulative binomial probability* refers to the probability that the binomial random variable falls within a specified range (e.g., is greater than or equal to a stated lower limit and less than or equal to a stated upper limit).

For example, we might be interested in the cumulative binomial probability of obtaining 45 or fewer heads in 100 tosses of a coin (see Example 1 below). This would be the sum of all these individual binomial probabilities.

$$b(x \leq 45; 100, 0.5) =$$

$$b(x = 0; 100, 0.5) + b(x = 1; 100, 0.5) + \ldots +$$

$$b(x = 44; 100, 0.5) + b(x = 45; 100, 0.5)$$

Example 1

What is the probability of obtaining 45 or fewer heads in 100 tosses of a coin?

Solution

To solve this problem, we compute 46 individual probabilities using the binomial formula. The sum of all these probabilities is the answer we seek. Thus,

$$b(x \leq 45; 100, 0.5) = b(x = 0; 100, 0.5) + b(x = 1; 100, 0.5) + \ldots +$$

$$b(x = 45; 100, 0.5) \, b(x \leq 45; 100, 0.5) = 0.184$$

Example 2

The probability that a student is accepted to a prestigious college is 0.3. If five students from the same school apply, what is the probability that at most two are accepted?

Solution

To solve this problem, we compute three individual probabilities using the binomial formula. The sum of all these probabilities is the answer we seek. Thus,

$$b(x \leq 2; 5, 0.3) = b(x = 0; 5, 0.3) + b(x = 1; 5, 0.3) + b(x = 2; 5, 0.3)$$

$$b(x \leq 2; 5, 0.3) = 0.1681 + 0.3601 + 0.3087 \, b(x \leq 2; 5, 0.3) = 0.8369$$

Example 3

What is the probability that the world series will last four games? Five games? Six games? Seven games? Assume that the teams are evenly matched.

Solution

This is a very tricky application of the binomial distribution. If you can follow the logic of this solution, you have a good understanding of the material covered up to this point.

In the world series, there are two baseball teams. The series ends when the winning team wins four games. Therefore, we define a success as a win by the team that ultimately becomes the world series champion.

For the purpose of this analysis, we assume that the teams are evenly matched. Therefore, the probability that a particular team wins a particular game is 0.5.

Let's look first at the simplest case. What is the probability that the series lasts only four games? This can occur if one team wins the first four games. The probability of the National League team winning four games in a row is

$$b(4; 4, 0.5) = {}_4C_4 * (0.5)^4 * (0.5)^0 = 0.0625$$

Similarly, when we compute the probability of the American League team winning four games in a row, we find that it is also 0.0625. Therefore, the probability that the series ends in four games would be $0.0625 + 0.0625 = 0.125$, since the series would end if either the American or the National League team won four games in a row.

Now let's tackle the question of finding the probability that the world series ends in five games. The trick in finding this solution is to recognize that the series can only end in five games if one team has won three out of the first four games. So let's first find the probability that the American League team wins exactly three of the first four games

$$b(3; 4, 0.5) = {}_4C_3 * (0.5)^3 * (0.5)^1 = 0.25$$

Okay, here comes some more tricky stuff, so pay attention to the logic. Given that the American League team has won three of the first four games, the American League team has a 50/50 chance of winning the fifth game to end the series. Therefore, the probability of the American League team winning the series in five games is $0.25 * 0.50 = 0.125$. Since the National League team could also win the series in five games, the probability that the series ends in five games would be $0.125 + 0.125 = 0.25$.

The rest of the problem would be solved in the same way. You should find that the probability of the series ending in six games is 0.3125, and the probability of the series ending in seven games is also 0.3125.

While this is statistically correct in theory, over the years the actual world series has turned out differently, with more series than expected lasting seven games. For an interesting discussion of why world series reality differs from theory, see Ben Stein's explanation of why seven-game world series are more common than expected.

Types of Acceptance Plans

Now that we have addressed the issue of replacement and no replacement samples, we are ready to provide a general summary of acceptance plans. LASPs fall into the following categories:

- **Single-sampling plans:** One sample of items is selected at random from a lot, and the disposition of the lot is determined from the resulting information. These plans are usually denoted as (n, c) plans for a sample size n, where the lot is rejected if there are more than c defectives. *These are the most common (and easiest) plans to use although not the most efficient in terms of average number of samples needed.*

- **Double-sampling plans:** After the first sample is tested, there are three possibilities:
  1. Accept the lot
  2. Reject the lot
  3. No decision

  If the outcome is point 3 above, and a second sample is taken, the procedure is to combine the results of both samples and make a final decision based on that information.

- **Multiple-sampling plans:** This is an extension of the double-sampling plans where more than two samples are needed to reach a conclusion. The advantage of multiple sampling is smaller sample sizes.

- **Sequential sampling plans:** This is the ultimate extension of multiple sampling where items are selected from a lot one at a time and after inspection of each item a decision is made to accept or reject the lot or select another unit.

- **Skip lot sampling plans:** Skip lot sampling means that only a fraction of the submitted lots are inspected.

Definitions of Basic Acceptance Plans

Deriving a plan within one of the categories listed above is discussed in the pages that follow. All derivations depend on the properties you want the plan to have. These are described using the following terms:

- **Acceptable Quality Level (AQL):** The AQL is a percent defective that is the baseline requirement for the quality of the producer's product. The producer would like to design a sampling plan such that there is a *high probability of accepting* a lot that has a defect level less than or equal to the AQL.

- **Lot Tolerance Percent Defective (LTPD):** The LTPD is a designated high defect level that would be unacceptable to the consumer. The consumer would like the sampling plan to have a *low probability of accepting* a lot with a defect level as high as the LTPD.

- **Type I Error (Producer's Risk):** This is the probability, for a given (n,c) sampling plan, of rejecting a lot that has a defect level equal to the AQL. The producer suffers when this occurs because a lot with acceptable quality is rejected. The symbol α is commonly used for Type I errors and typical values for α range from 0.2 to 0.01.

- **Type II Error (Consumer's Risk):** This is the probability, for a given (n,c) sampling plan, of accepting a lot with a defect level equal to the LTPD. The consumer suffers when this occurs because a lot with unacceptable quality is accepted. The symbol β is commonly used for Type II errors and typical values range from 0.2 to 0.01.

- **Operating Characteristic (OC) Curve:** This curve plots the probability of accepting the lot (Y-axis) versus the lot fraction or percent defectives (X-axis). *The OC curve is the primary tool for displaying and investigating the properties of a LASP.*

- **Average Outgoing Quality (AOQ):** A common procedure when sampling and testing are nondestructive is to inspect all of the pieces in all of the rejected lots and replace all defectives with good units. In this case, all rejected lots are made perfect and the only defects left are those in lots that were accepted. AOQs refer to the long-term defect level for this combined LASP and complete inspection of rejected lots process. If all lots come in with a defect level of exactly p, and the OC curve for the chosen (n,c) LASP indicates a probability p_a of accepting such a lot, over the long run the AOQ can easily be shown to be

$$AOQ = \frac{P_a P(N-n)}{N}$$

where N is the lot size.

- **Average Outgoing Quality Level (AOQL):** A plot of the AOQ (Y-axis) versus the incoming lot p (X-axis) will start at 0 for $p = 0$, and return to 0 for $p = 1$ (where every lot is completely inspected and rectified). In between, it will rise to a maximum. This maximum, which is the worst possible long-term AOQ, is called the AOQL.

- **Average Total Inspection (ATI):** When rejected lots are completely inspected, it is easy to calculate the ATI if the lots consistently yield a defect level of p. For a LASP (n, c) with a probability p_a of accepting a lot with defect level p, we have

$$\text{ATI} = n + (1 - p_a)(N - n)$$

where N is the lot size.

- **Average Sample Number (ASN):** For a single-sampling LASP (n,c) we know each and every lot has a sample of size n taken and inspected or tested. For double, multiple, and sequential LASPs, the amount of sampling varies depending on the number of defects observed. For any given double, multiple, or sequential plan, a long-term ASN can be calculated, assuming all lots come in with a defect level of p. A plot of the ASN, versus the incoming defect level p describes the sampling efficiency of a given LASP scheme.

The Final Choice Is a Trade-Off Analysis

Making a final choice between single- and multiple-sampling plans that have acceptable properties is a matter of deciding whether the average sampling savings gained by the various multiple-sampling plans justifies the additional complexity of these plans and the uncertainty of not knowing how much sampling and inspection will be done.

MIL STDs 105D and 114

These standards for sampling have been the work horse in industry for many years. They were developed by the U.S. government for defense and served a very important role in the evolution of improvement in many industries. The steps for using theses standards can be summarized as follows:

1. Decide on the AQL.
2. Decide on the inspection level.
3. Determine the lot size.
4. Enter the table to find sample size code letter.

5. Decide on type of sampling to be used.

6. Enter proper table to find the plan to be used.

Begin with normal inspection; follow the switching rules and the rule for stopping the inspection (if needed).

These standards, although very popular for many years, lately are not being used as much because of other techniques in sampling and the international standard of quality (ISO 9000). If the reader is interested in learning more about these standards, there are many sources in government publications, and the Internet, which one may find to their taste.

Stratified Sampling

Stratified random sampling refers to a sampling method that has the following properties:

- The population consists of N elements.
- The population is divided into H groups, called *strata*.
- Each element of the population can be assigned to one, and only one, stratum.
- The number of observations within each stratum N_h is known, and $N = N_1 + N_2 + N_3 + ... + N_{H-1} + N_H$.
- The researcher obtains a probability sample from each stratum.

In this discussion, we assume that the researcher draws a simple random sample from each stratum.

Advantages and Disadvantages

Stratified sampling offers several advantages over simple random sampling:

- A stratified sample can provide greater precision than a simple random sample of the same size.
- Because it provides greater precision, a stratified sample often requires a smaller sample, which saves money.
- A stratified sample can guard against an "unrepresentative" sample (e.g., an all-male sample from a mixed-gender population).
- We can ensure that we obtain sufficient sample points to support a separate analysis of any subgroup.

The main disadvantage of a stratified sample is that it may require more administrative effort than a simple random sample.

Proportionate versus Disproportionate Stratification

All stratified sampling designs fall into one of two categories, each of which has strengths and weaknesses as described below.

- **Proportionate stratification.** With proportionate stratification, the sample size of each stratum is proportionate to the population size of the stratum. This means that each stratum has the same sample fraction.
  - Proportionate stratification provides equal or better precision than a simple random sample of the same size.
  - Gains in precision are greatest when the values within the strata are homogeneous.
  - Gains in precision accrue to all survey measures.
- **Disproportionate stratification.** With disproportionate stratification, the sampling fraction may vary from one stratum to the next.
  - The precision of the design may be very good or very poor, depending on how the sample points are allocated to the strata. The way to maximize precision through disproportionate stratification is beyond the scope of this book.
  - If variances differ across strata, disproportionate stratification can provide better precision than proportionate stratification when sample points are correctly allocated to the strata.
  - With disproportionate stratification, the researcher can maximize the precision of a single important survey measure. However, gains in precision may not accrue to other survey measures.

Surveys

As we have mentioned before, *sampling* refers to the process of choosing a sample of elements from a total population of elements. In the statistical world, statisticians distinguish between two broad categories of sampling:

- **Probability sampling.** With probability sampling, every element of the population has a known probability of being included in the sample.
- **Nonprobability sampling.** With nonprobability sampling, we cannot specify the probability that each element will be included in the sample.

Each approach has advantages and disadvantages. The main advantages of nonprobability sampling are convenience and cost. However, with nonprobability samples, we cannot make probability statements about our sample statistics. For example, we cannot compute a confidence interval for an estimation problem or a region of acceptance for a hypothesis test.

Probability samples, in contrast, allow us to make probability statements about sample statistics. We can estimate the extent to which a sample statistic is likely to differ from a population parameter.

Quality of Survey Results

When researchers describe the quality of survey results, they may use one or more of the following terms:

- **Accuracy.** Accuracy refers to how close a sample statistic* is to a population parameter (a parameter is a measurable characteristic of a population, such as a mean or a standard deviation). Thus, if you know that a sample mean is 99 and the true population mean is 100, you can make a statement about the sample accuracy. For example, you might say the sample mean is accurate to within 1 unit.
- **Precision.** Precision refers to how close estimates from different samples are to each other. For example, the standard error is a measure of precision. When the standard error is small, estimates from different samples will be close in value, and vice versa. Precision is inversely related to standard error. When the standard error is small, sample estimates are more precise; when the standard error is large, sample estimates are less precise.

---

* a statistic is a characteristic of a sample. Generally, a statistic is used to estimate the value of a population parameter. For instance, suppose we selected a random sample of 100 students from a school with 1000 students. The average height of the sampled students would be an example of a statistic. So would the average grade point average. In fact, any measurable characteristic of the sample would be an example of a statistic.

- **Margin of error.** The margin of error expresses the maximum expected difference between the true population parameter and a sample estimate of that parameter. To be meaningful, the margin of error should be qualified by a probability statement. For example, a pollster might report that 50% of voters will choose the Democratic candidate. To indicate the quality of the survey result, the pollster might add that the margin of error is ±5%, with a confidence level of 90%. This means that if the same sampling method were applied to different samples, the true percentage of Democratic voters would fall within the margin of error 90% of the time.

The margin of error is equal to half of the width of the confidence interval. We already have addressed the issue of importance and the construction of confidence intervals.

Sample Design

A sample design can be described by two factors:

- **Sampling method.** Sampling method refers to the rules and procedures by which some elements of the population are included in the sample. Some common sampling methods are simple random sampling, stratified sampling, and cluster sampling.

- **Estimator.** The estimation process for calculating sample statistics is called the estimator. Different sampling methods may use different estimators. For example, the formula for computing a mean score with a simple random sample is different from the formula for computing a mean score with a stratified sample. Similarly, the formula for the standard error may vary from one sampling method to the next.

The "best" sample design depends on the survey objectives and on survey resources. For example, a researcher might select the most economical design that provides a desired level of precision. Or, if the budget is limited, a researcher might choose the design that provides the greatest precision without going over budget. Or other factors might guide the choice of sample design.

Simple Random Sampling

Simple random sampling refers to a sampling method that has the following properties:

- The population consists of N objects.
- The sample consists of n objects.
- All possible samples of n objects are equally likely to occur.

The main benefit of simple random sampling is that it guarantees that the sample chosen is representative of the population. This ensures that the statistical conclusions will be valid.

There are many ways to obtain a simple random sample. One way would be the lottery method. Each of the N population members is assigned a unique number. The numbers are placed in a bowl and thoroughly mixed. Then, a blind-folded researcher selects n numbers. Population members having the selected numbers are included in the sample. Yet another simple way is to generate a random table and follow it. This can be done with a random generator program (such as Excel, Minitab, or some other software) or see a table of random numbers in any basic statistics book.

Cluster Sampling

Cluster sampling refers to a sampling method that has the following properties:

- The population is divided into N groups, called *clusters*.
- The researcher randomly selects n clusters to include in the sample.
- The number of observations within each cluster M_i is known, and
  $M = M_1 + M_2 + M_3 + ... + M_{N-1} + M_N$.
- Each element of the population can be assigned to one, and only one, cluster.

Here we address the two common types of cluster sampling methods:

- **One-stage sampling.** All of the elements within selected clusters are included in the sample.

- **Two-stage sampling.** A subset of elements within selected clusters are randomly selected for inclusion in the sample.

Advantages and Disadvantages

Assuming the sample size is constant across sampling methods, cluster sampling generally provides less precision than either simple random sampling or stratified sampling. This is the main disadvantage of cluster sampling.

Given this disadvantage, it is natural to ask: Why use cluster sampling? Sometimes, the cost per sample point is less for cluster sampling than for other sampling methods. Given a fixed budget, the researcher may be able to use a bigger sample with cluster sampling than with the other methods. When the increased sample size is sufficient to offset the loss in precision, cluster sampling may be the best choice.

When to Use Cluster Sampling

Cluster sampling should be used only when it is economically justified—when reduced costs can be used to overcome losses in precision. This is most likely to occur in the following situations:

- Constructing a complete list of population elements is difficult, costly, or impossible. For example, it may not be possible to list all of the customers of a chain of hardware stores. However, it would be possible to randomly select a subset of stores (stage 1 of cluster sampling) and then interview a random sample of customers who visit those stores (stage 2 of cluster sampling).
- The population is concentrated in "natural" clusters (machines, facilities, city blocks, schools, hospitals, etc.). For example, to conduct personal interviews of operating room nurses, it might make sense to randomly select a sample of hospitals (stage 1 of cluster sampling) and then interview all of the operating room nurses at that hospital. Using cluster sampling, the interviewer could conduct many interviews in a single day at a single hospital. Simple random sampling, in contrast, might require the interviewer to spend all day traveling to conduct a single interview at a single hospital.

Even when the above situations exist, it is often unclear which sampling method should be used. Test different options, using hypothetical data if necessary. Choose the most cost-effective approach; that is, choose the sampling method that delivers the greatest precision for the least cost.

The Difference between Strata and Clusters

Although strata and clusters are both nonoverlapping subsets of the population, they differ in several ways:

- All strata are represented in the sample, but only a subset of clusters are in the sample.
- With stratified sampling, the best survey results occur when elements within the strata are internally homogeneous. However, with cluster sampling, the best results occur when elements within the clusters are internally heterogeneous.

How to Choose the Best Sampling Method

What is the best sampling method? The best sampling method is the sampling method that most effectively meets the particular goals of the study in question. The effectiveness of a sampling method depends on many factors. Because these factors interact in complex ways, the "best" sampling method is seldom obvious. Good researchers use the following strategies to identify the best sampling method:

- List the research goals (usually some combination of accuracy, precision, accessibility, and/or cost).
- Identify potential sampling methods that *might* effectively achieve those goals.
- Test the ability of each method to achieve each goal.
- Choose the method that does the best job of achieving the goals.

Obviously, there are many ways to chose a sampling method, and they all depend on the four items that we just discussed. There is no universal "best" method. These four items must be thoroughly understood and worked through so that efficient and good results may be reached. For example, we may work through each step thusly:

- List goals. This study has two main goals: (1) maximize precision and (2) stay within budget.
- Identify potential sampling methods. Because one of the main goals is to maximize precision, we can eliminate some of alternatives.

Sampling without replacement always provides equal or better precision than sampling with replacement, so we will focus only on sampling without replacement. Also, as long as the same clusters are sampled, one-stage cluster sampling always provides equal or better precision than two-stage cluster sampling, so we will focus only on one-stage cluster sampling.

This leaves us with four potential sampling methods—simple random sampling, proportionate stratified sampling, disproportionate stratified sampling, and one-stage cluster sampling. Each of these uses sampling without replacement. Because of the need to maximize precision, sometimes we may want to use the Neyman allocation with our disproportionate stratified sample. Neyman allocation is a sample allocation method that may be used with stratified samples. The purpose of the method is to maximize survey precision, given a fixed sample size. With Neyman allocation, the "best" sample size for stratum h would be

$$n_h = n * (N_h * S_h)/[\Sigma (N_i * S_i)]$$

where n_h is the sample size for stratum h, n is total sample size, N_h is the population size for stratum h, and S_h is the standard deviation of stratum h.

Neyman allocation is actually a special case of optimal allocation. Optimum allocation refers to a method of sample allocation used with stratified sampling. Optimum allocation is designed to provide the most precision for the least cost. Based on optimal allocation, the best sample size for stratum h is

$$n_h = n * [(N_h * S_h)/\text{sqrt}(c_h)]/[\Sigma (N_i * S_i)/\text{sqrt}(c_i)]$$

where n_h is the sample size for stratum h, n is the total sample size, N_h is the population size for stratum h, S_h is the standard deviation of stratum h, and c_h is the direct cost to sample an individual element from stratum h. (Note that c_h does not include indirect costs, such as overhead costs.)

- Test methods. A key part of the analysis is to test the ability of each potential sampling method to satisfy the research goals. Specifically, we will want to know the level of precision and the cost associated with each potential method. Quite often the test of choice here is the standard error to measure precision. The smaller the standard error, the greater the precision.

- Choose best method. Do not be misled here. Stay firm on the *a priori* decision plan. Sometimes the results may seem equal with several approaches, including cost. However, the methods differ drastically with respect to precision as measured by the standard error. So if the precision is an issue, then the appropriate method should be chosen over others. Cluster sampling provides the most precision (i.e., the smallest standard error).

We do not suggest that cluster sampling is the "best" in all cases. Other sampling methods may be best in other situations. Use the four-step process described above to determine which method is best in any situation.

Quality Control Charts and Sampling

We would be amiss if we did not at least mention the fact that when we do some type of quality control we invariably use some kind of statistical process control and some form of sampling. On one hand, quality-control charts are a graphical means for detecting systematic variation from the quality to be expected in a continuous production line, that is, variation greater than the random fluctuation, which is inevitable and allowable. The charts may be used as a simple method of performing statistical tests of the hypothesis that subsequently produced articles have essentially the same quality characteristics as previously produced articles.

Quality-control charts relate to a production line or process in action and indicate when a process should be examined for trouble, whereas acceptance sampling, as we saw earlier, relates to entire lots of product and enables a buyer to decide on the basis of a sample whether to accept a given lot from a supplier. Both quality-control and acceptance inspection procedures are based on judgment of the whole by a part; that is, they are based on sampling. These modern techniques have been found to be preferable to screening (complete inspection) in most cases because of lower cost. The work load in screening sometimes requires each inspection to be superficial or haphazard, so that a sampling plan involving careful inspection may actually give more reliable information about a whole process. Also, if the test destroys the item inspected (a flash bulb, for instance), screening is out of the question.

Quality-control charts may be used to keep a process up to a specified standard or simply to keep a process stable. A separate control chart should be used for each part produced or each different operation intended to produce different results. For instance, separate charts would be used for the operation of each of its parameters and they should be modified or changed completely as necessary when and if the process changes. That is any of the 5Ms and E change.

In quality control we use acceptance sampling that may be based on either attributes or variables. In the attributes case, the lot is rejected if the sample contains too many defectives. In the variables case, the criterion may be one sided or two sided, depending on specifications.

Attributes or Variables

The decision whether to use attributes or to use variables may be made on the basis of overall cost, keeping in mind the following points:

1. A larger sample is required for attributes sampling than for variables sampling to obtain equivalent discrimination between good and bad lots. Therefore, variables inspection is preferred if sample items are costly and inspection is destructive.

2. The actual measurements and computations required for variables inspection may be more costly than the yes-or-no decision and tallying required for attributes testing. This must be taken into account whenever the testing itself is difficult, expensive, or time consuming.

3. Variables methods produce as a byproduct information that can be valuable in diagnosing production ills.

4. Variables plans depend for exactness on an assumption of normality in the distribution of the variable measured, though the plans may be used as approximate methods when the distribution departs from normality. Attributes plans are not subject to such a restriction.

5. Attributes sampling is more widely known than variables sampling, and therefore, may require less training of inspectors.

Choice of Plan

There are many possible bases for choosing one sampling plan rather than another. In some cases a physical situation or the limitations of a budget may dictate the sample size or other aspects of the plan. The methods given in this chapter are basic and common. Furthermore they are based on the choice of a plan setting the parameters for the risks of rejecting a good lot or of accepting a bad one.

Having decided what is to constitute an inspection lot of the product (this may be a production lot, part of a production lot, or several production lots taken together), what is to constitute a unit or item in the lot, and whether to test by attributes or by variables, we must decide on (1) an acceptable quality level, (2) the number of groups of items to be sampled, and (3) the inspection level.

Inspection Level

After a decision has been made to use a single-, double-, or multiple-sampling plan, an inspection level must be chosen. The MIL STD 105 is a good choice to use for identifying the levels. Three levels of inspection, designated I, II, and III, correspond to the different levels of the importance of detecting defective items at the expense of large sample sizes. Inspection level III is appropriate if defectives must be rejected whenever possible, regardless of the size of the sample, as might be the case when safety item is a consideration. If, on the other hand, the cost of testing is unusually high and the acceptance of some defectives is not a serious matter, inspection level I is appropriate. Level II, a compromise, should be used unless there is a special need for one of the other levels.

Summary

In this chapter, we summarized some key items of understanding sampling techniques. The focus was to make the reader understand that it all starts with sampling and if the sampling is not appropriate and applicable the results will not be accurate or reflective of what is going on. There are many ways that the sample may be selected; however, here we discussed the lot acceptance sampling plan (LASP), which is a sampling scheme and a set of rules for making decisions. We also expanded the discussion into the MIL STD and quality control sampling as in statistical process control. Both attribute and variable samples were addressed. In the next chapter, we will focus on the understanding of computer programs for design and estimating design power. Part of the discussion will zero in the variations of ANOVA and how the modeling of the different setups may influence the results.

Selected Bibliography

American Statistical Association. (1950). *Acceptance Sampling*. Washington, DC: American Statistical Association.

Bowker, A. H. and Goode, H. P. (1952). *Sampling Inspection by Variables*. New York: McGraw-Hill.

Deming, W. E. *Some Theory of Sampling*. New York: Dover.

Dodge, H. F. and Romig, H. G. (1944). *Sampling Inspection Tables: Single and Double Sampling*. New York: Wiley.

Freeman, D., Pisani, R. and Purves, R. (2007). *Statistics*. 4th ed. (International Student Edition). New York: W. W. Norton.

Peach, P. (1947). *An Introduction to Industrial Statistics and Quality Control*. 2nd ed. Raleigh, NC: Edwards & Broughton Co.

Pearson, E. S. (1935). *The Application of Statistical Methods to Industrial Standardization and Quality Control*. London: British Standards Institution. (British Standard 600).

Scheffe, H. (1947). The relation of control charts to analysis of variance and chi-square tests. *American Statistical Association Journal*. 42:425–431 (corrections on p. 634).

Shewhart, W. A. (1931). *Economic Control of Quality of Manufactured Product*. New York: Van Nostrand.

Stamatis, D. H. (2003). *Six Sigma and Beyond. Statistical Process Control*. Boca Raton, FL: St. Lucie Press.

Tippett, L. H. C. (1950). *Technological Applications of Statistics*. New York: Wiley.

Tucker, H. (1998). *Mathematical Models in Sample Surveys: Multivariate Series*. Vol. 3. Hackensack, NJ: World Scientific.

Wald, A. (1947). *Sequential Analysis*. New York: Wiley.

10

Understanding Computer Programs for Design and Estimating Design Power

In the previous chapter, we discussed in a cursory manner the issue of acceptance sampling. In this chapter, we are focusing on understanding computer software programs for optimum design and estimating design power to our experiments.

Any statistical test of pattern requires a model against which to test the null hypothesis of no pattern. Models for ANOVA and ANCOVA take the form: Response (Y) = Factors (independent variables) + error, where the response refers to the data that require explaining, the factor or factors are the putative explanatory variables contributing to the observed pattern of variation in the response, and ε is the residual variation in the response left unexplained by the factor(s). (For the advanced reader please note that the model structure is the same as for regression analysis. In fact, at any time, one can conduct ANOVA and the same can be done with regression modeling.)

To illustrate the modeling characterization, generally we use standard notations and terms. For example, we use a standard notation to describe the full model and its testable terms. For example, the two-factor nested model in Section 2 is described by

(i) The full model, packed up into a single expression: $Y = B(A) + \varepsilon$

(ii) Its testable terms to be declared in a statistics package, unpacked from the full model: $A + B(A)$

A statistics package will require you to specify the model desired for a given dataset. You will need to declare which column contains the response variable Y, which column(s) contain the explanatory variable(s) to be tested, any nesting or cross-factoring of the explanatory variables, whether any of the variables are random rather than fixed factors, and whether any are covariates of the response.

Examples of ANOVA and ANCOVA Models

Each of the Sections 1 to 7 shows the anatomy of a full analysis. We have followed a standard notation and, hopefully, the reader will be able to use them with any software of his or her choice. Each section presents the setup of the experiment. However, we must also warn the reader that where appropriate, these include alternative restricted and unrestricted models (Searle, 1971), and Model 1 and Model 2 designs (Newman et al., 1997). Refer to the protocols in Doncaster (2007) and Doncaster and Davey (2007) to see which mean squares are used for the F-ratio denominators and, consequently, how many error degrees of freedom are available for testing significance. The examples have not used *post hoc pooling* though this may be an option or an alternative to some quasi F-ratios, and the underlying assumptions have not been evaluated though this would need to be done for real data sets.

A statistics package may give different F-ratios and p-values to those shown here, for a variety of reasons:

- In unbalanced and nonorthogonal designs and ANCOVA models, default use of Type-III adjusted SS for models that require Type II
- In balanced mixed models and ANCOVA models, default use of an unrestricted model when the design may suite a restricted model
- In designs with randomized blocks and split plots, default use of Model-1 analysis when the test hypothesis may require Model 2
- In ANCOVA mixed models, default use of residual error for all denominator mean squares when the test hypothesis may have a different error variance

1. One-factor designs
Analysis of differences between sample means of one categorical factor (including orthogonal contrasts) or trend with one covariate
  1.1   One-factor model $Y = A + \varepsilon$
   - Planned orthogonal contrasts

2. Nested designs
Analysis of two or more factors in a replicated hierarchy with levels of each nested in (belonging to) levels of the next
  2.1   Two-factor nested model $Y = B(A) + \varepsilon$
  2.2   Three-factor nested model $Y = C(B(A)) + \varepsilon$

3. Fully replicated factorial designs
Analysis of crossed combinations of factor levels randomly assigned to sampling units in replicated samples (including orthogonal contrasts) and/or of trends with cross-factored covariates
  3.1   Two-factor fully cross-factored model $Y = B|A + \varepsilon$
   - Planned orthogonal contrasts for levels of factors B and/or A, and contrasts for two-factor analysis missing one combination of levels

3.2   Three-factor fully cross-factored model $Y = C|B|A + \varepsilon$

3.3   Cross-factored with nesting model $Y = C|B(A) + \varepsilon$

3.4   Nested cross-factored model $Y = C(B|A) + \varepsilon$

4. Randomized block designs

Analysis of one or more categorical factors with levels, or combinations of levels, randomly assigned in blocking sampling units of plots within blocks, and replicated only across blocks (including orthogonal contrasts, balanced incomplete block, Latin squares, and Youden square variants on the one-factor complete block design).

4.1   One-factor randomized complete block model $Y = S'|A$ and with planned orthogonal contrasts

   –   Balanced incomplete block variant $Y = S'|A$

   –   Latin square variant $Y = C|B|A$ with replicate Latin squares in blocks and stacked squares for crossover designs. Some computer software programs have an option to allocate treatment levels at random

   –   Youden square variant $Y = C|B|A$

4.2   Two-factor randomized complete block model $Y = S'|B|A$

4.3   Three-factor randomized complete block model $Y = S'|C|B|A$

5. Split-plot designs

Analysis of two or more categorical cross-factors with levels randomly assigned in split-plot sampling units of sub-sub-plots nested in sub-plots and/or sub-plots nested in plots and/or plots nested in blocks and replicated only across levels of the nesting (repeated-measures) factor(s)

5.1   Two-factor split-plot model (a) $Y = B|P'(S'|A)$

5.2   Three-factor split-plot model (a) $Y = C|P'(S'|B|A)$

5.3   Three-factor split-plot model (b) $Y = C|B|P'(S'|A)$

5.4   Split split-plot model (a) $Y = C|Q'(B|P'(S'|A))$

5.5   Split split-plot model (b) $Y = C|P'(B|S'(A))$

5.6   Two-factor split-plot model (b) $Y = B|S'(A)$

5.7   Three-factor split-plot model (c) $Y = C|B|S'(A)$

5.8   Split-plot model with nesting $Y = C|S'(B(A))$

5.9   Three-factor split-plot model (d) $Y = Y = C|S'(B|A)$

6. Repeated-measures designs

Analysis of one or more categorical factors with levels, or combinations of levels, assigned in repeated-measures sampling units of subjects repeatedly tested in a temporal or spatial sequence, and replicated only across subjects

6.1 One-factor repeated-measures model $Y = S'|A$

6.2 Two-factor repeated-measures model $Y = S'|B|A$

6.3 Two-factor model with repeated measures on one cross-factor $Y = B|S'(A)$

6.4 Three-factor model with repeated measures on nested cross-factors $Y = C(B)|S'(A)$

6.5 Three-factor model with repeated measures on two cross-factors $Y = C|B|S'(A)$

6.6 Nested model with repeated measures on a cross-factor $Y = C|S'(B(A))$

6.7 Three-factor model with repeated measures on one factor $Y = C|S'(B|A)$

7. Unreplicated designs

Analysis of fully randomized factorial (crossed) combinations of factor levels without replication

7.1 Two-factor cross-factored unreplicated model $Y = B|A$

7.2 Three-factor cross-factored unreplicated model $Y = C|B|A$

There are many commercial software programs for planning designs and estimating design power. However, the basic structure of all these programs has the following in common:

1. *List all terms and degrees of freedom in any model for analysis of variance or covariance.* By doing so you specify the model as a hierarchical nesting of sampling units in factors, representing each variable by a single letter. Thus, for example, requesting: "$P(B|S(A))$" will yield testable terms for any of the models 3.3, 5.6, or 6.3, depending on the nature of the variables and the replication of the sampling unit, P.

2. *List testable terms, degrees of freedom, and critical F-values for any of the numbered designs above.* Many software programs allow you to specify your own sample sizes in the numbered ANOVA and ANCOVA designs above, assuming fixed treatment factors. For each estimable effect, it shows the test and error degrees of freedom, the critical F at $\alpha = 0.05$, and the effect size with 80% detection probability (the value of θ/σ that gives the test a power of 0.8).

   Use the program to evaluate alternative experimental designs for a given workload of data points, targeting a low critical F for treatment effects. The critical F will vary according to the distribution of data points between levels of sampling units and treatments. For a given total data points, critical F will be increased by the inclusion of nesting, covariates, blocking, split plots, or repeated measures. These may be desirable or intrinsic features of the experimental design, and they will increase power to detect treatment effects if they reduce error variances sufficiently to compensate for the higher critical F.

3. *Calculate statistical power for any balanced model.* Most software programs calculate statistical power prospectively for any balanced model with a proposed size of samples, given a threshold ratio of treatment effect size, θ (the standard deviation of the treatment variability) to error effect size, σ (the standard deviation of the random unmeasured variation). It can also calculate the value of θ/σ required to achieve a target power.

A pilot study may be needed to obtain an initial observed F from samples of size n. Then $[(F-1)/n]^{1/2}$ will provide an unbiased estimate of the population θ/σ, with which to evaluate the potential to gain power from more replication (e.g., Kirk, 1968). Freeware is available on the web to further explore the relationships between n, θ, σ, and power for specified designs (e.g., Lenth, 2006–2009).

4. *Find critical F-values for any number of test and error degrees of freedom and value of α.*

Types of Statistical Models

Use the following key to identify the appropriate section of model structures above, then look at your data and make appropriate decision.

1. Can you randomly sample from a population with independent observations?

   Yes → 2.

   No → the data may not suit statistical analysis of any sort.

2. Are you interested either in differences between sample averages or in relationships between covariates?

   Yes → 3.

   No → the data may not suit ANOVA or ANCOVA.

3. Does one or more of your explanatory factors vary on a continuous scale (e.g., distance, temperature, etc.) as opposed to a categorical scale (e.g., taxon, sex, etc.)?

   Yes → consider treating the continuous factor as a covariate and using ANCOVA designs in Sections 1 to 3; this will be the only option if each sampling unit takes a unique value of the factor. The response and/or covariate may require transformation to meet the assumption of linearity. Analyze with a general linear model (GLM) and, for nonorthogonal designs, consider using Type II-adjusted

SS if cross-factors are fixed, or Type III-adjusted SS if one or more cross-factors are random (and an unrestricted model, checking correct identification of the denominator MS to the covariate). Be aware that adjusted SS can increase or decrease the power to detect main effects.

No → 4.

4. Can all factor levels be randomly assigned to sampling units without stratifying any crossed factors and without taking repeated measures on plots or subjects?

    Yes → 5.

    No → 9.

5. Are all combinations of factor levels fully replicated?

    Yes → 6.

    No → use an unreplicated design (Section 7).

Fully randomized and fully replicated designs

6. Do your samples represent the levels of more than one explanatory factor?

    Yes → 7.

    No → use a one-factor design (Section 1), considering options for orthogonal contrasts (e.g., Model 1.1).

7. Is each level of one factor present in each level of another?

    Yes → 8.

    No → use a nested design with each level of one factor present in only one level of another (Section 2).

8. Use a fully replicated factorial design (Section 3), taking into account any nesting within the cross-factors (Model 3.3 or Model 3.4). For balanced and orthogonal designs with one or more random cross-factors, consider using a restricted model, and consider *post-hoc* pooling if an effect has no exact *F*-test. If cross-factors are not orthogonal (e.g., sample sizes are not balanced), use GLM and consider using Type II-adjusted SS if cross-factors are fixed, or Type III-adjusted SS if one or more cross-factors are random (an unrestricted model).

Stratified random designs

9. Are sampling units grouped spatially or temporally and are all treatment combinations randomly assigned to units within each group?

    Yes → use a design with randomized blocks (Section 4). If all factor combinations are fully replicated, analyze with Section 3 ANOVA tables; otherwise consider analysis by Model 1 (assumes treatment-by-block interactions) or Model 2 (assumes no treatment-by-block interactions). For a single treatment factor, consider options to use a balanced incomplete block or, with cross-factored blocks, a Latin square or Youden square.

No → 10.

10. Are different treatments applied at different spatial scales and their levels randomly assigned to blocks or to plots within blocks?

Yes → use a design with split plots (Section 5), taking into account nesting among sampling units.

No → use a repeated-measures design (Section 6) for repeated measurement of each sampling unit at treatment levels applied in a temporal or spatial sequence. If all factor combinations are fully replicated, analyze with Section 3 ANOVA tables.

Strong Recommendations

Here are *two very wrong things* that people try to do with any software:

- **Retrospective power** (also known as observed power, *post-hoc* power). You've got the data, did the analysis, and did not achieve "significance." So you compute power retrospectively to see whether the test was powerful enough or not. This is an empty question. Of course it wasn't powerful enough—that's why the result isn't significant. Power calculations are useful for design, not analysis. (Note: These comments refer to power computed based on the observed effect size and sample size. Considering a different sample size is obviously prospective in nature. Considering a different effect size might make sense, but probably what you really need to do instead is an equivalence test; see Hoenig and Heisey, 2001.)

- **Specify T-shirt effect sizes** ("small," "medium," and "large"). This is an elaborate way to arrive at the same sample size that has been used in past social science studies of large, medium, and small size (respectively). The method uses a standardized effect size as the goal. Think about it: For a "medium" effect size, you'll choose the same *n* regardless of the accuracy or reliability of your instrument, or the narrowness or diversity of your subjects. Clearly, important considerations are being ignored here. "Medium" is definitely not the message!

Here are *three very right things* you can do:

- **Use power prospectively** for planning future studies. Software programs such as Minitab are useful for determining an appropriate sample size, or for evaluating a planned study to see whether it is likely to yield useful information.

- **Put science before statistics**. It is easy to get caught up in statistical significance and such, but studies should be designed to meet scientific goals, and you need to keep those in sight at all times (in planning *and* analysis). The appropriate inputs to power/sample-size calculations are effect sizes that are deemed clinically important, based on careful considerations of the underlying scientific (not statistical) goals of the study. Statistical considerations are used to identify a plan that is effective in meeting scientific goals—not the other way around.

- **Do pilot studies**. Investigators tend to try to answer all questions with one study. However, you usually cannot do a definitive study in one step. It is far better to work incrementally. A pilot study (or screening experiment) helps you establish procedures, understand and protect against things that can go wrong, and obtain variance estimates needed in determining sample size. A pilot study with 20 to 30 degrees of freedom for error is generally quite adequate for obtaining reasonably reliable sample-size estimates.

Following the above guidelines is good for improving your chances of being successful. You will have established that you have thought through the scientific issues, that your procedures are sound, and that you have a defensible sample size based on realistic variance estimates and scientifically tenable effect-size goals.

Summary

In this chapter, we have addressed seven specific designs with their sub designs. The intent was to familiarize the reader with a variety of designs and power with any computer software program available so that an optimum design and estimated power be selected appropriately and applicably to our experiments.

References

Doncaster, C. P. (2007). Computer software for design of analysis of variance and covariance. Retrieved [date] from http://www.southampton.ac.uk/~cpd/anovas/

Doncaster, C. P. and Davey, A. J. H. (2007). *Analysis of Variance and Covariance: How to Choose and Construct Models for the Life Sciences*. Cambridge University Press, Cambridge, MA.

Hoenig, J. M. and Heisey, D. M. (2001). The abuse of power: The pervasive fallacy of power calculations for data analysis. *The American Statistician*. 55:19–24.

Kirk, R. E. (1968, 1982, 1994). *Experimental Design: Procedures for the Behavioral Sciences*. Brooks/Cole, Belmont, CA.

Lenth, R. V. (2006-2009). Java Applets for Power and Sample Size [Computer software]. Retrieved May, 25, 2010, from http://www.stat.uiowa.edu/~rlenth/Power

Newman, J. A., Bergelson, J. and Grafen, A. (1997). Blocking factors and hypothesis tests in ecology: Is your statistics text wrong? *Ecology*. 78:1312–1320.

Searle, S. R. (1971, 1997). *Linear Models*. John Wiley, New York.

Epilogue

Risk is a normal part of business and professional life. Nothing in the business world is straightforward or totally predictable. Change and external factors are always present to make us think better about the decisions we make.

Risk needs understanding and managing for at least three reasons:

1. The first is an obvious one. By definition professional life and business need to be somewhat conservative and defensive. After all, both are interested in protecting the current position in the market place and at the same time maximizing future opportunities.

2. The second is the need to embrace change and put in place strategic objectives for the development of the specific undertaking.

3. Third, positively addressing risks can identify opportunities. Businesses and individuals who understand their risks better than their competitors can create market advantage.

The challenge of these three reasons, however, has always been for owner managers to find a comprehensive and robust process with which to manage and monitor the influencing factors in their business. That is where statistics enters the picture.

A coherent risk management plan plays a major role in providing an answer through statistics and probability. It can help in both personal and business endeavors to avoid over extending themselves, while at the same time coordinating opportunities and identifying risks on an informed basis. The plan must be backed fully by personal commitment (if personal) and by all partners and directors, as appropriate and applicable (for business), and should detail the impact as well as the likelihood of specific events (both favorable and unfavorable) occurring. In this plan, it is important to include the responsibilities and time frames for managing such risks.

From a generic perspective, at least for business, the risk management plan should include the following five main approaches to risk. They are:

1. Remove the organization from the risk. For example, changing what the business does or withdrawing from a particular market. Statistics can help with the comparison of being good as we are or better with a new approach.

2. Sharing or transferring the risk. For example, insuring or perhaps subcontracting work out. Statistics can help with both the evaluation and selection of suppliers.

3. Mitigating the risk. This would include taking actions to reduce the likelihood or impact of isolated factors. Statistics can help evaluate the probability of success or failure in a given action taken.

4. Doing nothing but ensuring there are contingency plans in place. Statistics can help with alternative "what if" decisions and select the best or stay the current course.

5. Accepting the risk as being at a level of impact, which can be "lived with." We all must be cognizant and accept that risk management is not about avoiding risks, but identifying and managing it and taking chances on an informed basis. (This is intelligent strategic progression and it is most definitely a major contributor to overall success.) Statistics can help with the determination of Type I or Type II error and plan for the consequences of each.

To accomplish this plan all parties involved (including senior management) must understand and use some level of statistics. The statistical thinking will identify where an organization is, where it wants to go and if that direction is worth pursuing. In this book, we have identified some basic and advanced statistics that every quality professional should be exposed to and use at every opportunity.

When statistics is used as part of decision making, risks are minimized; efficiency of resources is maximized; and effectiveness for the benefit of the customer is always at forefront of the decision-making process.

Appendix A: Minitab Computer Usage

Ever since the proliferation of both lean and Six Sigma methodologies, the Minitab software is often used as the primary tool for statistical analysis by the quality professional. That does not mean that Minitab is the only software that can do the required analysis. Programs such as SAS, SPSS, Design Ease, Excel, and many others are just as good; however, some are more difficult to use than the others.

In this appendix, some of the important aspects that a practitioner using Minitab should know are discussed. The focus of the discussion is on the worksheet and the basic statistics screens. The appendix is presented in an outlined format because of the technical content and the art that is required. A worksheet has been chosen to elaborate because it is a very powerful screen and unless one understands its capability, one cannot take advantage of the software in its entirety. Basic statistics because of the ease and the convenience it presents. If you understand the flow of these, the rest of the guiding screens are not difficult.

In addition, we have included a short demonstration of the MSA discussion using ANOVA and the 6 panels.

Introduction to Minitab

What is Minitab?

- Minitab is a powerful statistical software package that removes the difficulties associated with analyzing data and using statistical tools.
- This module provides an introduction and overview of its use.

Section 1—Introducing the Software

Contents of this section are as follows:

- Introduction to the software
- Navigating using the Project Manager
- Data in Minitab
  - Data types
  - Changing data types

- Minitab overview
  - Opening worksheets
  - Worksheet folder
  - Worksheet description
  - Column description
  - Data structure and manipulation
  - Subset/split worksheets
  - Information folder

Minitab Worksheet

Opening Minitab

Menu bar

Tool bar

The session window

The worksheet/data window

Project manager window (minimised)

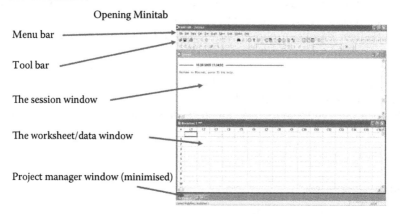

By default Minitab opens with the above

Everything one wants to do in Minitab starts with the worksheet. The worksheet looks like an Excel spreadsheet, but be aware of that:

- You cannot write formulas based on cells
- Most operations work on *whole* columns of data
- Any text in a cell will format the column as a Text Column, labeled Cn-T,

which prevents any mathematical analysis in that column

- The gray first row is for column names only, not for data
- Two columns cannot have the same name in any one worksheet

Minitab Files

There are two basic types of Minitab files

1) Minitab Projects;	"filename.MPJ"
Worksheets	Graphs
Session Window Output	Session Command History
2) Minitab Worksheets;	"filename.MTW"
Columns	Constants
Matrices	Column Descriptions
Worksheet Descriptions	

Save your work as a Minitab Project File to save all your data, graphs, and preferences together. VERY IMPORTANT: If you save your work as a worksheet file, you only save the data.

Minitab Windows

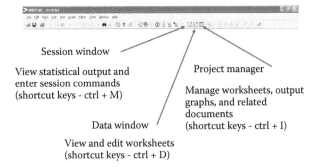

Session window

View statistical output and
enter session commands
(shortcut keys - ctrl + M)

Project manager

Manage worksheets, output
graphs, and related
documents
(shortcut keys - ctrl + I)

Data window

View and edit worksheets
(shortcut keys - ctrl + D)

Navigating with the Project Manager

Toolbar view ─────

Detached by clicking on
and dragging

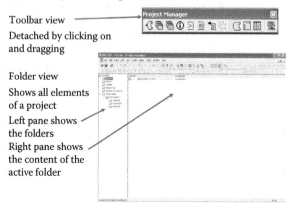

Folder view

Shows all elements
of a project

Left pane shows
the folders

Right pane shows
the content of the
active folder

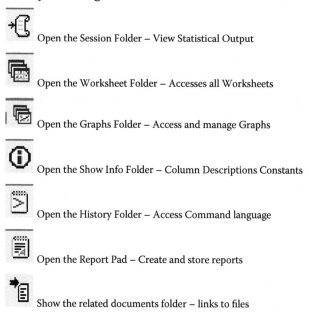

Minitab—The Project Manager Toolbar

Open the Session Folder – View Statistical Output

Open the Worksheet Folder – Accesses all Worksheets

Open the Graphs Folder – Access and manage Graphs

Open the Show Info Folder – Column Descriptions Constants

Open the History Folder – Access Command language

Open the Report Pad – Create and store reports

Show the related documents folder – links to files

Data Types in Minitab

Minitab operates using columns of data.

The arrow indicates the direction that the data is entered in.

Click on the arrow to change its direction.

The names of the columns are entered above the column (not in row 1)

Minitab (automatically) recognizes several different types of data:
Numeric
Text (T)
Date/Time (D)

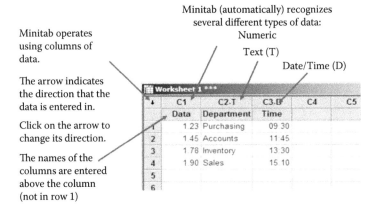

Entering Data in Minitab

Data can be entered into Minitab in a variety of ways:

1) Manually

2) Cutting and pasting from other application like excel

3) Opening a worksheet from another software application directly into Minitab

Because Minitab columns work "top-down", missing data is represented by an asterisk.

Changing Data Types in Minitab (1)

Sometimes Minitab makes mistakes in recognising the type of data in a column.

For example, when you cut and paste data, just one piece of text will cause the whole column to be labelled as text, as shown.

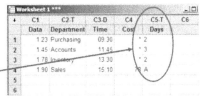

Unfortunately, deleting the text won't change the column type back automatically. This needs to be done manually; see next page.

Changing Data Types in Minitab (2)

Minitab: Data > Change Data Type > Text to Numeric

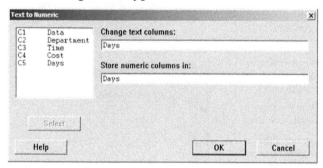

This function is set to change all the data in column C5 (Days) to Numeric data, and store it back in C5.

NB: You can delete the text data before using this function, but it is not necessary.

Importing Data from Excel

It is possible to import Excel worksheets directly into Minitab, but check the format of the data first. Remember that Minitab only allows one row for the column headings, and the Excel worksheet needs to reflect this.

File > Open > Minitab Worksheet

Under *files of type* choose excel (*.xls), as below.

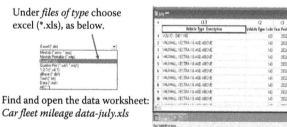

Find and open the data worksheet: *Car fleet mileage data-july.xls*

Minitab—Overview

The following example introduces the basic functions of Minitab and demonstrates the key navigation tools. The example includes three worksheets:

1. "stacked data"—data in a stacked format
2. "shift"—data unstacked by shift
3. "region"—data unstacked by region

Normally, Minitab files are saved as Minitab Projects, which have the file extension "MPJ".

Find and open the data file *Unavailability.mpj* by double clicking on it from within Windows Explorer.

Minitab—Show Worksheets Folder

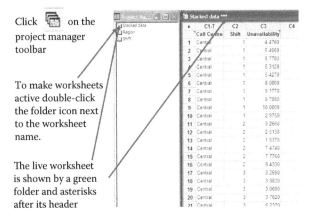

Click ▦ on the project manager toolbar

To make worksheets active double-click the folder icon next to the worksheet name.

The live worksheet is shown by a green folder and asterisks after its header

Worksheet Descriptions in Minitab

To add or edit an existing worksheet description click anywhere in the relevant worksheet, then select:

Editor > Worksheet > Description

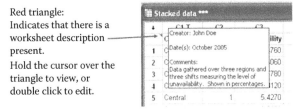

Red triangle:
Indicates that there is a worksheet description present.

Hold the cursor over the triangle to view, or double click to edit.

Column Descriptions in Minitab

To add or edit an existing column description, click the relevant column, then select:

Editor > Column > Description

Red triangle:
denotes an available column description.
Hover the cursor over the triangle to view or double click to edit.

Data Structure and Manipulation (1). Minitab operates using columns of data.

So, where
traditionally we
might structure data
like this:

	Jan	Feb	Mar	Apr
Location 1	289	295	300	301
Location 2	70	73	75	76
Location 3	168	174	180	189

In Minitab, the data
should be structured
in columns like
these:

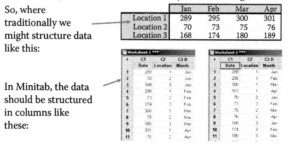

Data Structure and Manipulation (2)

File > Open Worksheet > OldversusNew.mpj

Open a new worksheet of raw
data

The next step is to include
categories against the data.

1. Sample number
2. Time of sample

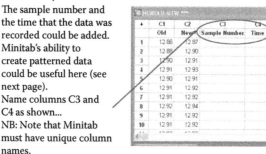

Data Structure and Manipulation (3)

The sample number and
the time that the data was
recorded could be added.
Minitab's ability to
create patterned data
could be useful here (see
next page).
Name columns C3 and
C4 as shown...
NB: Note that Minitab
must have unique column
names.

Data Structure and Manipulation (4)

Exercise: Use MINITAB's patterned data function as shown to create a set of
sample numbers in Column 3 (C3).

Calc > Make Patterned Data > Simple Set of Numbers

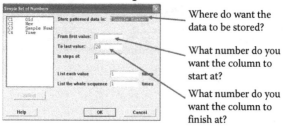

Where do want the
data to be stored?

What number do you
want the column to
start at?

What number do you
want the column to
finish at?

Data Structure and Manipulation (5)

Exercise: Use MINITAB's patterned data function as shown to create a set of sample numbers in Column 4 (C4).

Calc > Make Patterned Data > Simple Set of Date/Time Numbers

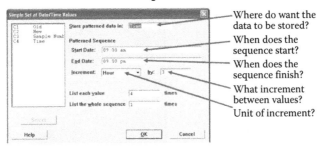

—Where do want the data to be stored?

—When does the sequence start?

—When does the sequence finish?

—What increment between values?

—Unit of increment?

Data Structure and Manipulation (6)

Your data should now look like this: Note that the heading of column 4 has a "D" suffix, because it is a date/time format.

Before analysing the data, you need to check all the different types of data have their own column....

Columns 1 and 2 contain the same type of data, so would be better placed in the same column. Minitab can stack data like this into one column;   ......see next page.

Data Structure and Manipulation (7)

What are we trying to do by stacking the data?
Answer: We would like to have one column containing the measured data (from C1 and C2).

Data > Stack > Columns

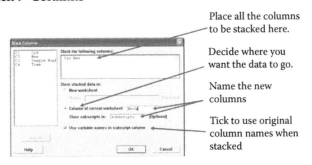

Place all the columns to be stacked here.

Decide where you want the data to go.

Name the new columns

Tick to use original column names when stacked

Data Structure and Manipulation (8)

The resulting data set looks like this:
The stacked data has been placed in C5, with a "subscript" in C6.

A subscript defines a characteristic of the data - in this case whether it was from the old or new process.
Exercise: Cut and paste the time data manually into C7, as shown here.

	C1 Old	C2 New	C3 Sample Number	C4 Time	C5 Both subscript	C6-T	C7 Times	C8
1	12.86	12.87	1	09:00 am	12.86 Old		09:00 am	
2	12.86	12.90	2	09:00 am	12.88 Old		09:00 am	
3	12.90	12.91	3	09:00 am	12.90 Old		09:00 am	
4	12.91	12.93	4	09:00 am	12.91 Old		09:00 am	
5	12.90	12.91	5	12:00 pm	12.90 Old		12:00 pm	
6	12.91	12.92	6	12:00 pm	12.91 Old		12:00 pm	
7	12.91	12.92	7	12:00 pm	12.91 Old		12:00 pm	
8	12.92	12.94	8	12:00 pm	12.92 Old		12:00 pm	
9	12.91	12.92	9	03:00 pm	12.91 Old		03:00 pm	
10	12.91	12.92	10	03:00 pm	12.91 Old		03:00 pm	
11	12.90	12.92	11	03:00 pm	12.90 Old		03:00 pm	
12	12.88	12.90	12	03:00 pm	12.88 Old		03:00 pm	
13	12.90	12.92	13	06:00 pm	12.90 Old		06:00 pm	
14	12.91	12.92	14	06:00 pm	12.91 Old		06:00 pm	
15	12.94	10.96	15	06:00 pm	12.94 Old		06:00 pm	

Subset/splitting out data

Subset Worksheet Function: Used to copy specified rows from the active worksheet to a new worksheet. You can specify the subset based on row numbers, brushed points on a graph, or a condition such as unmarried males younger than 50 years.

Split Worksheet Function: Splits, or unstacks, the active worksheet into two or more new worksheets based on one or more "By" variables.

Subset Worksheet and *Split Worksheet* always copy data to new worksheets. You can use *Copy Columns* to **replace data in the current worksheet with a subset.**

Subset Worksheet (1)

Using the file again:

We want to separate out the data at point 25 where it changes from central to western

A key point here is that we should always retain the original raw data file intact. Therefore, we could use the *subset worksheet* function.

	C1-T Call Centre	C2 Shift	C3 Unavailability	C4	C5
6	Central	1	8.0000		
7	Central	1	9.1770		
8	Central	1	9.7580		
9	Central	1	10.0000		
10	Central	1	2.9750		
11	Central	2	3.2660		
12	Central	2	2.5130		
13	Central	2	1.6370		
14	Central	2	7.4740		
15	Central	2	7.7760		
16	Central	2	8.4330		
17	Central	3	3.2690		
18	Central	3	3.8830		
19	Central	3	3.0690		
20	Central	3	3.7620		
21	Central	3	6.2370		
22	Central	3	5.4350		
23	Central	3	6.4160		
24	Central	3	7.4090		
25	Western	1	5.0970		
26	Western	1	3.2200		
27	Western	1	2.6080		

Subset Worksheet (2)

Data > Subset Worksheet

In name enter *central*
click 'specify which rows to
include'

Click row numbers and enter
the row numbers in the box.

You can enter either every
number – 1 2 3 4 5 etc or
enter the range – 1:24

NB: The colon is essential

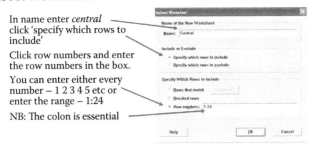

The result is shown on the next page.

Subset Worksheet (3)

A separate worksheet for the Central data has been created and the original
raw data file for all centers has been retained.

	C1-T	C2	C3	C4	C5	C6	C7	C8	C
	Call Centre	Shift	Unavailability						
1	Central	1	4.4760						
2	Central	1	5.4060						
3	Central	1	5.7780						
4	Central	1	5.3120						
5	Central	1	5.4270						
6	Central	1	8.0000						
7	Central	1	9.1770						
8	Central	1	9.7580						
9	Central	1	10.0000						
10	Central	1	2.9750						
11	Central	2	3.2660						
12	Central	2	2.5130						
13	Central	2	1.5370						
14	Central	2	7.4740						

This can now be repeated for Western and Eastern data.

Subset Worksheet (4)

If we click on the
'show worksheets' icon
project manager shows
the original worksheet
and the newly created
ones for central,
western and eastern.
The green folder shows
which folder is currently
active.

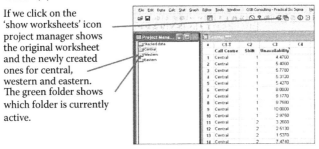

If we want to separate the worksheet by shift, we would then use Split Worksheet

Split Worksheet (1)

The "Split worksheet" function will create two or more worksheets from an original data file

Example: Split stacked data into three separate worksheets, one for each shift

Data > Split Worksheet

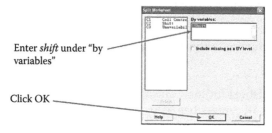

Enter *shift* under "by variables"

Click OK

Split Worksheet (2)

Minitab will then create three new worksheets—one for each *Shift*

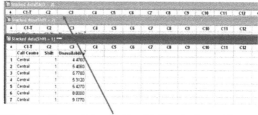

NB. The active worksheet is identified by the 3 asterisks

Split Worksheet (3)

If we click on the *show worksheets* icon, the project manager shows the original worksheet and the newly created ones for shift 1, shift 2, and shift 3.

The green folder shows which folder is currently active.

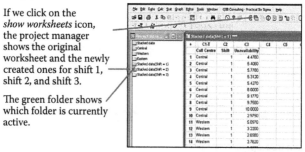

The Info Folder

Minitab'sproject manager can be used to provide a summary of the worksheet. Click here for information...

The *info* window can be useful for identifying missing data, particularly on large worksheets.

Section 2—Analyzing the Data

Contents of this section are as follows:

- Minitab's functions—statistical and graphical
- Accessing the *Stat* guide
- Minitab graphs—updating
- Displaying basic statistics
- Graphical summary
- Scatterplot and brushing
- Bar chart
  - Changing colors of bars
  - Adding reference lines
- Time series plot
- Histogram
- Box plot
- Dot plot

Analyzing the Data using Graphs

Graphing (or "plotting") is one of the most powerful tools of the Six Sigma process. "Reading" graphs will often tell you the story of the process. The statistics can then be used to provide mathematical confidence in the conclusions you draw from the graphs.

In the following pages, we look at the following graphs in MINITAB:

Basic statistics	Scatterplots
Bar chart	Histograms
Time series plots	Box plot
Marginal plot	Normality testing

Graphing is *always* the first step of analyzing data.

Minitab's Statistical Functions

Minitab's "stat" menu
contains all of the
statistical functions
that you will need
within six sigma.

"Basic statistics"
contains the most
frequently used
commands.

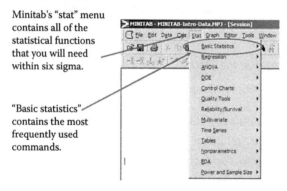

Minitab's Graphical Functions (1)

Minitab's "Graph" menu contains all of the graphical functions that you will need within Six Sigma.

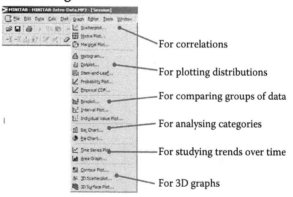

For correlations

For plotting distributions

For comparing groups of data

For analysing categories

For studying trends over time

For 3D graphs

Minitab's Graphical Functions (2)

Almost all of Minitab's graphs can be accessed through several different menu locations. For example,

Stat > Basic Stats > Display Descriptive Stats

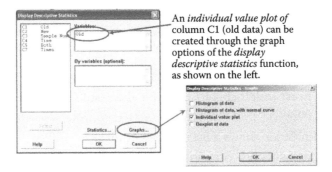

An *individual value plot of* column C1 (old data) can be created through the graph options of the *display descriptive statistics* function, as shown on the left.

Minitab's Graphical Functions (3)

An *Individual Value Plot* of column C1 (Old data) can also be created through the Graph menu, as shown below:

Graph > Individual Value Plot

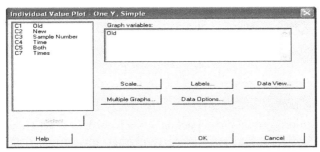

Exercise: **Create Individual Value plots for "Old" data (C1) using the two methods shown on the last two slides.**

Minitab's Graphical Functions (4)

The *Individual Value Plots* show the same data, but look different because they were created in different ways in Minitab.

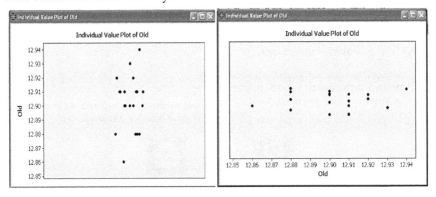

Exercise: Take a closer look at the graphics in the top left-hand corner of the graph windows you've created. Now go back to the data in the worksheet and change one of the values in the "Old" Column. What happens to the graphics?

Accessing the Stat Guide

MINITAB's *StatGuide* helps to interpret the results of the analyses. The *StatGuide* contains two windows:

1. The MiniGuide shows a list of topics available.
2. The main StatGuide contains specific interpretations of data.

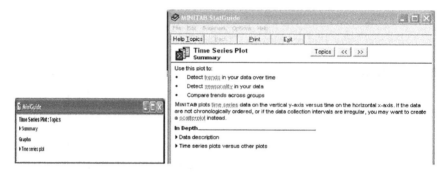

MINITAB's Graphs—updating (1)

Unlike common spreadsheet packages such as Excel, when you create a graph in MINITAB, and subsequently change the source data, the graph does not *necessarily* update itself. Some graphs do not have the capability to be updated. In this case:

A green cross indicates the graph is up to date

And a white cross indicates the graph is not up to date, and cannot be updated.

MINITAB's Graphs—updating (2)

Some graphs Do have the capability to be updated. In this case... A green cross on white indicates the graph is up to date...

...and a yellow circle indicates the graph is not up to date, but can be updated.

Graphs can be updated by right-clicking the graph and selecting:
"Update graph automatically"
or "Update graph now"

MINITAB'S Basic Statistics (1)

MINITAB has a wide variety of ways of displaying data. The following is a useful tool.

Stat > Basic Statistics > Descriptive Statistics

Double click on old and new to place them in the variables box.

Click statistics

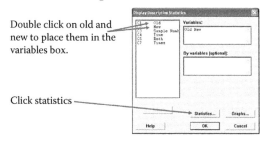

MINITAB's Basic Statistics (2)

Minitab gives the option to select a number of different statistics for example, check *trimmed mean*

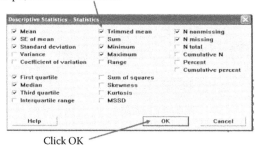

Click OK

Trimmed mean: The top and bottom 5% of data is ignored in calculating this version of the mean (average).

MINITAB's Basic Statistics (3)

The output from this function appears in the Session Window.

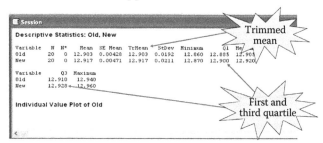

Trimmed mean—When the trimmed mean is very close to the standard mean, it shows the absence of significant outliers.

MINITAB's Basic Statistics (4)

This information can also be shown graphically. Select:

Stat > Basic Statistics > Graphical Summary

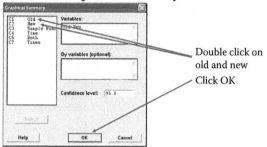

Double click on
old and new
Click OK

MINITAB's Basic Statistics (5)

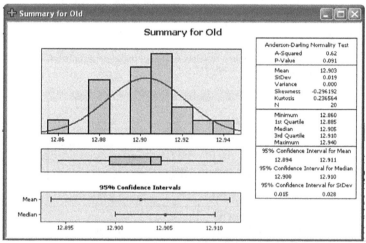

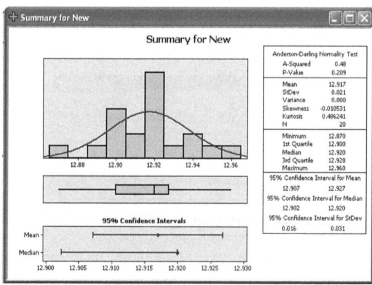

Scatter Plots (1)

What do we want to know?

How does the output from the Old process compare to the output from the New process?

Graph > Scatter Plot > Simple

The first step is to select the type of scatter plot you require.
In this case, select *simple*

Click simple

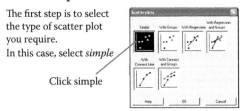

Scatter Plots (2)

Firstly, double left click on *old* to place it in the Y box

Then double left click on *new* to place it in the X box.

Click OK

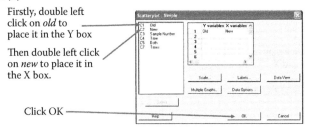

The scatter plot will graph point 1 of the *Old* data against point 1 of the *New* data, then point 2 of the *Old* against point 2 of the *New* and so on.

Scatter Plots (3)

If the two processes are producing the same outputs, we would expect a scatter plot to show a 45-degree line through the origin. The output should look like this:

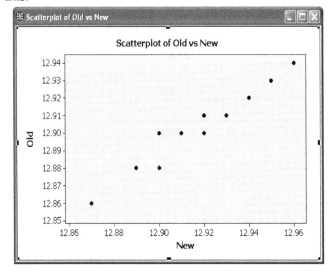

Analysis: The alignment isn't perfect, but it looks like the Old and New processes produce about the same output—except for one small detail: The scales are not the same!

To see a true representation, the *X* axis and the *Y* axis must be scaled the same.

Scatter Plots (4)

To adjust the scatter
plot with axes of the
same scale, then
follow this process:

Click on Y-axis

Click on editor

Click on edit Y-axis

Click same axis for
Y and X

Click OK

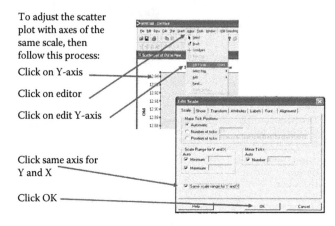

Scatter Plots (5)

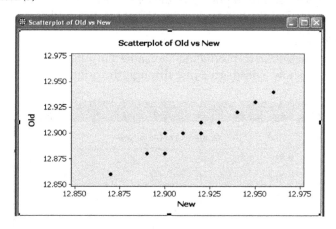

MINITAB Graphs—Brushing

Some MINITAB graphs can also be "interrogated" using the Brush function, found under the menu: **Editor > Brush**

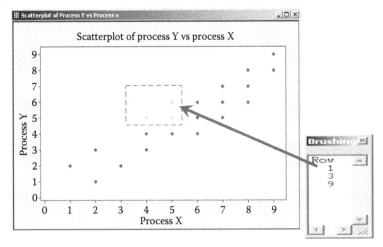

The brush function allows you to brush over specific points on a graph, and a smaller window then appears to give you the row numbers of those data points.

Note that the Brush function can only be used on graphs that are "up to date."

Bar Charts (1)

What do we want to know? How do we graphically summarize frequency data where there are several categorical variables? Open the file:
Graph > Bar Chart

Unavailability.MPJ
Under bars represent choose
"A function of a variable"

Under one Y choose cluster

Click ok

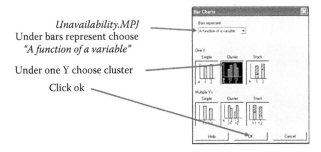

Bar Charts (2)

In the graph variables box enter *unavailability*.

Under categorical variables enter, *call centre then shift.*

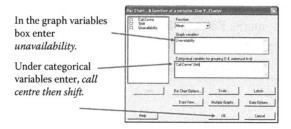

Bar Charts (3)

Interpreting the results.

Why does the level of unavailability over the three shifts decline in central and increase in western?

Why is the level of unavailability on shift 2 so much higher in eastern than the other two regions?

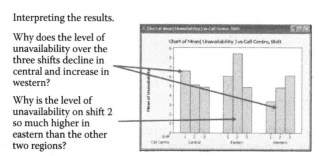

MINITAB can cluster the data by regions within the shifts. In MINITAB press the shortcut keys "CTRL-E" to bring back your last dialog box. Now enter the Categorical variables in the order *Shift,* then *Call Centre.*

Bar Charts (4)

Changing colors of the bars: Click any bar on the chart once and MINITAB will highlight all the bars. Once the bars are highlighted, select:
Editor > Edit bars

In the dialogue box click *groups*

Enter shift in categorical variables

Click ok

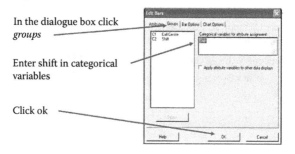

Bar Charts (5)

Changing colors of the bars…

Changing the colours of the bars...

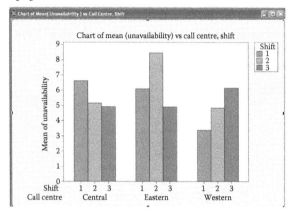

Bar Charts—Adding Reference Lines

Management may want to know where each region performed against targets of 3%, 4%, and 5%. We can add these reference lines to the chart. First, right click the graph and select:
Add > Reference Lines

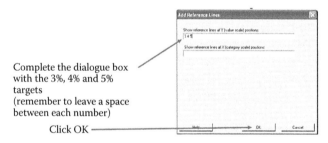

Complete the dialogue box with the 3%, 4% and 5% targets
(remember to leave a space between each number)

Click OK

Bar Charts—Adding Reference Lines

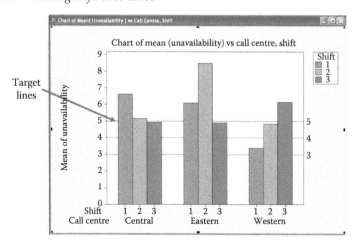

Target lines

Time Series Plots

What do we want to know? Because our data are time ordered, we should see whether there are any patterns over time. Open the file *OldversusNew.mpj*

Graph > Time Series Plot > Simple

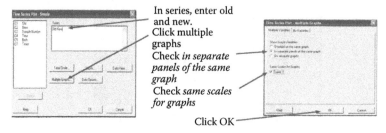

In series, enter old and new.
Click multiple graphs
Check *in separate panels of the same graph*
Check *same scales for graphs*

Click OK

Time Series Plots

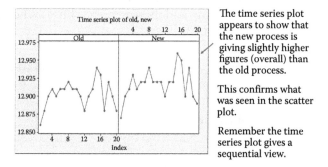

The time series plot appears to show that the new process is giving slightly higher figures (overall) than the old process.

This confirms what was seen in the scatter plot.

Remember the time series plot gives a sequential view.

We can now superimpose the New process onto the Old process to confirm this difference between New and Old.

Time Series Plots

To superimpose one Time Series Plot directly on top of the other...
Graph > Time Series Plot > Multiple

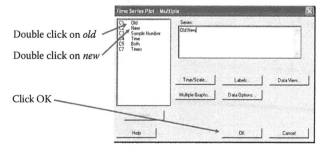

Double click on *old*

Double click on *new*

Click OK

Time Series Plots

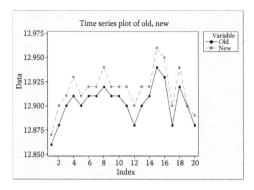

This version shows clearly that not only does the *new* process take longer than the old, but also both sets of data are following virtually the same pattern over the twenty readings

Histograms (1)

Continuing with graphical analysis: The two processes in the previous graphs appear to be different, and so it may be worth investigating the distributions within each process to understand their differences.

Graph > Histogram > Simple

Double click on *old* to place it in graph 1

Double click on *new* to place it in graph 2

Click multiple graphs

Click OK

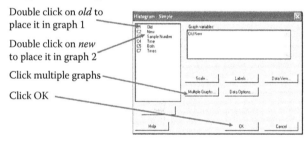

Histograms (2)

The *Multiple Graphs* command enables you to plot two graphs side by side, using the same scale for comparison.

Click "*in separate panels of the same graph*"

Tick same Y and same X

Click OK

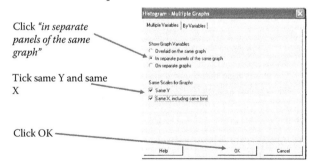

Histograms (3)

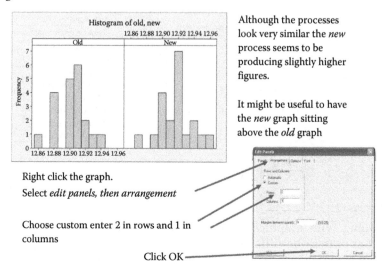

Although the processes look very similar the *new* process seems to be producing slightly higher figures.

It might be useful to have the *new* graph sitting above the *old* graph

Right click the graph.

Select *edit panels, then arrangement*

Choose custom enter 2 in rows and 1 in columns

Click OK

Histograms (4)

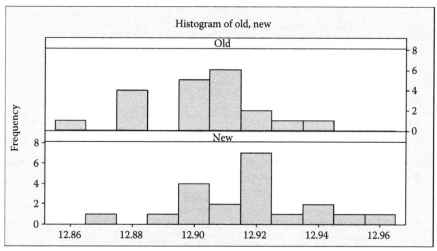

The observation that the new process appears to be producing slightly higher figure than the old process is now much easier to see.

Histograms (5)

We now get two separate histograms; one for Existing and one for New. To view them side by side we need to follow the command we used earlier: **Editor > Layout Tool**

The histograms can then be arranged as follows:

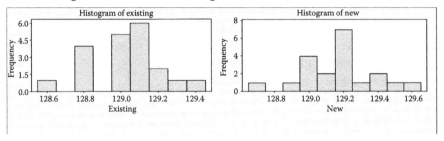

So, what do these histograms tell us?

- The "average" output for the new process looks slightly higher than the 'average' for the existing process.
- Both distributions appear fairly normal—but only a statistical Normality test as performed later can confirm this.

Histograms (6)

We can also get MINITAB to fit a distribution line on to the histogram.
Graph > Histogram > With Fit

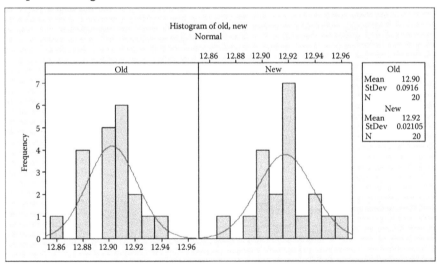

Box plot (1)

- A box plot is a tool for observing differences in position and/or spread between data groups.
- They are often used before more advanced statistical "hypothesis" tests such as *t*-tests, *F*-tests, ANOVA, etc.

- They are particularly useful during the Analyse phase, to divide data by rational subgroups. This can help identify potentially critical X's and sources of nonnormality.
- If data have been well collected then it can be stratified into rational subgroups, by logical factors such as:

Shift Number

Machine Number

   Order Number

     Day of the Week

      Operator

Box plots (2)

Graph > Box plot > Multiple Y's—Simple

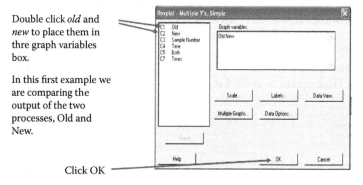

Double click *old* and *new* to place them in thre graph variables box.

In this first example we are comparing the output of the two processes, Old and New.

Click OK

Box plots (3)

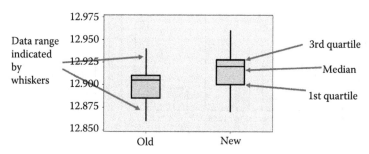

Data range indicated by whiskers

3rd quartile

Median

1st quartile

Old   New

Structure of a Box Plot:

The vertical lines above and below the box are "whiskers" and represent the tails of the distribution. The Box represents the middle 50% of the data, and the horizontal line in the box is the Median or the midpoint of the data. A quartile represents one quarter of the data.

Box plots (4)

Graph > Box plot > One Y—With Groups

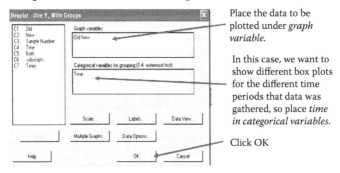

Place the data to be plotted under *graph variable*.

In this case, we want to show different box plots for the different time periods that data was gathered, so place *time in categorical variables.*

Click OK

Box plots (5)

Editor > Layout Tool

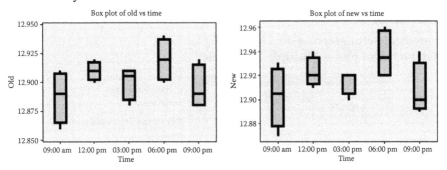

From these box plots, immediate points for investigation stand out:

1. Why a wider spread of data on both processes at 9:00 am?
2. Why are the results at 6:00 pm higher in both processes?

Box plots (6)

We can, if we wish, merge both of these charts into one.

Graph > Box plot > Multiple Y's—With Groups

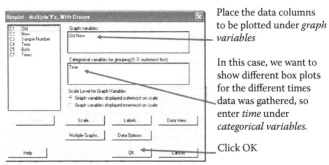

Place the data columns to be plotted under *graph variables*

In this case, we want to show different box plots for the different times data was gathered, so enter *time* under *categorical variables.*

Click OK

Box plots (7)

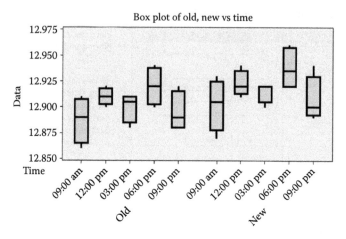

Dot plots (1)

Another useful graphical tool is the dot plot.
Graph > Dot plot > Multiple Y's Simple

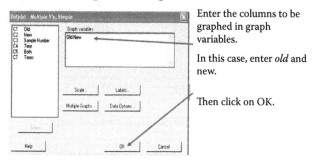

Enter the columns to be graphed in graph variables.

In this case, enter *old* and new.

Then click on OK.

Dot plots (2)

The dot plot is similar to a histogram because it helps to show the distribution of data groups. In this case, the two dot plots enable the variations in two or more groups to be examined side by side.

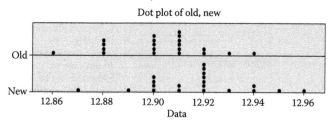

Graph Editing Tools (1)—Editing toolbar

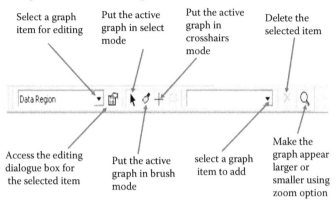

Graph Editing Tools (2)—Annotation toolbar

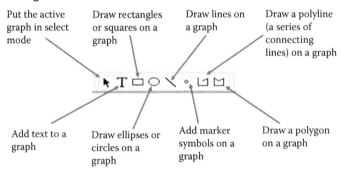

Analyzing the Data Using Graphs—Summary

The **scatterplot** was used to compare the *New* and *Old* processes.

Basic statistics were used to confirm differences in key statistics such as the mean (average) and standard deviation (variation).

The **histogram** provided a visual check of the shape of the distribution of each process (data set).

Times Series Plots were used to check for trends over time.

A **box plot** was used to compare the new and old processes against each other, and the results were also stratified into different time sections.

The **dot plot** was used to confirm information from previous graphs.

Finally, the **Normality Test** was used to create a **probability plot**, and to investigate whether the data sets are Normally distributed.

Variable MSA—Using Minitab

As we already discussed in the previous section, we begin by opening the worksheet and place the data in the columns as follows.

USL = 1.0   Replicate 1   Replicate 2
LSL = 0.5   (Randomized order)

Part Trial 1	Operator Trial 1	Response Trial 1	Parts Trial 2	Operator Trial 2	Response Trial 2
1	1	0.60000	8	1	0.85000
2	1	1.00000	2	1	1.00000
3	1	0.80000	10	1	0.60000
4	1	0.95000	6	1	1.00000
5	1	0.45000	5	1	0.55000
6	1	1.00000	4	1	0.85000
7	1	0.95000	7	1	0.95000
8	1	0.80000	1	1	0.65000
9	1	1.00000	9	1	1.00000
10	1	0.70000	3	1	0.85000
1	2	0.55000	1	2	0.55000
2	2	0.95000	2	2	1.05000
3	2	0.75000	3	2	0.80000
4	2	0.75000	4	2	0.80000
5	2	0.40000	5	2	0.40000
6	2	1.05000	6	2	1.00000
7	2	0.90000	7	2	0.95000
8	2	0.70000	8	2	0.75000
9	2	0.95000	9	2	1.00000
10	2	0.50000	10	2	0.55000
1	3	0.55000	1	3	0.50000
2	3	1.00000	2	3	1.05000
3	3	0.80000	3	3	0.80000
4	3	0.80000	4	3	0.80000
5	3	0.50000	5	3	0.45000
6	3	1.05000	6	3	1.00000
7	3	0.95000	7	3	0.95000
8	3	0.80000	8	3	0.80000

10 process outputs
3 operators
2 replicates

- Have operator 1 measure all samples once (as shown in the outlined block)

- Then, have operator 2 measure all samples once

- Continue until all operators have measured samples once (this is Replicate 1)

- Repeat these steps for the required number of replicates, (parts in random order)

- Enter data into Minitab columns as shown

Replicate 1   Replicate 2
                                                  (Randomized order)

Part Trial 1	Operator Trial 1	Response Trial 1	Parts Trial 2	Operator Trial 2	Response Trial 2
1	1	0.60000	8	1	0.85000
2	1	1.00000	2	1	1.00000
3	1	0.80000	10	1	0.60000
4	1	0.95000	6	1	1.00000
5	1	0.45000	5	1	0.55000
6	1	1.00000	4	1	0.85000
7	1	0.95000	7	1	0.95000
8	1	0.80000	1	1	0.65000
9	1	1.00000	9	1	1.00000
10	1	0.70000	3	1	0.85000
1	2	0.55000	1	2	0.55000
2	2	0.95000	2	2	1.05000
3	2	0.75000	3	2	0.80000
4	2	0.75000	4	2	0.80000
5	2	0.40000	5	2	0.40000
6	2	1.05000	6	2	1.00000
7	2	0.90000	7	2	0.95000
8	2	0.70000	8	2	0.75000
9	2	0.95000	9	2	1.00000
10	2	0.50000	10	2	0.55000
1	3	0.55000	1	3	0.50000
2	3	1.00000	2	3	1.05000
3	3	0.80000	3	3	0.80000
4	3	0.80000	4	3	0.80000
5	3	0.50000	5	3	0.45000
6	3	1.05000	6	3	1.00000
7	3	0.95000	7	3	0.95000
8	3	0.80000	8	3	0.80000

Manipulate the Data

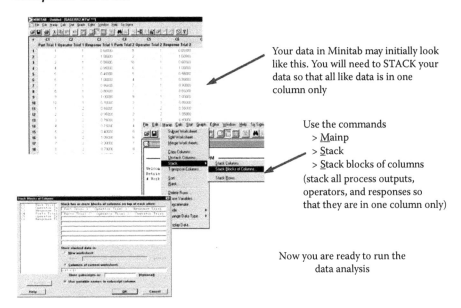

Your data in Minitab may initially look like this. You will need to STACK your data so that all like data is in one column only

Use the commands
> Manip
> Stack
> Stack blocks of columns
(stack all process outputs, operators, and responses so that they are in one column only)

Now you are ready to run the data analysis

Stacked and Ready for Analysis

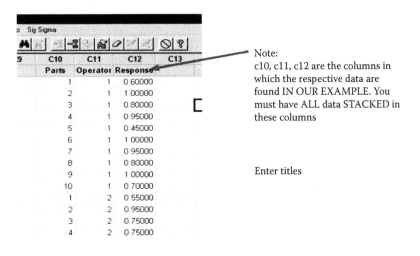

Note:
c10, c11, c12 are the columns in which the respective data are found IN OUR EXAMPLE. You must have ALL data STACKED in these columns

Enter titles

Prepare the Analysis

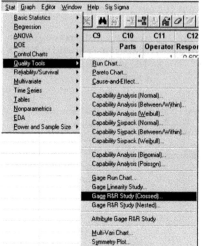

Use the commands
> Stat > quality tools
> gage R&R Study (Crossed)

   Each process output
   measured by each
   operator

   OR

> Gage R&R study (Nested)
   for "destructive tests"
   where each process output
   is measured uniquely by
   each operator

(Crossed & Nested: Defined on next page)

Choose Method of Analysis

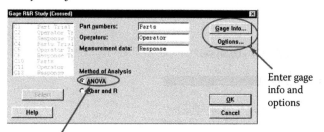

Enter gage
info and
options

ANOVA method is preferred
• Gives more information

Adding Tolerance (Optional)

Upper Specification
Limit (USL)
Minus
Lower Specification
Limit (LSL)

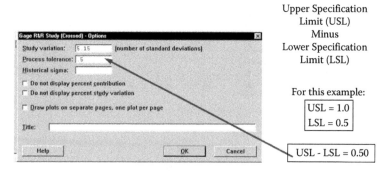

For this example:

USL = 1.0
LSL = 0.5

USL - LSL = 0.50

(Note: Industry standard set at 5.15 standard deviations, if supplier uses 6.0,
comparisons will show larger error than with industry.)

MSA Graphical 6 Panel

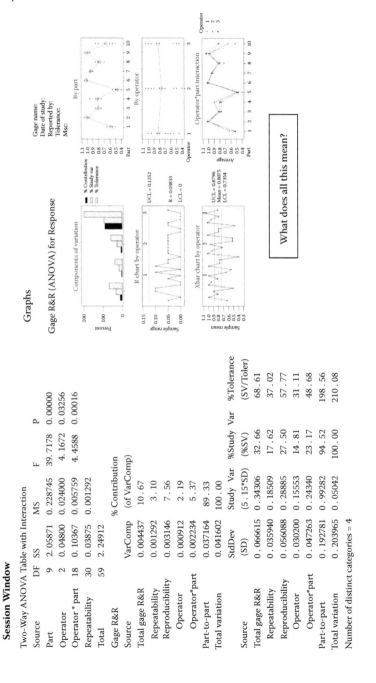

Session Window

Two-Way ANOVA Table with Interaction

Source	DF	SS	MS	F	P
Part	9	2.05871	0.228745	39.7178	0.00000
Operator	2	0.04800	0.024000	4.1672	0.03256
Operator * part	18	0.10367	0.005759	4.4588	0.00016
Repeatability	30	0.03875	0.001292		
Total	59	2.24912			

Gage R&R

Source	VarComp	% Contribution (of VarComp)
Total gage R&R	0.004437	10.67
Repeatability	0.001292	3.10
Reproducibility	0.003146	7.56
Operator	0.000912	2.19
Operator*part	0.002234	5.37
Part-to-part	0.037164	89.33
Total variation	0.041602	100.00

Source	StdDev (SD)	Study Var (5.15*SD)	%Study Var (%SV)	%Tolerance (SV/Toler)
Total gage R&R	0.066615	0.34306	32.66	68.61
Repeatability	0.035940	0.18509	17.62	37.02
Reproducibility	0.056088	0.28885	27.50	57.77
Operator	0.030200	0.15553	14.81	31.11
Operator*part	0.047263	0.24340	23.17	48.68
Part-to-part	0.192781	0.99282	94.52	198.56
Total variation	0.203965	1.05042	100.00	210.08

Number of distinct categories = 4

Mini-Tab Graphical Output

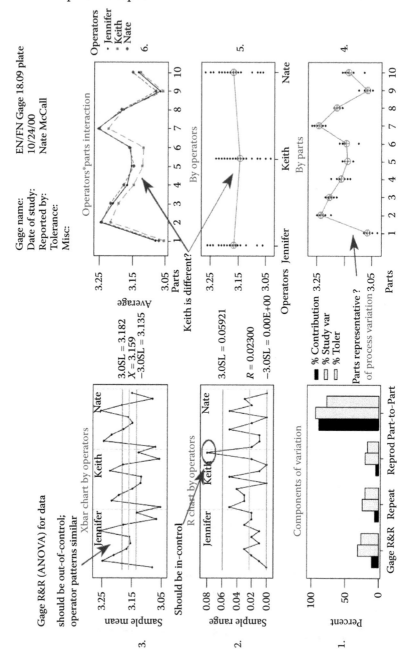

What do we look for in these charts:

Total Gage R&R < 30% (% Study Variation), and *Part-to-Part to be the biggest contributor* (similar as shown).

R-Chart must be *In-Control*. An *Out-of-Control Range Chart* indicates *Poor Repeatability*. Also want to see five or more levels (look across the points) of range within the control limits.

Xbar Chart must be *50% Out-Of-Control* or more (indicating the measurement system can tell a good part from a bad part) and similar patterns between operators.

Want spread of the 10 MSA parts to represent the actual process variation.

Want the *Operator Means* to be the same (straight red line).

Want *lines to be parallel & close to each other*. This indicates *Good Reproducibility*.

Graphical Output—6 Graphs In All

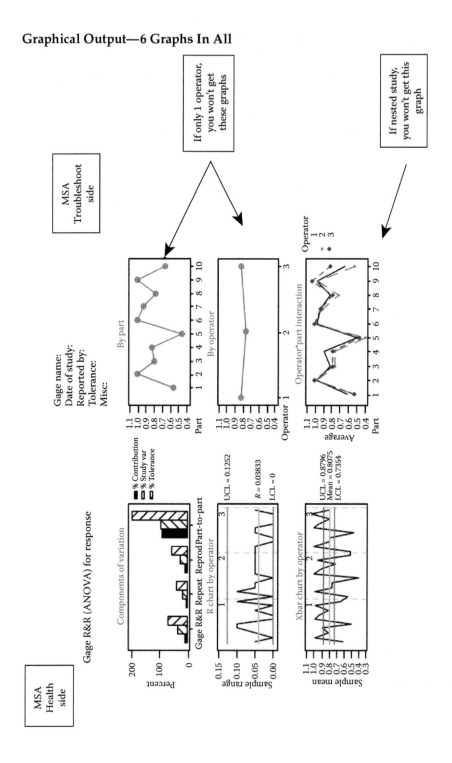

Destructive Test

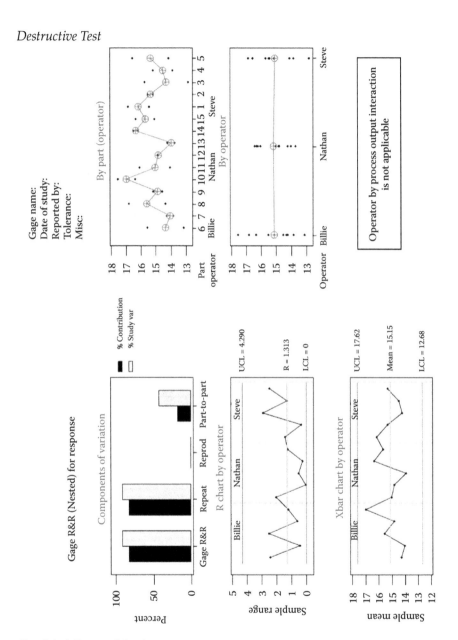

Graphical Output Metrics

Chart Output

- *Tall Bar Charts*: Distinguishes the components of Variation in percentages (%).
  - Repeatability, Reproducibility and Parts
(want low gage R&R, high part-to-part variation)

- *R Chart*: Helps identify unusual measurements
  - Repeatability/Resolution
(no outliers permitted)
- *Xbar Chart*: Shows sampled process output variety
  - Reproducibility/sensitivity
(want similar patterns for each operator)
These are your leading graphical indicators

Bar Charts for Components

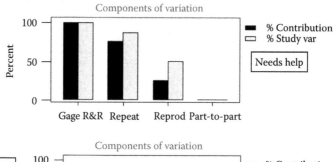

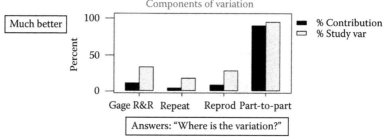

Answers: "Where is the variation?"

Look at the R Chart first, then Xbar Chart

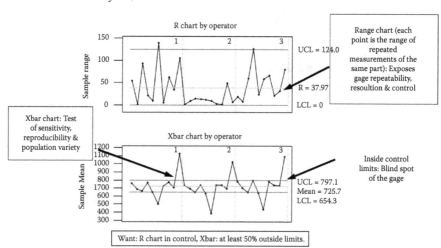

Want: R chart in control, Xbar: at least 50% outside limits.

More R Chart Indicators

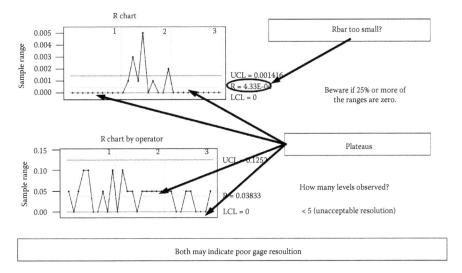

ANOVA Tabular Output %

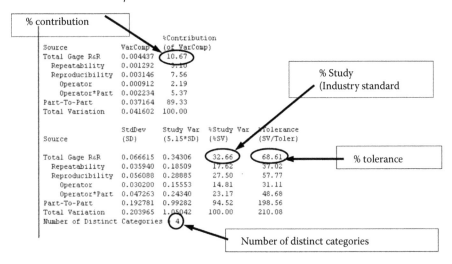

Appendix B: Formulas Based on Statistical Categories

Parameters

- Population mean = $\mu = (\Sigma\, X_i)/N$
- Population standard deviation = $\sigma = \text{sqrt}\,[\Sigma\,(X_i - \mu)^2/N]$
- Population variance = $\sigma^2 = \Sigma\,(X_i - \mu)^2/N$
- Variance of population proportion = $\sigma_P^2 = PQ/n$
- Standardized score = $Z = (X - \mu)/\sigma$
- Population correlation coefficient = $\rho = [1/N]^*\Sigma\{[(X_i - \mu_X)/\sigma_x]^* [(Y_i - \mu_Y)/\sigma_y]\}$

Statistics: Unless otherwise noted, these formulas assume simple random sampling.

- Sample mean = $x = (\Sigma\, x_i)/n$
- Sample standard deviation = $s = \text{sqrt}\,[\Sigma\,(x_i - x)^2/(n - 1)]$
- Sample variance = $s^2 = \Sigma\,(x_i - x)^2/(n - 1)$
- Variance of sample proportion = $s_P^2 = pq/(n - 1)$
- Pooled sample proportion = $p = (p_1 * n_1 + p_2 * n_2)/(n_1 + n_2)$
- Pooled sample standard deviation = $s_p = \text{sqrt}\,[(n_1 - 1)^*s_1^2 + (n_2 - 1)^*s_2^2]/(n_1 + n_2 - 2)]$
- Sample correlation coefficient = $r = [1/(n - 1)] * \Sigma\{[(x_i - x)/s_x]^*[(y_i - y)/s_y]\}$

Correlation

- Pearson product-moment correlation = $r = \Sigma\,(xy)/\text{sqrt}\,[(\Sigma\, x^2) * (\Sigma\, y^2)]$
- Linear correlation (sample data) = $r = [1/(n - 1)]^*\Sigma\{[(x_i - x)/s_x] * [(y_i - y)/s_y]\}$
- Linear correlation (population data) = $\rho = [1/N]^* \Sigma\{[(X_i - \mu_X)/\sigma_x]^* [(Y_i - \mu_Y)/y]\}$

Simple Linear Regression

- Simple linear regression line: $\hat{y} = b_0 + b_1 x$
- Regression coefficient = $b_1 = \Sigma\, [(x_i - x)\,(y_i - y)]/\Sigma\, [(x_i - x)^2]$
- Regression slope intercept = $b_0 = y - b_1 * x$
- Regression coefficient = $b_1 = r * (s_y/s_x)$
- Standard error of regression slope = $s_{b1} = \text{sqrt}\, [\Sigma(y_i - \hat{y}_i)^2/(n-2)]/\text{sqrt}\,[\Sigma(x_i - x)^2]$

Counting

- n factorial: $n! = n * (n-1) * (n-2) * \ldots * 3 * 2 * 1$. By convention, $0! = 1$.
- Permutations of n things, taken r at a time: $_nC_r = n!/(n-r)!$
- Combinations of n things, taken r at a time: $_nC_r = n!/r!(n-r)! = {_nP_r}/r!$

Probability

- Rule of addition: $P(A \cup B) = P(A) + P(B) - P(A \cap B)$
- Rule of multiplication: $P(A \cap B) = P(A)\,P(B|A)$
- Rule of subtraction: $P(A') = 1 - P(A)$

Random Variables

In the following formulas, X and Y are random variables, and a and b are constants.

- Expected value of $X = E(X) = \mu_x = \Sigma\, [x_i * P(x_i)]$
- Variance of $X = \text{Var}(X) = \sigma^2 = \Sigma\, [x_i - E(x)]^2 * P(x_i) = \Sigma\, [x_i - \mu_x]^2 * P(x_i)$
- Normal random variable = z-score = $z = (X - \mu)/\sigma$
- Chi-square statistic = $X^2 = [(n-1) * s^2]/\sigma^2$
- F statistic = $f = [s_1^2/\sigma_1^2]/[s_2^2/\sigma_2^2]$

- Expected value of sum of random variables $= E(X + Y) = E(X) + E(Y)$
- Expected value of difference between random variables $= E(X - Y) = E(X) - E(Y)$
- Variance of the sum of *independent* random variables $= \text{Var}(X + Y) = \text{Var}(X) + \text{Var}(Y)$
- Variance of the difference between *independent* random variables $= \text{Var}(X - Y) = E(X) + E(Y)$

Sampling Distributions

- Mean of sampling distribution of the mean $= \mu_x = \mu$
- Mean of sampling distribution of the proportion $= \mu_p = P$
- Standard deviation of proportion $= \sigma_p = \text{sqrt}[P^*(1-P)/n] = \text{sqrt}(PQ/n)$
- Standard deviation of the mean $= \sigma_x = \sigma/\text{sqrt}(n)$
- Standard deviation of difference of sample means $= \sigma_d = \text{sqrt}[(\sigma_1^2/n_1) + (\sigma_2^2/n_2)]$
- Standard deviation of difference of sample proportions $= \sigma_d = \text{sqrt}\{[P_1(1 - P_1)/n_1] + [P_2(1 - P_2)/n_2]\}$

Standard Error

- Standard error of proportion $= \text{SE}_p = s_p = \text{sqrt}[p * (1 - p)/n] = \text{sqrt}(pq/n)$
- Standard error of difference for proportions $= \text{SE}_p = s_p = \text{sqrt}\{p * (1 - p) * [(1/n_1) + (1/n_2)]\}$
- Standard error of the mean $= \text{SE}_x = s_x = s/\text{sqrt}(n)$
- Standard error of difference of sample means $= \text{SE}_d = s_d = \text{sqrt}[(s_1^2/n_1) + (s_2^2/n_2)]$
- Standard error of difference of paired sample means $= \text{SE}_d = s_d = \{\text{sqrt}[(\Sigma(d_i - d)^2/(n - 1)]\}/\text{sqrt}(n)$
- Pooled sample standard error $= s_{\text{pooled}} = \text{sqrt}[(n_1 - 1) * s_1^2 + (n_2 - 1) * s_2^2/(n_1 + n_2 - 2)]$
- Standard error of difference of sample proportions $= s_d = \text{sqrt}\{[p_1(1 - p_1)/n_1] + [p_2(1 - p_2)/n_2]\}$

Discrete Probability Distributions

- Binomial formula: $P(X = x) = b(x; n, P) = {}_nC_x * P^x * (1 - P)^{n-x} = {}_nC_x * P^{x} * Q^{n-xs}$
- Mean of binomial distribution $= \mu_x = n * P$
- Variance of binomial distribution $= \sigma_x^2 = n * P * (1 - P)$
- Negative Binomial formula: $P(X = x) = b^*(x; r, P) = {}_{x-1}C_{r-1} * P^r * (1 - P)^{x-r}$
- Mean of negative binomial distribution $= \mu_x = rQ/P$
- Variance of negative binomial distribution $= \sigma_x^2 = r * Q/P^2$
- Geometric formula: $P(X = x) = g(x; P) = P * Q^{x-1}$
- Mean of geometric distribution $= \mu_x = Q/P$
- Variance of geometric distribution $= \sigma_x^2 = Q/P^2$
- Hypergeometric formula: $P(X = x) = h(x; N, n, k) = [{}_kC_x] [{}_{N-k}C_{n-x}]/[{}_NC_n]$
- Mean of hypergeometric distribution $= \mu_x = n * k/N$
- Variance of hypergeometric distribution $= \sigma_x^2 = n * k * (N - k) * (N - n)/[N^2 * (N - 1)]$
- Poisson formula: $P(x; \mu) = (e^{-\mu}) (\mu^x)/x!$
- Mean of Poisson distribution $= \mu_x = \mu$
- Variance of Poisson distribution $= \sigma_x^2 = \mu$
- Multinomial formula: $P = [n!/(n_1! * n_2! * \ldots n_k!)] * (p_1^{n_1} * p_2^{n_2} * \ldots * p_k^{n_k})$

Linear Transformations

For the following formulas, assume that Y is a linear transformation of the random variable X, defined by the equation: $Y = aX + b$.

- Mean of a linear transformation $= E(Y) = Y = aX + b$.
- Variance of a linear transformation $= \mathrm{Var}(Y) = a^2 * \mathrm{Var}(X)$.
- Standardized score $= z = (x - \mu_x)/\sigma_x$.
- t-score $= t = (x - \mu_x)/[s/\mathrm{sqrt}(n)]$.

Estimation

- Confidence interval: Sample statistic ± Critical value * Standard error of statistic

- Margin of error = (Critical value) * (Standard deviation of statistic)
- Margin of error = (Critical value) * (Standard error of statistic)

Hypothesis Testing

- Standardized test statistic = (Statistic − Parameter)/(Standard deviation of statistic)
- One-sample z-test for proportions: z-score $= z = (p - P_0)/\text{sqrt}(p * q/n)$
- Two-sample z-test for proportions: z-score $= z = z = [(p_1 - p_2) - d]/\text{SE}$
- One-sample t-test for means: t-score $= t = (x - \mu)/\text{SE}$
- Two-sample t-test for means: t-score $= t = [(x_1 - x_2) - d]/\text{SE}$
- Matched-sample t-test for means: t-score $= t = [(x_1 - x_2) - D]/\text{SE} = (d - D)/\text{SE}$
- Chi-square test statistic $= X^2 = \Sigma[(\text{Observed} - \text{Expected})^2/\text{Expected}]$

Degrees of Freedom

The correct formula for degrees of freedom (DF) depends on the situation (the nature of the test statistic, the number of samples, underlying assumptions, etc.).

- One-sample t-test: $\text{DF} = n - 1$
- Two-sample t-test: $\text{DF} = (s_1^2/n_1 + s_2^2/n_2)^2 / \{[(s_1^2/n_1)^2/(n_1 - 1)] + [(s_2^2/n_2)^2/(n_2 - 1)]\}$
- Two-sample t-test, pooled standard error: $\text{DF} = n_1 + n_2 - 2$
- Simple linear regression, test slope: $\text{DF} = n - 2$
- Chi-square goodness of fit test: $\text{DF} = k - 1$
- Chi-square test for homogeneity: $\text{DF} = (r - 1) * (c - 1)$
- Chi-square test for independence: $\text{DF} = (r - 1) * (c - 1)$

Sample Size

Below, the first two formulas find the smallest sample sizes required to achieve a fixed margin of error, using simple random sampling. The third formula assigns sample to strata, based on a proportionate design. The fourth formula, Neyman allocation, uses stratified sampling to minimize variance, given a fixed sample size. And the last formula, optimum allocation, uses stratified sampling to minimize variance, given a fixed budget.

- Mean (simple random sampling): $n = \{z^2 * \sigma^2 * [N/(N-1)]\}/\{ME^2 + [z^2 * \sigma^2/(N-1)]\}$
- Proportion (simple random sampling): $n = [(z^2 * p * q) + ME^2]/[ME^2 + z^2 * p * q/N]$
- Proportionate stratified sampling: $n_h = (N_h/N) * n$
- Neyman allocation (stratified sampling): $n_h = n * (N_h * \sigma_h)/[\Sigma (N_i * \sigma_i)]$

Appendix C: General Statistical Formulas

This appendix provides the reader with some common statistical formulas that are used in the pursuit of quality.

1. Coefficient of variation

   For a population: $v = \dfrac{\sigma}{\mu}$

   For a sample: $v = \dfrac{s}{\overline{X}}$

   where

   μ = the population mean
   $\overline{X}$ = the sample mean
   σ = population standard deviation
   S = sample standard deviation

2. Pearson's coefficient of skewness

   For a population: $sk = \dfrac{\mu - MO}{\sigma}$

   For a sample: $sk = \dfrac{\overline{X} - mo}{s}$

   where

   sk = skewness
   MO = population mode
   mo = sample mode

3. Coefficient of kurtosis

   From ungrouped data:

   For a populaton: $K = \dfrac{\left[\sum (X - \mu)^4\right]/N}{\sigma^4}$

   For a sample: $K = \dfrac{\left[\sum (x - \overline{X})^4\right]/n}{s^4}$

   where

   K = kurtosis

X = observed population values

x = observed sample values

N = population size

n = sample size

σ^4 = squared population variance

s^4 = squared sample variance

From grouped data:

For a population: $K = \dfrac{\left[\sum f(X-\mu)^4\right]/\sum f}{\sigma^4}$

For a sample: $K = \dfrac{\left[\sum f(x-\bar{X})^4\right]/\sum f}{s^4}$

where

X = class midpoints

f = class frequencies

4. The probability of event A: classical approach

$$p(A) = \frac{P(A)_{\text{fav}}}{p(A)_{\text{pos}}} = \frac{n}{N}$$

where

$p(A)_{\text{fav}}$= number of equally likely basic outcomes favorable to the occurrence of event A

$p(A)_{\text{pos}}$ = number of equally likely basic outcomes possible

5. The probability of event A: empirical approach

$$p(A) = \frac{p(A)_{\text{past}}}{p(A)_{\text{max}}} = \frac{k}{M}$$

where

$p(A)_{\text{past}}$= number of times event A occurred in the past during a large number of experiments

$p(A)_{\text{max}}$= maximum number of times that event A could have occurred during these experiments

6. The factorial

$n! = n \times (n-1) \times (n-2) \times \cdots \times 3 \times 2 \times 1$ where 0! And 1! are equal to 1 by definition.

7. Permutations for x at a time out of n distinct items (with no repetition among the x items)

$$P_x^n = \frac{n!}{(n-x)!} \text{ where } x \leq n$$

8. Permutations for x out of n items, given k distinct types and $k < n$

$$P^n_{x=1, x=2 \cdots x=k} = \frac{n!}{x_1! x_2! \cdots x_k!}$$

where x_1 items are of one kind, x_2 items are of a second kind, ... and x_k items are of a kth kind, and where $x_1 + x_2 + \ldots + x_k = n$.

9. Combinations for x at a time out of n distinct items (with no repetitions allowed among the x items)

$$C^n_x = \frac{n!}{x!(n-x)!}$$

where $x \leq n$; note that this formula collapses to 1 for the limiting case of $x = n$:

$$C^n_n = \frac{n!}{n!(n-n)!} = \frac{n!}{n!0!} = \frac{n!}{n!(1)!} = 1$$

10. General addition law

$$p(A\,\text{or}\,B) = p(A) + p(B) - p(A\,\text{and}\,B)$$

11. Special addition law for mutually exclusive events

$$p(A\,\text{or}\,B) = p(A) + p(B)$$

12. General multiplication law

a. $p(A\,\text{and}\,B) = p(A) \times p(B|A)$

b. $p(A\,\text{and}\,B) = p(A) \times p(A|B)$

13. Special multiplication law for independent events

$$p(A\,\text{and}\,B) = p(A) \times p(B)$$

14. Bayes's theorem

$$p(E|R) = \frac{p(E)p(R|E)}{p(E) \times p(R|E) + p(E) \times p(R|E)}$$

15. Bayes's theorem rewritten

$$p(E|R) = \frac{p(E\,\text{and}\,R)}{p(R)}$$

16. Summary measures for the probability distribution of a random variable, R

$$\mu_r = E_R = \sum p(R=x) \times x$$

$$\sigma_R^2 = \sum p(R=x) \times (x - \mu_R)^2$$

$$\sigma_R = \sqrt{\sigma_R^2}$$

where

μ_r = arithmetic mean

σ_R^2 = variance

σ_R = standard deviation of the probability distribution of random variable, R

$p(R = x)$ = probability of the random variable = x

x = an observed value of that variable

17. The binomial formula

$$p(R = x|n, \Pi) = c_x^n \times \Pi a \times (1 - \Pi)^{n-x}$$

where

$x = 0, 1, 2,...$

n = number of successes in n trails

$c_x^n = n! / [x!(n - x)!]$

Π = probability of success in any one trial

18. Summary measures for the binomial probability distribution of random variable, R

$$\mu_R = E_R = n \times \Pi$$

$$\sigma_R^2 = n \times \Pi(1 - \Pi)$$

$$\sigma_R = \sqrt{\sigma_R^2}$$

Skewness: zero for $\Pi = 0.5$, positive for $\Pi < 0.5$, and negative for $\Pi > 0.5$

19. The hypergeometric formula

$$p(R = x|n, N, S) = \frac{C_x^S C_{n-x}^N - S}{C_x^N}$$

where

x = number of "successes" in a sample

n = sample size

N = population size

S = number of population with success characteristic

Clearly, $x = 0, 1, 2,...$, n or N (whichever is smaller); $n < N$; $S < N$.

20. Summary measures for the hypergeometric probability distribution of random variable, R

$$\mu_R = E_R = n \times \Pi = n(S|N)$$

$$\sigma_R^2 = n \times \Pi(1 - \Pi)\left(\frac{N - n}{N - 1}\right)$$

$$\sigma_R = \sqrt{\sigma_R^2}$$

where $\prod = S/N$

21. The Poisson formula

$$p(R = x|\mu) = \frac{e^{-\mu}\mu^x}{x!}$$

where

x = number of occurrences (x = 0, 1, 2,...,∞)

μ = mean number of occurrences within the examined units of time or space

e = 2.71828

22. Summary measures for the Poisson probability distribution

$$\mu_R = E_R = \lambda t$$

$$\sigma_R^2 = \lambda t$$

$$\sigma_R = \sqrt{\sigma_R^2}$$

where

σ_R = standard deviation of Poisson random variable, R

λ = Poisson process rate

t = number of units of time or space examined

23. The normal probability density function

$$f(x) = \frac{1}{\sigma_R\sqrt{2\pi}} e^{-1/2\left(\frac{x-\mu_R}{\sigma_R}\right)}$$

where

$f(x)$ = function of observed value of x

π = 3.14159

24. The standard normal deviate

$$Z = \frac{x-\mu_R}{\sigma_R}$$ where x = observed value of random variable, R

25. The standard normal probability density function

$$f(z) = \frac{1}{\sqrt{2\pi}e^{-z^2/2}}$$ where $z = (x-\mu_R)/\sigma_R$ and $-\infty \leq z \leq \infty$

26. The exponential probability density function

$$f(x) = \lambda e^{-\lambda x}$$ where $x > 0, \lambda > 0$

27. Greater than cumulative exponential probabilities

$$p(R > x) = e^{-\lambda x} = e^{-x/\mu_R}$$

Appendix C: General Statistical Formulas

28. Less than cumulative exponential probabilities

$$p(R < x) = 1 - e^{-\lambda x} = 1 - e^{-x/\mu_R}$$

29. Combination formula for exponential probabilities

$$p(x_1 < R < x_2) = 1 - \left[e^{-\lambda x_2} + \left(1 - e^{-\lambda x_1} \right) \right] e^{-\lambda x} = e^{-\lambda x_1} - e^{-\lambda x_2}$$

30. The uniform probability density function

$$f(x) = \frac{1}{b - a} \text{ if } a \le x \le b \text{ otherwise } f(x) = 0$$

31. Less than cumulative uniform probabilities

$$p(R \le x) = \frac{x - a}{b - a} \text{ If } a \le x \le b \text{ otherwise } p(R \le x) = 0$$

32. Summary measures for the uniform probability density

$$\mu_R = \frac{a + b}{2}$$

$$\sigma_R^2 = \frac{(b - a)^2}{12}$$

$$\sigma_R = \sqrt{\frac{(b - a)^2}{12}}$$

33. Summary measures of the sampling distribution of $\bar{X}$.

When selections of sample elements are statistically independent events, typically referred to as "the large-population case," because $n < 0.05N$

$$\mu_{\bar{x}} = \mu$$

$$\sigma_{\bar{x}}^2 = \frac{\sigma^2}{n}$$

$$\sigma_{\bar{x}} = \frac{\sigma}{\sqrt{n}}$$

When selections of sample elements are statistically dependent events, typically referred to as "the small-population case," because $n \ge 0.05N$

$$\sigma_{\bar{x}}^2 = \frac{\sigma^2}{n} \left(\frac{N - n}{N - 1} \right)$$

$$\sigma_{\bar{x}} = \frac{\sigma}{\sqrt{n}} \sqrt{\frac{N - n}{N - 1}}$$

where

$\mu_{\bar{x}}$ = mean of the sampling distribution of the sample mean

$\sigma_{\bar{X}}^2$ = variance of sampling distribution of the sample mean

$\bar{X}$ = sampling distribution of the sample mean

μ = population mean

σ^2 = population variance

σ = population standard deviation

34. Summary measures of the sampling distribution of P.

When selections of sample elements are statistically independent events, typically referred to as "the large-population case," because $n < 0.05N$

$$\mu_p = \Pi$$

$$\sigma_p^2 = \frac{\Pi(1-\Pi)}{n}$$

$$\sigma_p = \frac{\Pi(1-\Pi)}{n}$$

When selections of sample elements are statistically dependent events, typically referred to as "the small-population case," because $n \geq 0.05N$

$$\sigma_p^2 = \frac{\Pi(1-\Pi)}{n}\left(\frac{N-n}{N-1}\right)$$

$$\sigma_p = \sqrt{\frac{\Pi(1-\Pi)}{n}}\sqrt{\frac{N-n}{N-1}}$$

where

μ_p = mean of the sampling distribution of the sample proportion

σ_p^2 = variance of sampling distribution of the sample proportion

σ_p = standard deviation of the sampling distribution of the sample proportion, p

Π = population proportion

35. Upper and lower limits of confidence intervals for the population mean, μ and the proportion, Π (using the normal distribution).

$$\mu = \bar{X} \pm \left(z\sigma_{\bar{X}}\right)$$

$$\Pi = p \pm \left(z\sigma_p\right)$$

where

$\bar{X}$ = sample mean

p = sample proportion

$\sigma_{\bar{X}}$ = standard deviation of sample

σ_p = standard deviation of the sampling distribution of the sample proportion, p

Assumptions: sampling distribution of $\bar{X}$ or p is normal because: (a) the population distribution is normal or (b) the central limit theorem applies [for the mean, $n \geq 30$, $n < 0.05N$; for the proportion, $n\Pi \geq 5$ and $n(1-\Pi) \geq 5$]

36. Upper and lower limits of confidence interval for the population mean (using the t distribution)

$$\mu = \bar{X} \pm \left(t \frac{s}{\sqrt{n}} \right)$$

Assumption: $n < 30$ and the population distribution is normal.

37. Upper and lower limits of confidence interval for the difference between two population means (large and independent samples from large or normal populations)

$$\mu_A - \mu_B > \bar{X}_A - \bar{X}_B \pm \left(\sqrt{\frac{S_A^2}{n_A} + \frac{S_B^2}{N_b}} \right)$$

Assumptions: $n_A \geq 30$ and $n_B \geq 30$ and sampling distribution $\bar{X}_A - \bar{X}_B$ is normal because $n_A \leq 0.05N_A$ and $n_B \leq 0.05N_B$ or because population distributions are normal. Note: for small and independent samples, the same formula applies except that t replaces z. In that case, $n_A < 30$ and $n_B < 30$, but the two populations are assumed to be normally distributed and to have equal variances.

38. Upper and lower limits of confidence interval for the difference between two population means (large and matched-pairs samples from large or normal populations)

$$\mu_A - \mu_B > \bar{D} \pm \left(z \frac{S_D}{\sqrt{n}} \right)$$

where

$\bar{D}$ = mean of sample matched-pairs differences

S_D = standard deviation of sample matched-pairs differences

n = number of matched sample pairs

Assumptions: $n > 30$ and the sampling distribution of $\bar{D}$ is normal because $n < 0.05N$ or because the populations are normally distributed. Note: for small and matched-pairs sample, the same formula applies except that t replaces z. In that case, $n < 30$ but populations are assumed to be normally distributed and to have equal variances.

39. Upper and lower limits of confidence interval for the difference between two population proportions (large and independent samples from large populations)

$$\Pi_A - \Pi_B > p_a - p_b \pm \left(\sqrt{\frac{p_A(1-p_A)}{n_A} + \frac{p_B(1-p_b)}{n_b}} \right)$$

Assumptions: $n_A \geq 30$ and $n_B \geq 30$ and sampling distribution $p_A - p_B$ is normal because $n\Pi \geq 5$ and $n(1-\Pi) \geq 5$ for both samples.

40. Required sample size for specified tolerable error and confidence levels when estimating the population mean

$$n = \left(\frac{z\sigma}{e} \right)^2$$

where

z = standard normal deviate appropriate for the desired confidence level

σ = population standard deviation

e = tolerable error level

41. Required sample size for specified tolerable error and confidence levels when estimating the population proportion

$$n = \frac{z^2 p(1-p)}{e^2}$$

where p is an estimate of the population proportion.

42. A chi-squire statistic

$$x^2 = \sum \frac{(f_o - f_e)^2}{f_e}$$

where

f_o = observed frequency

f_e = expected frequency

43. Alternative chi-square statistic

$$x^2 = \frac{s^2(n-1)}{\sigma^2}$$

44. Expected value and standard deviation of the sampling distribution of W (given: H_0: the sampled populations are identical)

$$\mu_W = \frac{n_A(n_A + n_B + 1)}{2}$$

$$\sigma_W = \sqrt{\frac{n_A n_B(n_A + n_B + 1)}{12}}$$

Note: if W were defined as the rank sum of sample B, one would have to interchange subscripts A and B in the μ_W formula.

45. Normal deviate for the Wilcoxon rank-sum test

    $$z = \frac{w - \mu_W}{\sigma_W}$$

    Assumption: $n_A \geq 10$ and $n_B \geq 10$

46. Mann–Whiney test statistic

    $$U = \left[(n_A n_B) + \frac{n_A(n_A + 1)}{2}\right] - W$$

    Note: W was defined as the rank sum of sample A. If it were defined as the rank sum of sample B, one would have to interchange the subscripts A and B in this formula.

47. Normal deviate for Mann–Whiney test

    $$s = \frac{U - \mu_u}{\sigma_u}$$

    Assumption: $n_A \geq 10$ and $n_B \geq 10$

48. Expected value and standard deviation of the sampling distribution of U (given: H_0: the sampled populations are identical)

    $$\mu_u = \frac{n_A n_B}{2}$$

    $$\sigma_u = \sqrt{\frac{n_A n_B (n_A + n_B + 1) susp}{12}} = \sigma_W$$

49. The Wilcoxon signed-rank test statistic. It concerns itself with the direction of the difference as well as the size. The distinction of the difference is made by ranking them.

    $$z = \frac{T - \dfrac{N(N+1)}{4}}{\sqrt{\dfrac{N(N+1)(2N+1)}{24}}}$$

    where T = is the smaller value for either ΣR^+ or ΣR^-

    N = is the number of matched pairs, excluding those with a deviation of zero.

    R = the ranking

50. Expected value and standard deviation of the sampling distribution of T (given: H_0: there are no population differences)

    $$\mu_r = 0$$

    $$\sigma_r = \sqrt{\frac{n(n+1)(2n+1)}{6}}$$

51. Normal deviate for Wilcoxon singed-rank test.

$$z = \frac{T - \mu_T}{\sigma_T} = \frac{T}{\sigma_T}$$

Assumption: $n \geq 10$

Miscellaneous Statistical Formulas

1. Expected value and standard deviation of the sampling distribution of R_H (given H_0: Sampling is random.).

$$\mu_{RH} = \frac{n_H(n_T + 1)}{n_H + n_T}$$

$$\sigma_{RH} = \sqrt{\frac{n_H(n_T + 1)(n_H - 1)}{(n_H + n_T)^2} \left(\frac{n_T}{n_H + n_T - 1} \right)}$$

2. Normal deviate for the number-of-runs test.

$$z = \frac{R_H - \mu_{RH}}{\sigma_{RH}}$$

3. 11.P The Kruskal-Wallis test statistic.

$$K = \left[\frac{12}{n(n+1)} \left(\Sigma \frac{W_i^2}{n_i} \right) \right] - [3(n+1)]$$

4. The limits of the prediction interval for the average value of Y, given X (small-sample case).

$$\mu_{YX} = \hat{Y}_X \pm t_{\frac{\alpha}{2}} \cdot S_{YX} \sqrt{\frac{1}{n} + \frac{(X - \bar{X})^2}{\Sigma X^2 - n\bar{X}^2}}$$

where Xs are observed values of the independent variable ($\bar{X}$ being their mean), while $\hat{Y}_X$ is the estimated value of dependent variable Y, S_{YX} is the sample standard error of the estimate of Y, n is the sample size, and the t-statistic.

5. The limits of the prediction interval for the average value of Y, given X (large-sample case).

$$\mu_{YX} = \hat{Y}_X \pm \left(z_{a/2} \cdot \frac{S_{YX}}{\sqrt{n}} \right)$$

where $\hat{Y}_X$ is the estimated value of Y, the dependent variable, S_{YX} is the sample standard error of the estimate of Y, n is the sample size, and z is the standard normal deviate.

6. The limits of the prediction interval for an individual value of Y, given X (small-sample case).

$$I_{Y.X} = \hat{Y}_X \pm \left(t_{\frac{\alpha}{2}} \cdot S_{Y.X} \sqrt{\frac{1}{n} + \frac{(X - \bar{X})^2}{\Sigma X^2 - n\bar{X}^2} + 1} \right)$$

where Xs are observed values of the independent variable ($\bar{X}$ being their mean), While $\hat{Y}_X$ is the estimated value of dependent variable Y, S_{YX} is the sample standard error of the estimate of Y, n is the sample size, and the t-statistic.

7. The limits of the confidence interval for β (small-sample case).

$$\beta = b \pm \left(t_{a/2} \cdot \frac{S_{Y.X}}{\sqrt{\Sigma X^2 - n\bar{X}^2}} \right)$$

where b is the estimated (slope) regression coefficient, Xs are observed values of the independent variable ($\bar{X}$ being their mean), while, S_{YX} is the sample standard error of the estimate of dependent variable Y, n is the sample size, and the t-statistic.

Note: For large samples, the normal deviate $z_{a/2}$ replaces $t_{a/2}$.

8. The limits of the confidence interval for α (small-sample case).

$$\alpha = a \pm \left(t_{a/2} \frac{S_{Y.X}}{\sqrt{\Sigma X^2 - n\bar{X}^2}} \sqrt{\frac{\Sigma X^2}{n}} \right)$$

where a is the estimated (intercept) regression coefficient, Xs are observed values of the independent variable ($\bar{X}$ being their mean), while S_{YX} is the sample standard error of the estimate of dependent variable Y, n is the sample size, and the t-statistic.

Note: For large samples, the normal deviate $z_{a/2}$ replaces $t_{a/2}$

9. Components of total variation.

Total Variation = Explained Variation + Unexplained Variation

or

Total Sum of Squares (Total SS) = Regression Sum of Squares (RSS) + Error Sum of Squares (ESS)

$$\Sigma(Y - \bar{Y})^2 = \Sigma(\hat{Y}_X - \bar{Y})^2 + \Sigma(Y - \hat{Y}_X)^2$$

where Y_s are observed values of the dependent variable ($\bar{Y}$ being their mean) and $\hat{Y}_X$ is the estimated value of Y, given a value of independent variable X.

10. Sample coefficient of determination.

$$r^2 = \frac{\text{explained variation}}{\text{total variation}} = \frac{\text{RSS}}{\text{total SS}} = \frac{\Sigma(\hat{Y}_x - \bar{Y})^2}{\Sigma(Y - \bar{Y})^2}$$

Note: Because total SS − ESS = RSS, we can also write

$$r^2 = \frac{\text{total SS} - \text{ESS}}{\text{total SS}} = 1 - \frac{\text{ESS}}{\text{total SS}} = 1 - \frac{\Sigma(Y - \hat{Y}_x)^2}{\Sigma(Y - \bar{Y})^2}$$

where Ys are observed values of the dependent variable ($\bar{Y}$ being their mean) and $\hat{Y}_x$ is the estimated value of Y, given a value of independent variable X.

11. The limits of the prediction interval for an individual value of Y, given X (large-sample case).

$$I_{Y.X} = \hat{Y}_X \pm (Z_{a/2} \cdot S_{Y.X})$$

where $\hat{Y}_X$ is the estimated value of Y, the dependent variable, S_{YX} is the sample standard error of the estimate of Y, and z is the standard normal deviate.

12. Sample coefficient of determination: Alternative formulation.

$$r^2 = \frac{a\Sigma Y + b\Sigma XY - n\bar{Y}^2}{\Sigma Y^2 - n\bar{Y}^2}$$

where Xs are observed values of the independent variable, Ys are observed associated values of the dependent variable ($\bar{Y}$ being their mean), while n is sample size, and a and b are the estimated (intercept and slope) regression coefficients, respectively.

13. Pearson's sample coefficient of correlation.

$$r = \frac{\Sigma XY - n\overline{XY}}{\sqrt{(\Sigma X^2 - n\bar{X}^2)(\Sigma Y^2 - n\bar{Y}^2)}}$$

where Xs are observed values of the independent variable ($\bar{X}$ being their mean), Ys are observed associated values of the dependent variable ($\bar{Y}$ being their mean), and n is the sample size.

14. Sample coefficient of nondetermination.

$$k^2 = \frac{\text{unexplained variation}}{\text{total variation}} = \frac{\text{ESS}}{\text{total SS}} = \frac{\Sigma(Y - \hat{Y}_x)^2}{\Sigma(Y - \bar{Y})^2} = 1 - r^2$$

15. Sample coefficient of multiple determination.

$$R^2 = \frac{\text{explained variation}}{\text{total variation}} = \frac{\text{RSS}}{\text{total SS}} = \frac{\Sigma(\hat{Y} - \bar{Y})^2}{\Sigma(Y - \bar{Y})^2}$$

where Ys are observed values of the dependent variable ($\bar{Y}$ being their mean), while $\hat{Y}_s$ are estimated values of Ys.

16. Sample coefficient of multiple determinations: Alternative formulation.

$$R^2 = \frac{a\Sigma Y + b_1 \Sigma X_1 Y + b_2 \Sigma X_2 Y - n\bar{Y}^2}{\Sigma Y^2 - n\bar{Y}^2}$$

where a and b's are estimated partial regression coefficients, Xs are observed values of the independent variables, Ys are associated observed values of the dependent variable ($\bar{Y}$ being their mean), and n is the sample size.

Formulas Used in the Process of Constructing Control Charts

This section provides some typical formulas used in the control charting process. The list by no means is an exhaustive one; however, it provides the quality professional with some basic approaches to statistical understanding for the everyday application of some common charts. Specifically, Part I provides the most common descriptive statistical formulas, and Part II provides the formulas for calculating control chart limits and capability.

1. General Formulas

1. Arithmetic Mean (average) from ungrouped data.

For a population: $\mu = \dfrac{\Sigma X}{N}$

For a sample: $\bar{X} = \dfrac{\Sigma X}{n}$

For the average of the average $\bar{\bar{X}} = \dfrac{\Sigma \bar{X}}{n}$

where ΣX is the sum of all observed population (or sample) values, N is the number of observations in the population, and n is the number of observations in the sample. The $\Sigma \bar{X}$ is the average of the samples.

2. Arithmetic Mean (Average) from grouped data.

For a population: $\mu = \dfrac{\Sigma fX}{N}$

For a sample: $\bar{X} = \dfrac{\Sigma fX}{n}$

where ΣfX is the sum of all class-frequency(f) times class-midpoint (X) products, N is the number of observations in the population, and n is the number of observations in the sample.

3. Median from ungrouped data.

For a population: $M = X_{\frac{N+1}{2}}$ in an ascending ordered array

For a sample: $m = X_{\frac{n+1}{2}}$ in an ascending ordered array

where X is an observed population (or sample) value, N is the number of observations in the population, and n is the number of observations in the sample.

4. Median from grouped data.

For a population: $M = L + \dfrac{\left(\dfrac{N}{2}\right) - F}{f} w$

For a sample: $m = L + \dfrac{\left(\dfrac{n}{2}\right) - F}{f} w$

where L is the lower limit of the median class, f is its absolute frequency, and w is its width, while F is the sum of frequencies up to (but not including) the median class, N is the number of observations in the population, and n is the number of observations in the sample.

5. Mode from grouped data.

For a population or a sample: MO or $mo = L + \dfrac{d_1}{d_1 + d_2} w$

where L is the lower limit of the modal class, w is its width, and d_1 and d_2, respectively, are the differences between the modal class frequency density and that of the preceding or following class.

6. Weighted mean from ungrouped data

For a population: $\mu_w = \dfrac{\Sigma wX}{\Sigma w}$

For a sample: $\bar{X}_w = \dfrac{\Sigma wX}{\Sigma w}$

where $\Sigma w1X$ is the sum of all weight (w) times observed-value (x) products, while Σw equals N (the number of observations in the population) or n (the number of observations in the sample).

7. Mean absolute deviation

   From ungrouped data

   For population: $\text{MAD} = \dfrac{\Sigma|X - \mu|}{N}$

   For a sample: $\text{MAD} = \dfrac{\Sigma|X - \bar{X}|}{n}$

   where $\Sigma|X - \mu|$ is the sum of the absolute differences between each observed population value, X, and the population mean, μ, while N is the number of observations in the population, and where $\Sigma|X - \bar{x}|$ is the sum of the absolute differences between each observed sample value, X, and the sample mean, $\bar{X}$, while n is the number of observations in the sample.

   From grouped data

   Denoting absolute class frequencies by f and class midpoints by X, substitute $\Sigma f|X - \mu|$ or $\Sigma f|X - \bar{X}|$ for the numerators given here.

   Note: Occasionally, absolute deviations from the median rather than from the mean are calculated; in which case μ is replaced by M, and $\bar{X}$ is replaced by m.

8. Variance from ungrouped data.

   For a population: $\sigma^2 \dfrac{\Sigma(x - \mu)^2}{N}$

   For a sample: $s^2 \dfrac{\Sigma(x - \bar{x})^2}{n - 1}$

   where $\Sigma(x - \mu)^2$ is the sum of squared deviations between each population value, X, and the population mean, μ, with N being the number of observations in the population, while $\Sigma(x - \bar{x})^2$ is the sum of squared deviations between each sample value, X, and the sample mean, $\bar{x}$, with n being the number of observations in the sample.

9. Variance from grouped data.

   For a population: $\sigma^2 \dfrac{\Sigma f(x - \mu)^2}{N}$

   For a sample: $s^2 \dfrac{\Sigma f(x - \bar{x})^2}{n - 1}$

   where absolute class frequencies are denoted by f, class midpoints of grouped population (or sample) values by x, the population (or sample) mean byμ (or $\bar{x}$), and the number of observations in the population (or sample) by N (or n).

10. Standard deviation from ungrouped data.

For a population: $\sigma = \sqrt{\dfrac{\Sigma(x-\mu)^2}{N}}$

For a sample: $s = \sqrt{\dfrac{\Sigma(x-\bar{x})^2}{n-1}}$

where $\Sigma(x-\mu)^2$ is the sum of squared deviations between each population value, x, and the population mean, μ, with N being the number of observations in the population, while $\Sigma(x-\bar{x})^2$ is the sum of squared deviations between each sample value, x, and the sample mean, $\bar{x}$, with n being the number of observations in the sample.

11. Standard deviation from grouped data.

For a population: $\sigma = \sqrt{\dfrac{\Sigma f(x-\mu)^2}{N}}$

For a sample: $s = \sqrt{\dfrac{\Sigma f(x-\bar{x})^2}{n-1}}$

where absolute class frequencies are denoted by f, class midpoints of grouped population (or sample) values by x, the population (or sample) mean by μ (or $\bar{x}$), and the number of observations in the population (or sample) by N (or n).

12. Variance from ungrouped data—shortcut method.

For a population: $\sigma^2 = \dfrac{\Sigma x^2 - N\mu^2}{N}$

For a sample: $s^2 = \dfrac{\Sigma x^2 - n\bar{x}^2}{n-1}$

where Σx^2 is the sum of squared population (or sample) values, μ^2 is the squared population mean and $\bar{x}^2$ the squared sample mean, N is the number of observations in the population, and n is the number of observations in the sample.

13. Variance from grouped data—shortcut method.

For a population: $\sigma^2 = \dfrac{\Sigma fx^2 - N\mu^2}{N}$

For a sample: $s^2 = \dfrac{\Sigma fx^2 - n\bar{x}^2}{n-1}$

Where ΣfX^2 is the sum of absolute-class-frequency (f) times squared-class-midpoint (x) products, μ^2 is the squared population mean and

$\bar{x}^2$ is the squared sample mean, N is the number of observations in the population and n is the number of observations in the sample.

2. Control Chart Limits

a. X-bar and R Chart

Process average $= \text{Centerline} = \overline{\overline{X}}$

$$\text{UCL} = \overline{\overline{X}} + A_2\,\bar{R}$$

$$\text{LCL} = \overline{\overline{X}} - A_2\,\bar{R}$$

Process variation

$$\text{UCL} = D_4\,\bar{R}$$

$$\text{Centerline} = \bar{R}$$

$$\text{LCL} = D_3\,\bar{R}$$

where UPC and LCL denote the upper and lower control limits of the process, A_2, D_4, and D_3 are constants, $\overline{\overline{X}}$ is the process average and $\bar{R}$ is the average range.

b. Individual (X) and Moving Range Chart

Process average

$$\text{UCL} = \overline{\overline{X}} + E_2\bar{R}$$

Note that in this case, $\overline{X}$ and $\overline{\overline{X}}$ are the same.

$$\text{Centerline} = \overline{\overline{X}}$$

$$\text{LCL} = \overline{\overline{X}} - E_2\bar{R}$$

Process variation

$$\text{UCL} = D_4\bar{R}$$

$$\text{Centerline} = \bar{R}$$

$$\text{LCL} = D_3\,\bar{R}$$

where UPC and LCL denote the upper and lower control limits of the process, E_2, D_4, and D_3 are constants, $\overline{\overline{X}}$ is the process average, and $\bar{R}$ is the average range. To calculate the range, the experimenter must couple the individual data points in groups. The most sensitive (and recommended) grouping is a sample of two observations. More may be grouped together, however, much of the sensitivity will be lost.

c. X-bar and s chart

Process average

$$UCL = \overline{\overline{X}} + A_3\, \overline{s}$$

$$\text{Centerline} = \overline{\overline{X}}$$

$$LCL = \overline{\overline{X}} - A_3\, \overline{s}$$

Process variation

$$LCL = B_4 \overline{s}$$

$$\text{Centerline} = \overline{s}$$

$$LCL = B_3\, \overline{s}$$

where UPC and LCL denote the upper and lower control limits of the process, A_3, B_4, and B_3 are constants, $\overline{\overline{X}}$ is the process average, and $\overline{s}$ is the average standard deviation.

d. Median Chart

Process average

$$UCL = \tilde{\overline{X}} + \tilde{A}_2\, \overline{R}$$

$$\text{Centerline} = \tilde{\overline{X}}$$

$$LCL = \tilde{\overline{X}} - \tilde{A}_2\, \overline{R}$$

Process variation

$$UCL = D_4\, \overline{R}$$

$$\text{Centerline} = \overline{R}$$

$$LCL = D_3\, \overline{R}$$

where UPC and LCL denote the upper and lower control limits of the process, $\tilde{A}_2$, D_4, and D_3 are constants, $\tilde{\overline{X}}$ is the process median (median of all the sample medians), and $\overline{R}$ is the average range.

e. p-chart

$$UCL = \overline{p} + 3\sqrt{\frac{\overline{p}(1 - \overline{p})}{n}}$$

$$\text{Centerline} = \overline{p}$$

$$LCL = \bar{p} - 3\sqrt{\frac{\bar{p}(1-\bar{p})}{n}}$$

where UCL and LCL are the limits of the process and $\bar{p}$ is the average proportion defective.

f. np-chart

$$UCL = \bar{n}p + 3\sqrt{\frac{\bar{n}p(1-\bar{p})}{n}}$$

$$\text{Centerline} = \bar{p}$$

$$LCL = \bar{n}p - 3\sqrt{\frac{\bar{n}p(1-\bar{p})}{n}}$$

where UCL and LCL are the limits of the process and $\bar{n}p$ is the number of average proportion defective.

g. Standardized-value p-chart

$$P_s = \frac{(p-\bar{p})}{\sqrt{\frac{\bar{p}\,\bar{q}}{n}}}$$

Where P_s is the standard p-value
p is the observed sample proportion defective
q is the $(1-p)$ yield
n is the sample size
$\bar{p}$ is the process average proportion defective

h. c-chart

$$UCL = \bar{c} + 3\sqrt{\bar{c}}$$

$$\text{Centerline} = \bar{c}$$

$$LCL = \bar{c} - 3\sqrt{\bar{c}}$$

where UCL and LCL are the limits of the process and $\bar{c}$ is the average defect.

i. u-chart

$$UCL = \bar{u} + 3\sqrt{\frac{\bar{u}}{n}}$$

$$\text{Centerline} = \bar{u}$$

$$LCL = \bar{u} - 3\sqrt{\frac{\bar{u}}{n}}$$

where UCL and LCL are the limits of the process and $\bar{u}$ is the average of the number defects.

3. Capability

a. *Process Capability*

$$C_p = \frac{USL - LSL}{6\,\sigma_x}$$

where USL and LSL are the upper and lower specifications and σ_x is the standard deviation.

b. Capability Ratio

$$C_r = \frac{6\,\sigma_x}{USL - LSL}$$

where USL and LSL are the upper and lower specifications and σ_x is the standard deviation.

c. Capability Index

$$C_{pk} = \frac{Z_{min}}{3}$$

where Z_{min} is the less value of (USL-Xbar)/3σ and (Xbar-LSL)/3σ

d. *Target Ratio Percent*

$$TR_p = \frac{3\sigma}{Z_{min}} \times 100$$

Appendix D: Hypothesis Testing Roadmap

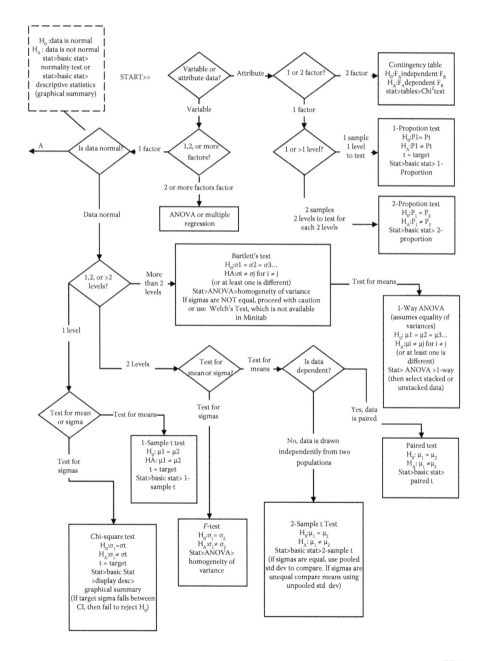

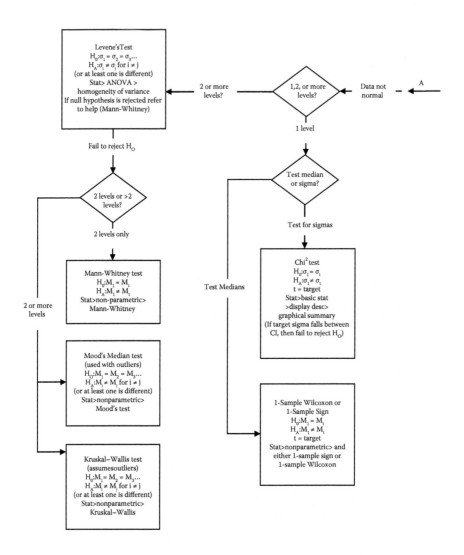

Appendix E: Test Indicators

Statistical test indicator

Types of difference		Nominal	Ordinal	Metric
Between 1 set of answers and a standard		X test, X² test	Kolmogorov Smirnov test	t-test
Between 2 sets of answers from	One sample	Mcnemar test	Sign test, Wilcoxon matched pairs	Dependent samples t-test
	Two independent samples	X test, X² test	Mann–Whitney (Wilcoxon) median test (x^2)	Independent samples t-test
Between k(k≥3) sets of answers from	One sample	Cochran's Q test	Friedman or Kendall coefficient of concordance	Repeated measures ANOVA
	k Independent Samples	X² test	Kruskal–Wallis test	Oneway ANOVA
Number of variables 2 variables		Pei coefficient contingency coefficient	Spearman rank correlation Kendall's ton	Pearson' S V
> 2 variables Analysis of interdependence		Latent structure analysis	Nonmetric multidimensional scaling	Factor analysis Cluster analysis
Analysis of dependence using single criterion & multiple predictors or independent variables (level of criterion on top, level of predictors on row).	Nominal	Search MNA	Mononova (Kruskal)	Mca search ANOVA regression
	Ordinal	Nonparametric discriminant analysis (Kendall)	Guttman-lingoes CM-2 regression	Regression (after transformations)
	Metric	Multiple discriminant analysis	Monotone regression (Carroll)	Multiple regression

Level of measurement

253

Selection path for control charts

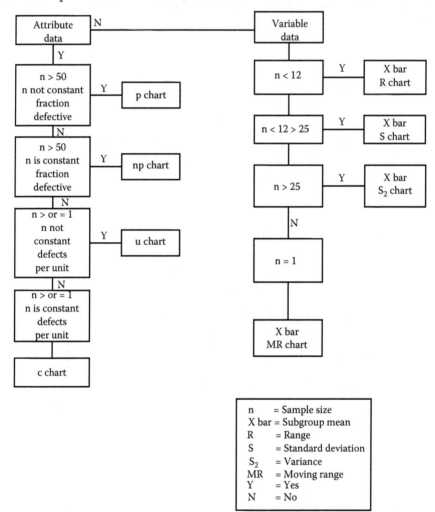

A selection guide for some common multivariate techniques

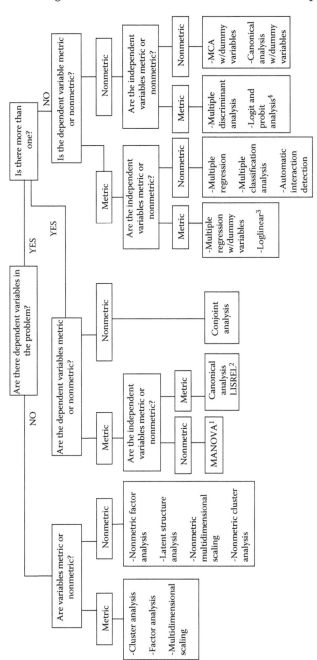

1. Factor may be considered nonmetric independent variable in that it organizes the data into groups. The reader will notice that MANOVA is used here as the umbrella of several analyses, rather than the individual classifications of MANOVA and other multivariate analyses of variance.
2. LISREL refers to a linear structural equation model for latent variables. It is a family of models appropriate for conformity factor analysis, path analysis, time series analysis, recursive and nonrecursive models, and covariance structure models. Because it may handle dependence and interdependence, metric and nonmetric, it is arbitrarily placed in this diagram.
3. The dependent variable is metric only when we consider the number of cases in the cross tabulation cell used to calculate the logs.
4. The independent variable is metric only in the sense that a transformed proportion is used.

Appendix F: Hypothesis Testing—Selected Explanations and Examples

Chapters 7, 8, 9, and 10 present some theoretical and conceptual explanations, appendix attempts to formalize these explanations with a succinct summary of sorts and with several examples.

In all statistical designs, there are five attributes that must be adhered to. They are:

1. Simplicity (kiss): Always keep it as simple as possible.
2. Flexibility: Be willing to be flexible in both designing and implementing the design of choice.
3. Reliability: Make sure that the design is reliable for the goal of the study.
4. Economics: Unless the study is cost-effective, it will not be followed. Therefore, make sure you chose a design that is easy to use, appropriate for the application at hand, and within the set budget.
5. Acceptance: A successful design is one that has been accepted by all stakeholders. Do not underestimate the influence of managers and operators. Acceptance does not mean that only the individuals who are responsible for the study or doing the study have to accept it, rather it should be accepted by everyone directly and indirectly involved.

Decision Making

Decisions are made based on a hypothesis. There are two alternatives available for decision making:

1. Based on Type I error
2. Based on Type II error

	H is True	H is False
Accept H	Correct decision	Type II error
Reject H	Type I error	Correct decision

$$\text{Error} = Z_{a/2} \frac{\sigma}{\sqrt{n}}$$

Confidence interval

$$\bar{x} - Z_{a/s}\frac{s}{n} < \mu < \bar{x} + Z_{a/2}\frac{s}{\sqrt{n}}$$

_t_–distribution

$$t = \frac{\bar{x} - \mu}{s/\sqrt{n}} \text{ (small samples: less than 30)}$$

Steps in testing hypothesis

1. State hypotheses (preferably in equation form).
2. State whether test is to be one-tail or two-tail.
3. State assumptions.
4. State the statistic to be used; compute its value and state degrees of freedom.
5. State value of *a* to be used and why it was selected to be high or low.
6. State limits; these must be consistent with #2.
7. State results and/or decision or action.

Test concerning means

1. $H_0 : \mu = 40$
   $H_1 : \mu \neq 40$ $\alpha = 0.05$ $Z = -1.96 - 1.96$

   $$z = \frac{\bar{x} - \mu_0}{\sigma/\sqrt{n}}$$

   $\bar{x} = 42.80$ $\sigma = 12.20$ and $n = 100$

   $$z = \frac{42.80 - 40.00}{12.20/\sqrt{100}} = 2.30$$

   Since 2.30 > 1.96, we reject H_0
   if $\alpha = .01$ or $z = -2.58 - 2.58$, then accept H_0

2. $H_0: \mu = 190$
   $H_1: \mu = > 190$
   $\alpha = .05$ $n = 200$ $x = 192.01$ $\sigma = 18.85$

   $$z = \frac{192.01 - 190}{18.85/\sqrt{200}} = 1.51$$

   Since 1.51 < 1.96, we accept H_0

3. $H_0: \mu = 3000$ $\alpha = 0.05$ (small samples)
   $H_1: \mu < 3000$

   $x = 2960$ $s = 46.8$ $n = 6$ $df = 5$

   $$t = \frac{2960 - 3000}{46.8/\sqrt{6}} = -2.09$$

   Since $|-2.09| < |-2.015|$, reject H_0 [We deal with a t-test, we use absolute values].

Differences between means

$$n_1 = 80 \qquad n_2 = 100$$

1. $H_0 : \mu_1 - \mu_2 = 0$
   $H_1 : \mu_1 - \mu_2 \neq 0$ $\alpha = .05$ $\overline{x_1} = 64.20$ $\overline{x_2} = 71.41$

   $$s_1 = 16.00 \qquad s_2 = 22.13$$

   $$z = \frac{\overline{x_1} - \overline{x_2}}{\sqrt{\dfrac{s_1^2}{n_1} + \dfrac{s_1^2}{n_2}}}$$

   $$z = \frac{64.20 - 71.41}{\sqrt{\dfrac{(16)^2}{80} + \dfrac{22.13}{100}}} = -2.53$$

   Since $|-2.53| < 1.96$, we reject H_0

2.

Brand 1	72	68	76	64	$n_1 = n_2 = 4$	$df = 8$	$\alpha = .05$
Brand 2	75	74	80	83	$x_1 = 70$	$x_2 = 78$	

$$t = \frac{\overline{x_1} - \overline{x_2}}{\sqrt{\dfrac{\sum(x_1 - \overline{x_1})^2 + \sum(x_2 - \overline{x_2})^2}{n_1 + n_2 - 2}} \cdot \left(\dfrac{1}{n_1} + \dfrac{1}{n_2}\right)} = \frac{70 - 78}{\sqrt{\dfrac{80 + 54}{4 + 4 - 2}\left(\dfrac{1}{4} + \dfrac{1}{4}\right)}}$$

$$= -2.39$$

Since $|-2.39| < |-2.447|$, H_0 is accepted

$$\sum\left(x_1 - \overline{x_1}\right)^2 = 2^2 + (-2)^2 + 6^2 + (-6)^2 = 80$$

$$\sum\left(x_2 - \overline{x_2}\right)^2 = (-3)^2 + (-4)^2 + 2^2 + 5^2 = 54$$

Use t-test for small samples.

Tests concerning proportion

1. $H_0 : p = .80$
   $H_1 : p \neq .80$ $x = 245$ $\alpha = .05$ $n = 320$ $p_0 = .80$

$$z = \frac{x - np_0}{\sqrt{np_0(1 - p_0)}} = \frac{245 - (320(.80))}{\sqrt{320(.80)(.20)}} = -1.54$$

Since $1.54 < 1.96$, accept H_0

Differences among proportion

$$H_0 : p_1 = p_2 = p_3$$
$$H_1 : p_1 \neq p_2 \neq p_3$$

Use chi-square.
  If cells have less than five observations, then use contingency coefficient.

$$c = \sqrt{\frac{x^2}{x^2 + n}}$$

The larger the c, the stronger the relationship.

The estimation of proportions
Binomial distribution assumption

$$\therefore \mu = np \quad \text{and} \quad \sigma = \sqrt{np(1 - p)}$$

$$\text{Error} = z_{a/2} \sqrt{\frac{p(1 - p)}{n}}$$

if we substitute x/n for p, we can assert with a probability $\approx 1 - \alpha$ that the error will be less than

$$E = Z_{a/2} \sqrt{\frac{\frac{x}{n}\left(1 - \frac{x}{n}\right)}{n}} \quad \text{then} \quad n = p\,(1 - p)\left[\frac{z_{a/2}}{E}\right]^2$$

The estimation of σ

$$\sigma_e = \frac{\sigma}{\sqrt{2n}} \quad \text{for large samples.}$$

Chi-Square
PROBLEM:

An inspection campaign yields a 10% response, while another, more expensive one, yields a 12% response. We want to decide whether the observed 2% difference is significant or whether it is merely a chance.

Example:

A frozen orange juice processor wants to determine whether the proportions of people who prefer (in taste tests) an acid-type juice to a non-acid type juice are the same in three distinct geographical regions. The processor has taken random samples of 200, 150, and 100 in Region A, B, and C and found 124, 75, and 120 people in these three samples who prefer the acid-type juice to the nonacid type. The results can be shown in the table form as:

Rekey

The question is whether the observed differences among the three sample proportions $124/200 = .62$, $75/150 = .50$, and $120/200 = .60$, for regions A, B, and C, are too large to be accounted for by chance or whether they may reasonably be attributed to chance.

$$\text{Hypothesis} = H_0\text{: } p_1 = p_2 = p_3.$$

$$\text{Alternate} = H_1\text{: } p_1, p_2, \text{ and } p_3 \text{ are not equal.}$$

Solution:

If the null hypothesis is true, we can combine the three samples and estimate the common proportion of people who prefer the acid type as:

$$\frac{124+75+120}{200+150+200} = \frac{319}{550} = .58$$

With this estimate of the common population proportion, we can expect to find $200(.58) = 116$, $150(.58) = 87$, and $200(.58) = 116$ for each of the regions, respectively.

Now if you subtract these totals from the respective sample totals, we would expect:

Region A	Region B	Region C
$200 - 116 = 84$	$150 - 87 = 63$	$200 - 116 = 84$

These results then are called expected and are shown in parenthesis below the actual observed frequencies.

Rekey

To test the hypothesis $p_1 = p_2 = p_3$, we must compare the frequencies of the actual with those we could expect if the null hypothesis were true.

$\therefore$ H_0 is accepted if the two frequencies are alike; if the discrepancies between the two sets are large, then this suggests that H_0 must be false and H_1 is accepted.

Using them in the formula

$$X^2 = \sum \frac{(f-e)^2}{e}$$

where e = expected and f = observed

$$X^2 = \frac{(124-116)^2}{116} + \frac{(75-87)^2}{87} + \frac{(120-116)^2}{116} + \frac{(76-84)^2}{84}$$

$$+ \frac{(75-63)^2}{63} + \frac{(80-84)^2}{84} = 5.583$$

if X^2 = small, this implies that the agreement is close.

If X^2 = large, this implies that the agreement is poor.

The exact criterion for this decision depends on the calculated table value and degrees of freedom. In this case, the $df = 2 = 3 - 1$ and table value is 5.991.

Since $X^2_{calc.} = 5.583 < X^2_{table} = 5.991$, we accept H_0.

Correlation

- When one wants to find out the degree to which two variables are related, he or she is asking a correlation question.
- The strength or closeness of the relationship is described when one talks about the correlation coefficient; it measures the degree of that relationship.
- The correlation coefficients vary from –1 to 0 to = 1

What Is Correlation?

The π or correlation coefficient (quite often it is denoted by the letter r) is simply the mean of the z-score products or

$$\pi = \frac{\sum z_x z_y}{N}$$

where N = number of pairs.

First, you convert every x-score and y-score value to a z-score, and second, you multiply each z-score of x by its z-score of y.

To find the value of π from the raw scores you utilize the following formula

$$\pi = \frac{N\sum XY - \left(\sum X\right)\left(\sum Y\right)}{\sqrt{\left[N\sum X^2 - \left(\sum X\right)^2\right]\left[N\sum Y^2 - \left(\sum Y\right)^2\right]}}$$

where

$\sum XY$ = the product of each x by its y, then sum the results
N = Number of pairs
$\sum X$ = Sum of x scores
$\sum Y$ = Sum of y scores
$\sum X^2$ = Square the x scores, then sum the results
$\sum Y^2$ = Square the y scores, then sum the results

Assumptions for Correlation

1. Relationship is linear.
2. Scores of the population form a normal distribution curve.
3. Scattergram is homoscedastic (the distance from the raw data to the regression line is equal).
4. Data are at interval level of measurement

Level of measurement	Correlation technique	Kind
Nominal	Contingency	Nonparametric
Ordinal	Spearman rank	Nonparametric
Interval	Pearson-product-moment	Parametric

$$\text{Contingency coefficient} = c = \sqrt{\frac{x^2}{x^2 + N}}$$

$$\text{Spearman rank} = \text{rho} (\rho) = 1 - \frac{6\sum D^2}{N(N^2 - 1)}$$

where x^2 is the chi square; D is the difference between the ranks in the column labeled; N is the number of individuals in the group.

Example:

INDIVIDUAL	X	Y	XY	X^2	Y^2
A	3	5	15	9	25
B	4	5	20	16	25
C	4	4	16	16	16
D	5	3	15	25	9
E	7	2	14	49	4
F	8	2	16	64	4
G	10	1	10	100	1
$N = 7$	$\Sigma X = 41$	$\Sigma Y = 22$	$\Sigma XY = 106$	$\Sigma X^2 = 279$	$\Sigma Y^2 = 84$

Using the formula

$$\pi = \frac{N\sum XY - \left(\sum X\right)\left(\sum Y\right)}{\sqrt{\left[N\sum X^2 - \left(\sum X\right)^2\right]\left[N\sum Y^2 - \left(\sum Y\right)^2\right]}}$$

Once substituting the values, we get

$$\pi = \frac{(7)106 - (41)(22)}{\sqrt{[7(279) - (41)^2][7(84) - (22)^2]}} = -\frac{160}{168.19} = -0.95$$

Gauge for Interpreting Coefficient of Correlation

A. 0.85 TO 1.00 (OR –0.85 TO – 1.00)   High

B. 0.50 TO 0.84 (OR –0.50 TO –0.84)   Moderate

C. 0 TO 0.49 (OR 0.0 TO –0.49)   Low

NOTE: The stronger the relationship between the two variables, the closer the coefficient is to +1.00 or –1.00. However, it must be emphasized that this relationship in no way implies causality.

Regression Analysis

While correlation analysis is concerned with the strength of the relationship between two variables, regression analysis describes the way in which one variable is related to another.

Regression analysis derives an equation, which can be used to estimate the unknown value of one variable on the basis of the known value of the other variable.

Aims of Regression and Correlation Analysis

1. Regression analysis provides estimates of the dependent variable for given values of the independent variable.

2. Regression analysis provides measures of the errors that are likely to be involved in using the regression line to estimate the dependent variable.

3. Regression analysis provides an estimate of the effect on the mean value of Y of a one-unit change in X.

4. Correlation analysis provides estimates of how strong the relationship is between the two variables. The coefficient of correlation (π value or r^2) and the coefficient of determination $\pi^2 = ((\text{Variation explained by} - \text{regression})/(\text{Total variation}))$ are two measures generally used for this purpose.

5. The linear regression model is of the form

$$Y = a + bx + e$$

where
  $Y = $ Dependent variable
  $A = Y$ Intercept
  $b = $ Coefficient of correlation
  $x = $ independent variable
  $e = $ Error

Hazards and Problems in Regression and Correlation

1. It is by no means true that a high coefficient of determination or correlation between two variables means that one variable causes the other variable to vary.

2. Even if an observed correlation is due to a causal relationship, the direction of causation may be the reverse of that implied by the regression.

3. Regression sometimes used to forecast values of the dependent variable corresponding to values of the independent variable lying beyond the sample range.

4. It is important to recognize that a regression based on past data may not be a good predictor, due to shifts in the regression line.

5. When carrying out a regression, it is important to try to make sure you follow the assumptions you have set for your problem as well as the general assumption of regression analysis.

Multiple Regression and Correlation

Whereas a simple regression includes only one independent variable a multiple regression includes two or more independent variables. Basically, there are two important reasons why a multiple regression must often be used instead of a simple regression: (1) one frequently can predict the dependent variable more accurately if more than one independent variable is used and (2) if the dependent variable depends on more than one independent variable, a simple regression of the dependent variable on a single independent variable may result in a biased estimate of the effect of this independent variable on the dependent variable. The general model for multiple regression is:

$$Y = a + b_1 x_{1i} + b_2 x_{2i} + \dots + e$$

where Y = dependent variable; X_1 and X_2 = independent variables; a = intercept constant; b_1 and b_2 = correlation coefficient; e = error term.

Regression

If $y = a + bx$ explains, then

$$y' = a + bx \quad \text{predicts.}$$

The least squares criterion requires that we minimize the sum of the squares of the differences between the observed values of y and the corresponding predicted values of y'.

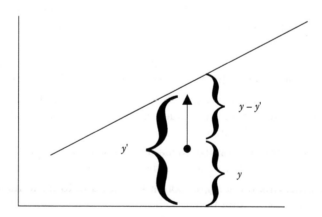

To solve the predicted equation, we must find the numerical values of a and b appearing in the equation $y' = a + bx$ for which $\Sigma(y - y')^2$ is as small as possible. The minimum values (using calculus) of the $\Sigma(y - y')^2$ yields the following two equations for a and b:

1. $\Sigma y = na + b(\Sigma x)$

    and

2. $\Sigma xy = a(\Sigma x) + b(\Sigma x^2)$

    where N = number of observations; Σx, Σy = sums of the given x's and y's; Σx^2 = sum of squares of the x's and, Σxy = sum of the products of x's and y's

    For example,

Time	Efficiency	x^2	xy
14	55	196	770
34	60	1156	2040
18	44	324	792
40	78	1600	3120
48	75	2304	3600
18	50	324	900
44	62	1936	2728
30	65	900	1950
26	55	676	1430
38	70	1444	2660
$\Sigma x = 310$	$\Sigma y = 614$	$\Sigma x^2 = 10860$	$\Sigma xy = 19990$

Now substitute in (1) and (2).

3. $614 = 10a + 310b$

4. $19990 = 310a + 10860b$

    to solve (1) and (2) equations using algebra we obtain.

5. $a = \dfrac{\Sigma(y)(\Sigma x^2) - (\Sigma(y)(\Sigma xy))}{n(\Sigma x^2) - (\Sigma x)^2}$

6. $b = \dfrac{n\Sigma xy - (\Sigma x)(\Sigma y)}{n(\Sigma x^2) - (\Sigma x)^2}$

    In our example, we have

7. $a = \dfrac{(614)(10860) - (310)(19990)}{10(10860) - (310)^2} = 37.69$

8. $b = \dfrac{10(19990) - (310)(614)}{10(10860) - (310)^2} = 0.7648$

    Alternate Way:

$a = \dfrac{\Sigma y - b\Sigma x}{n} \dfrac{614 - 0.7648(310)}{10} = 37.69$

The prediction equation we want then is

A. $y' = 37.69 + 0.7648x$: Given this equation we can now predict what happens to the y when the x is changed.

Confidence Limits for Regression

Standard error estimate

9. $s_e = \sqrt{\dfrac{\Sigma(y - y')^2}{n - 2}}$

where $n - 2 = df$.

This measures that portion of the total variation of the y's, which can be attributed to chance. Equation (9) can be rewritten as

10. $s_e = \sqrt{\dfrac{\Sigma y^2 - a\Sigma y - b\Sigma xy}{n - 2}}$

If we substitute our values, we have

$$s_e = \sqrt{\dfrac{38764 - 37.69(614) - 0.7648(19990)}{10 - 2}} = 6.5$$

To predict the confidence for a value y_0' for a given value x_0 of x, we can assert with a probability of $1 - a$ that the value we will actually obtain will lie between the limits $y_0' - A$ and $y_0' + A$. This is called the limits of prediction.

11. $A = t_{a/2}\ S_e\sqrt{\dfrac{n+1}{n} + \dfrac{(x_0 - \bar{x})^2}{\Sigma x^2 - n\bar{x}^2}}$

where

$t = t - $distribution; $a/2 = $probability; $x_0 = $the value we want to make a prediction; $y_0' = $the new value after x_0 substitution; n, $\bar{x}$, and Σx^2 come from original data.

In our example, let us assume $x_0 = 30$ when substituting this value in equation (A) we get:

$$y_0' = 37.69 + 0.07648(30) = 61$$

To find 95% limits of prediction we substitute in (11)
 $S_e = 6.5$; $n = 10$; $\bar{x} = 310/10 = 31$; $\Sigma x^2 = 10860$ and $t_{.025} = 2.306$ (for eight degrees of freedom, we lost two in calculating a and b).

$$A = 2.306(6.5)\sqrt{\dfrac{11}{10} + \dfrac{(30 - 31)^2}{10860 - 10(31)^2}} = 16$$

Therefore, the limits of prediction are

$$61 - 16 = 45 \text{ and } 61 + 16 = 77$$

$$R^2 = \frac{\text{Variation explained by regression}}{\text{Total variation}}$$

$$R^2 = \frac{b_1 \sum\limits_{i=1}^{n} (x_1 - \bar{x}_1)(y - \bar{y}) + b_2 \sum\limits_{i=1}^{n} (x_2 - \bar{x}_2)(y - \bar{y})}{\sum\limits_{i=1}^{n} Y^2 - \dfrac{\left(\sum\limits_{i=1}^{n} Y\right)^2}{n}}$$

∴ R^2 measures the proportion of the total variation in the dependent variable that is explained by the regression equation.

ADJ – RSQ, the amount of shrinkage of R^2 because of the ratio of the number of independent variables to the size of the sample. (Other things being equal, the larger the ratio the greater the overestimation of R)

$$\hat{R}^2 = 1 - (1 - R^2) \frac{N-1}{N-K-1}$$

where $\hat{R}^2$ = ADJ – RSQ; R^2 = squared multiple correlation; N = size of sample; K = number of predictors.

Variance

$$S_x^2 = \frac{\sum (X - \bar{X})^2}{N-1} = \frac{\sum X^2}{N-1}$$

When dealing with raw scores

$$\sum x^2 = \sum x^2 - \frac{\left(\sum x\right)^2}{N}$$

$$\sum y^2 = \sum y^2 - \frac{\left(\sum y\right)^2}{N}$$

Covariance

$$S_{xy} = \frac{\sum (X - \bar{X})(Y - \bar{Y})}{N-1} = \frac{\sum XY}{N-1}$$

$$\sum{}_{xy} = \frac{\left(\sum X\right)\left(\sum Y\right)}{N} + \sum XY \text{ (sum of cross products)}$$

In short, the variance indicates the variation of a set of scores from their mean, whereas the covariance indicates the covariation of two sets of scores from their respective means.

In regression, we said that the values of a and b are of critical importance; they can be calculated by:

$$b = \frac{\sum XY}{\sum X^2}$$

$$a = \bar{Y} - b\bar{X}$$

And the regression equation becomes

$$Y' = a + bX$$

Note that there is no e because $e = Y - Y'$ and in this case $\Sigma(Y - Y')^2 = $ Sum of squared residuals.

Partitioning the Sum of Squares

$$\sum y^2 = \underset{(1)}{\sum \left(Y' - Y\right)^2} + \underset{(2)}{\sum \left(Y - Y'\right)^2}$$

$$\sum y^2 = SS_{reg} + SS_{res} \quad \text{or} \quad 1 = \frac{SS_{reg}}{\sum y^2} + \frac{SS_{res}}{\sum y^2}$$

where SS_{reg} = regression sum of squares; SS_{res} = residual sum of squares.

The aforementioned principle in regression analysis states that the deviation sum of squares of the dependent variable Σy^2 is partitioned into two parts; $\Sigma(Y' - Y)^2$ is the sum of squares due to regression, or the regression sum of squares; and $\Sigma(Y - Y')^2$ is the sum of squares due to residuals, or the residual sum of squares.

NOTE: When (1) = 0 then Σy^2 = to the residual sum of squares, or the error, i.e., nothing has been gained by resorting information from X. When (2) = 0 then all the variability in Y is explained by regression, or by the information provided by X.

Testing the Regression

$$t = \sqrt{F} \quad \text{or} \quad F = \frac{SS_{reg}/df_1}{SS_{reg}/df_2} = \frac{SS_{reg}/K}{SS_{res}/(N-K-1)}$$

where
  N = sample size; K = number of independent variables.

Standardized Weights

$$\beta = b\frac{S_x}{S_y}$$

where S_x and S_y = standard deviation of x and y.

$$b = \beta\frac{S_y}{S_x}$$

NOTE:

$$\beta = b\frac{S_x}{S_y} = \frac{\sum xy}{\sum x^2 \sqrt{N-1}}\frac{\sqrt{\sum x^2}\sqrt{N-1}}{\sqrt{\sum y^2}} = \frac{\sum xy}{\sqrt{\sum x^2}\sqrt{\sum y^2}} = r_{xy}$$

$a = 0$ when using standard scores.
  For two variables:

$$\beta_1 = \frac{r_{y_1} - r_{y_2}r_{12}}{1 - r_{12}^2} \qquad \beta_2 = \frac{r_{y_2} - r_{y_1}r_{12}}{1 - r_{12}^2}$$

NOTE: When the independent variables are not correlated, i.e., $r_{12} = 0$; $\beta_1 = r_{y1}$, and $\beta_2 = r_{y2}$ as in the case of linear regression.

ANOVA (One-way)

One of the most powerful tools of statistical analysis is what is known as *analysis of variance*. Basically, it consists of classifying and cross-classifying statistical results and testing whether the means of a specified classification differ significantly.

For example, the output of a given process might be cross-classified by machines and operators (each operator having worked on each machine). From this cross-classification, it could be determined whether the mean qualities of the outputs of the various operators differed significantly. Also it could independently be determined whether the mean qualities of the outputs of the various machines differed significantly. Such a study would determine, for example, whether uniformity in quality of output could be increased by standardizing the procedures of the operators (say through training) and similarly whether it could be increased by standardizing the machines (say through resetting).

NOTE:

1. The ANOVA's paramount feature over the control charts is the fact that it is capable of giving you maximum information with limited amount of data.

2. A better applicability exists in laboratory environment rather than a process control.

ANOVA Requirements

1. Population normally distributed.
2. Variances equal.
3. Three or more independent groups.
4. Samples drawn randomly from the population.
5. At least interval level of measurement.

Formula

$$F = \frac{MS_B}{MS_W}$$

where
MS_B = The variance between the groups is called the mean square between.
MS_W = The variance within the groups is called the mean square within.

Computational Procedures for the One-Way ANOVA

Source of Variation	Sum of Squares (SS)	Degrees of Freedom (df)	Mean Square (MS)	F
Between groups	$\frac{(\Sigma x_1)^2}{n_1} + \frac{(\Sigma x_2)^2}{n_2} + \ldots$ $+ \frac{(\Sigma x_k)^2}{n_k} - \frac{(\Sigma x)^2}{N}$	$K-1$	$\frac{\text{Between SS}}{\text{Between df}}$	$\frac{\text{MS Between}}{\text{MS Within}}$
Within groups	(Total SS) – Between SS	(Total df) – (Between df) or $N-k$	$\frac{\text{SS Within}}{\text{df Within}}$	
Total	$\Sigma x^2 - \frac{(\Sigma x)^2}{N}$	$N-1$		

ANOVA

	Route 1	Route 2	Route	Route 4
	22	25	25	27
	26	27	29	29
	25	28	33	28
	25	26	30	30
	32	29	33	31
ΣX	26	27	30	29

$$H_0 : \mu_1 = \mu_2 = \mu_3 = \mu_4$$

$$H_1 : \mu_1 \neq \mu_2 \neq \mu_3 \neq \mu_4$$

$$\sum \bar{x} = \frac{26+27+30+29}{4} = \frac{112}{4} = 28$$

Assume that the population from which we are sampling can be approximated closely with normal distribution having the same standard deviation.

If, then our H_0 is true, this implies that our four samples are from one and the same population, hence:

$$s_{\bar{x}}^2 = \frac{10}{3}$$

since

$$s_{\bar{x}}^2 = \frac{(26-28)^2 + (27-28)^2 + (30-28)^2 + (29-28)^2}{4-1} = \frac{10}{3}$$

If our assumption about normality and population exists, then we can use:

$$\sigma\bar{x} = \frac{\sigma}{\sqrt{n}}$$

for samples from infinite population

$$s_{\bar{x}}^2 = \frac{10}{3} \quad \text{as an estimate} \quad \sigma^2 x = \left(\frac{\sigma}{n}\right)^2 = \frac{\sigma^2}{n}$$

Becomes

$$s_{\bar{x}}^2 = \sigma_{\bar{x}}^2 = \frac{\sigma^2}{n}$$

now substituting, we have:

$$n \cdot s_x^2 = \sigma^2 = 5 \cdot \frac{10}{3} = \frac{50}{3}$$

note that the 50/3 is the common variance of all four populations. If s^2 were known, we could compare $n \cdot s_x^2$ with σ^2 and reject the H_0 if $n \cdot s_x^2 > \sigma^2$.

In our example, since σ^2 is unknown we have no choice but to estimate it based on our data. This estimation comes from the sample variance i.e. s_1^2, s_2^2, s_3^2, s_4^2 as an estimate of σ^2 and as a consequence we can use their mean

$$\frac{s_1^2 + s_2^2 + s_3^2 + s_4^2}{4}$$

$$\frac{1}{4} \left[\frac{(22-26)^2 + (26-26)^2 + (25-26)^2 + (25-26)^2 + (32-26)^2}{5-1} \right.$$

$$+ \frac{(25-27)^2 + (27-27)^2 + (28-27)^2 + (26-27)^2 + (29-27)^2}{5-1}$$

$$+ \frac{(25-30)^2 + (29-30)^2 + (33-30)^2 + (30-32)^2 + (33-30)^2}{5-1}$$

$$\left. + \frac{(27-29)^2 + (29-29)^2 + (28-29)^2 + (30-29)^2 + (31-29)^2}{5-1} \right]$$

$$= \frac{118}{16}$$

With this result we have obtained two estimates of σ^2

$$n \cdot s_x^2 = \frac{50}{3} = 16\frac{2}{3} \quad \text{and} \quad \frac{s_1^2 + s_2^2 + s_3^2 + s_4^2}{4} = \frac{118}{15} = 7\frac{3}{8}$$

The first estimate is based on the variation among the sample means and the second estimate is based on the variation within the samples or the variation that is due to chance.

If the first estimate is *much* larger than the second one, then the H_0 should be rejected. Formally, the test or criteria for rejection in the F-test in the form of:

$$F = \frac{\text{Estimate of } \sigma^2 \text{ based on the variation among the } \overline{x}', s}{\text{Estimate of } \sigma^2 \text{ based on the variation with the samples}}$$

Substituting our values we have:

$$F = \frac{16\ 2/3}{7\ 3/8} = 2.26$$

From the table of the F-distribution based on the df the $a = .05$ we have a value of 3.24.

Since this value *does not* exceed the value of the calculated F we cannot reject the H_0 even though the first estimate it seems to be quite a bit larger than the second one. These differences may very well be attributed to chance.

One-way Analysis

Fundamental analysis of variance expresses a measure of the variation (total) of a set of data as a sum of terms each of which can be attributed to a specific source, or cause, of variation. The measure of the total variation of a set of data which we use in analysis of variance is the Total Sum of Squares.

$$SST = \sum_{i=1}^{k}\sum_{j=1}^{n}(x_{ij} - \bar{x}..)^2$$

where i and j are the individual observations and x, the grand mean.

If we divide the SST by $(n, k-1)$ we obtain the variance of the combined data. The total sum of squares of a set of data is interpreted in much the same way as its variance.

Letting $\bar{x}_i$ denote the mean of the ith sample, we rewrite the SST as follows:

$$SST = n \cdot \sum_{i=1}^{k}(\bar{x}_{i.} - \bar{x}..)^2 + \sum_{i=1}^{k}\sum_{j=1}^{n}(x_{ij} - \bar{x}_{i.})^2$$

Note that the two terms of the SST have been partitioned and that the first part measures the variation among the sample means. (If you divide it by $k-1$ we get indeed $n \cdot s_{\bar{x}}^2$). The second term measures the variation within the individual samples. (If you divide it by $k(n-1)$ we would get the mean of the variances of the individual samples, or the denominator of F).

The variation among the sample means is referred to as the treatment sum of squares SS(Tr) and the variation within the individual samples is referred to as the error sum of squares SSE.

To summarize:

$$SST = SS(Tr) + SSE$$

To check this identity let us substitute the various sum of squares from the preceding section, we get:

$$SST = (22-28)^2 + (26-28)^2 + (25-28)^2 + (25-28)^2 + (32-28)^2$$
$$+ (25-28)^2 + (27-28)^2 + (28-28)^2 + (26-28)^2 + (29-28)^2$$
$$+ (25-28)^2 + (29-28)^2 + (33-28)^2 + (30-28)^2 + (33-28)^2$$
$$+ (27-28)^2 + (29-28)^2 + (28-28)^2 + (30-28)^2 + (31-28)^2$$
$$= 168$$

$$SS(Tr) = 5[(26-28)^2 + (27-28)^2 + (30-28)^2 + (29-28)^2] = 50$$

$$SSE = SST - SS(Tr) = 168 - 50 = 118$$

or

$$SSE = (22-26)^2 + (26-26)^2 + (25-26)^2 + (25-26)^2 + (32-26)^2$$
$$+ (25-27)^2 + (27-27)^2 + (28-27)^2 + (26-27)^2 + (29-27)^2$$
$$+ (25-30)^2 + (29-30)^2 + (33-30)^2 + (30-30)^2 + (33-30)^2$$
$$+ (27-29)^2 + (29-29)^2 + (28-29)^2 + (30-29)^2 + (31-29)^2$$
$$= 118$$

To test the hypothesis of $\mu_1 = \mu_2 = \mu_3 = \mu_4$ against the Alternative hypothesis that he treatment mean are not equal we compare SS(Tr) and SSE with the appropriate F statistic.

To finalize the testing we construct a table with the following information:

Source of Variation	Degrees of Freedom	Sum of Squares	Mean Square	F
Treatments	$k-1$	SS(Tr)	$MS(Tr) = \dfrac{SS(Tr)}{k-1}$	$\dfrac{MS(Tr)}{MSE}$
Error	$k(n-1)$	SSE	$MSE = \dfrac{SSE}{k(n-1)}$	
Total	$nk-1$	SST		

Substituting our values into the table we have:

Source of Variation	Degrees of Freedom	Sum of Squares	Mean Square	F
Treatments	3	50	$\dfrac{50}{3}$	$\dfrac{50/3}{118/16} = 2.26$
Error	16	118	$\dfrac{118}{16}$	
Total	19	168		

$\therefore$ Since $2.26 < 3.24$ the hypothesis is accepted.

Short Cut Formulas

1. $SST = \displaystyle\sum_{i=1}^{k}\sum_{j=1}^{n} X^2{}_{ij} -_k \dfrac{1}{n} \cdot T..$

2. $SS(Tr) = \dfrac{1}{n}\displaystyle\sum_{i=1}^{k} T^2{}_{i.} -_k \dfrac{1}{n} .T^2.. \text{ and}$

3. $SSE = SST - SS(Tr)$ where
T_i = The total of the observations corresponding to the ith treatment (i.e. the sum of the values in the ith sample); $T..$ = The grand total of all the data: substituting, we find that:

$$SST = 15848 - \frac{1}{20}(560)^2 = 168$$

$$SS(Tr) = \frac{1}{5}(130^2 + 135^2 + 150^2 + 145^2) - \frac{1}{20}(560)^2 = 50$$

$$SSE = 168 - 50 = 118$$

Example

GROUP I	GROUP II	GROUP III
4	2	2
2	1	1
1	4	6
5	7	4
2	5	7

Solution: Part I

1. Find $\Sigma x_1, \Sigma x_2$ & Σx_3
2. Find $\Sigma x = \Sigma x_1 + \Sigma x_2 + \Sigma x_3$,
3. Find n_1, n_2, and n_3
4. Find $N = n_1 + n_2 + n_3$
5. Find the correction term $(\Sigma x)^2/N$
6. Find the between-group sum of squares (between SS)

$$\frac{\left(\sum X_1\right)^2}{n_1} + \frac{\left(\sum X_2\right)^2}{n_2} + \frac{\left(\sum X_3\right)^2}{n_3} - \frac{\left(\sum X\right)^2}{N}$$

7. Find degree of freedom $k - 1$ where $k = $ # of groups

Solution: Part II

1. Rewrite the scores for each group
2. Form the X_1^2, X_2^2 & X_3^2 columns
3. Find the $\Sigma X_1^2, \Sigma X_2^2$ & ΣX_3^2
4. Find $\Sigma X^2 = \Sigma X_1^2 + \Sigma X_2^2 + \Sigma X$
5. Find the total sum of squares $\Sigma X^2 - (\Sigma X)^2/N$
6. Find the total degrees of freedom $N - 1$

Solution: Part III

1. Find the within-groups sum of squares. Subtract the between-groups sum of squares from the total sum of squares and put the difference in the table.
2. Find the within-groups degrees of freedom. Subtract the between-groups degree of freedom from the total degrees of freedom and put the difference in the table.
3. Find the between-groups mean square. Divide the between-groups sum of squares by the between-groups degrees of freedom.
4. Find the within-groups mean square. Divide the within-groups sum of squares by the within-groups degrees of freedom.
5. Find F. Divide the between-groups mean square by the within-groups mean square.
6. Find the table value (within groups-down and between groups-across. Bold face usually is denoted for .01 significance and regular print is for .05 significance.)
7. If the F $\geq$ to table value reject hypothesis; if F $<$ table value accept hypothesis

Example

GROUP I	GROUP II	GROUP III
X_1	X_2	X_3
4	2	2
2	1	1
1	4	6
5	7	4
2	5	7

Solution: Part I

1. $4+2+1+5+2 = 14 = \sum X_1$

   $2+1+4+7+5 = 19 = \sum X_2$

   $2+1+6+4+7 = 20 = \sum X_3$

2. $14+19+20 = 53 = \sum X$

3. $n_1 = 5 \quad n_2 = 5 \quad n_3 = 5$

4. $5+5+5 = 15 = N$

5. $(53)^2/15 - 187.27$

6. $\dfrac{(14)^2}{5} + \dfrac{(19)^2}{5} + \dfrac{(20)^2}{5} - 187.27 = 4.13$

7. $k-1 = 3-1 = 2$

Solution: Part II

1 & 2	GROUP I		GROUP II		GROUP III	
	X_1	X_1^2	X_2	X_2^2	X_3	X_3^2
	4	16	2	4	2	4
	2	4	1	1	1	1
	1	1	4	16	6	36
	5	25	7	49	4	16
	2	4	5	25	7	49

3. $16+4+1+25+4 = 50 = \sum X_1^2$

   $4+1+16+49+25 = 95 = \sum X_2^2$

   $4+1+36+16+49 = 106 = \sum X_3^2$

4. $\sum X^2 = 50+95+106 = 251$

5. $251 - 187.27 = 63.73$

6. $15-1 = 14$

Solution: Part III
Summary Table

Source of Variation	Sum of Squares	Degrees of Freedom	Mean Square	F	Probability
Between	4.13	2	2.07		
				0.42	(n, s)
Within Groups	59.60	12	4.97		
Total	63.73	14			

1. $63.73 - 4.13 - 59.60$
2. $14 - 2 = 12$
   Or, use $N - k = 15 - 3 = 12$

3. $\dfrac{4.13}{2} = 2.07$

4. $\dfrac{59.60}{12} = 4.97$

5. $\dfrac{2.07}{4.97} = .42$

6. Table value is 3.88
7. Since 0.42 is less than 3.88 we do not reject the hypothesis.

EXAMPLE # 2

A manager wants to see if there is a significant difference between three machines.
   The data are as follows:

I	II	III
50	41	49
51	40	47
51	39	45
52	40	47

Use a one-way analysis to test the mean difference of the three machines at $\alpha = .05$ significance.

Solution:

I		II		III	
X_1	X_1^2	X_2	X_2^2	X_3	X_3^2
50	2500	41	1681	49	2401
51	2601	40	1600	47	2209
51	2601	39	1521	45	2025
52	2704	40	1600	47	2209

$$\sum X_1 = 204 \qquad \sum X_2 = 160 \qquad \sum X_3 = 188$$

$$\sum X_1^2 = 10406 \qquad \sum X_2^2 = 6402 \qquad \sum X_3^2 = 8844$$

$$n_1 = 4 \qquad n_2 = 4 \qquad n_3 = 4$$

$$N = n_1 + n_2 + n_3 = 4 + 4 + 4 = 12$$

$$\sum X = 552 \quad \text{and} \quad \frac{(\sum X)^2}{N} = 25392$$

Between-groups sum of squares

$$\frac{(204)^2}{4} + \frac{(160)^2}{4} + \frac{(188)^2}{4} - 25392$$

$$= 10404 + 6400 + 8836 - 25392$$

$$= 25640 - 25392 = +248$$

Between-groups *df*

$$k - 1 = 3 - 1 = 2$$

$$N = 12$$

$$\sum X^2 = 25652 \quad \text{and} \quad \frac{(\sum X)^2}{N} = 25392$$

Total sum of squares = 25652 − 25392 = + 260
   Total *df* = $N - 1 = 12 - 1 = 11$

Summary Table

Sources of Variation	Sum of Squares	df	Mean Squares	F
Between Groups	248	2	$\frac{248}{2} = 124$	
Within Groups	260-2148=12	11-2=9 OR N-k =12-3=9	$\frac{12}{9} = 1.33$	$\frac{124}{1.33} = 93$
Total	260	11		

Table value at $\alpha = .05$ is $F_{.05} = 4.26$

Therefore, the hypothesis is rejected and is concluded that the three machines are not different from each other.

Two-way Analysis of Variance

In the examples we have used thus far, we were unable to show that there really is a difference in the average time it takes, say, the manager or the truck driver along the different routes even though the sample mean varied from 26 min. to 30 min. Of course, there may actually be no difference, but it is also possible that he SSE, which served as an estimate of chance variation, may actually have been "inflated" by identifiable services of variation. For example, it is probable that every time he took Route 3 it rained or with Route 2 he left for work late thereby missing rush-hour congestion, etc.

To avoid this situation, we identify the first observation as taken on Monday, the second on Tuesday etc. Our hope, of course, lies with the fact that the variability which we have been ascribing to chance or at least put of it may well be due to differences in traffic conditions on different days of the week.

This situation suggests a two-way analysis of variance, in which the total variability of the data is partitioned into one component which we assemble to possible differences due to one variable (the difference treatments), a second component which we ascribe to possible differences due to second variable (referred to as blocks), while the remainder of the variability is ascribed to chance.

In our example, we will be looking at the different treatments (routes) and different blocks (days).

If we let $\bar{x}._j$ stand for the mean of all the values obtained for the jth day, the answer is given by the block sum of squares.

$$SSB = k \cdot \sum_{j=1}^{n} \left(\overline{x}_{\bullet j} - \overline{x}_{\bullet \bullet} \right)^2$$

which measures the variability of the average times obtained for the different days. This formula is very similar to the SS(Tr) provided we substitute the mean obtained for the different days for the mean obtained for the different routes, and correspondingly sum on j instead of i and interchange n and k. relating all this information to the short cut formula we have:

$$SSB = \frac{1}{k} \sum_{i=1}^{n} T^2_{\bullet j} - _k \frac{1}{n} . T^2_{\bullet \bullet}$$

where T_j is the total of the observations for the jth block (in our situation, the total of the values obtained for the jth day).

In two-way analysis we compute the SST and SS(Tr) the same way as 1 and 2, SSB in the same say as 4 and the SSE by subtraction, namely

$$SSE = SST - SS(Tr) - SSB$$

To proceed with the analysis we utilize a table as follows:

Sources of Variation	Degrees of Freedom	Sum of Squares	Mean squares	F
Treatments	k–1	SS(Tr)	$MS(Tr)=\dfrac{SS(Tr)}{k-1}$	$\dfrac{MS(Tr)}{MSE}$
Blocks	n–1	SSB	$MSB=\dfrac{SSB}{k(n-1)}$	$\dfrac{MSB}{MSE}$
Error	(n–1) (k–1)	SSE	$MSE=\dfrac{SSE}{(n-1)(k-1)}$	
Total	nk–1	SST		

Note that the two-way analysis of variance leads to two tests of significance:

1. The F statistic MS(Tr)/MSE tests the effects of the k treatments that are all equal.
2. The F statistic MSB/MSE tests the effects of the n blocks that are all equal.

In our example, we find that the totals for the five days are respectively 99, 111, 114, 111, and 125

$$(22 + 25 + 25 + 27) = 99, (26 + 27 + 29 + 29) = 111 \text{ etc.}$$

also

$$(99 + 111 + 114 + 111 + 125) = 560$$

Substituting these totals in the SSB formula we have:

$$\text{SSB} = \frac{1}{4}(99^2 + 111^2 + 114^2 + 111^2 + 125^2) - \frac{1}{20}(560)^2 = 86$$

the New error term is

$$\text{SSE} = 168 - 50 - 86 = 32$$

These two numbers are the same as before, thus, the table becomes:

Sources of Variation	Degrees of Freedom	Sum of Squares	Mean squares	F
Treatments	3	50	$\dfrac{50}{3}$	$\dfrac{50/3}{31/12} = 6.25$
Blocks	4	86	$\dfrac{86}{4}$	$\dfrac{86/4}{32/12} = 8.06$
Error	12	32	$\dfrac{32}{12}$	
Total	19	168		

Using the $a = .05$ we find that for 3 and 12 df $F = 3.49$ and that for 4 and 12 df $F = 3.26$

Since $6.24 > 3.49$ we reject the H_0 for treatments; Since $8.06 > 3.26$ we reject the H_0 for blocks. In other words, the difference between the sample means obtained in the four routes are significant, and so are the differences obtained for the different days of the week.

Factorial Designs

Factorial designs involve the simultaneous application of two or more different treatments within a single experiment.

Advantages

1. Since the same groups of subjects are utilized in estimating effects from the two or more treatment dimensions, there is an economy of both time and personnel.
2. The simultaneous application of treatments makes possible the elucidation of interactive effects among the treatment dimensions themselves (that is, one or some combinations of treatments may have an accelerating effect or a moderating effect upon the action of some other treatment).

Characteristics

Specific designs are labeled in terms of the number of levels of the treatments involved. The simplest possible is a design in which only two treatments, each occurring at only two levels and it would be called a 2×2 factorial design.

2x2

	1	A	2
1 B	11		12
2	21		22

Contrast

	(11)	(12)	(21)	(22)
A	+1	−1	1	−1
B	+1	+1	−1	−1
AB	+1	−1	−1	+1

2^k Design*

For complete factorial designs, there will be, in general

$$2^m - 1 \text{ main and interaction effects}$$

$$(\text{e.g., } 2^2 \text{ or } 2 \times 2 \text{ where } 2^2 - 1 = 3)$$

for a design occurring at m levels, there will be m main effects.

$$m(m-1)/2 \text{ first-order interactions}$$

$$m(m-1)(m-2)/6 \text{ second-order interactions}$$

RESPONSE GRAPHS

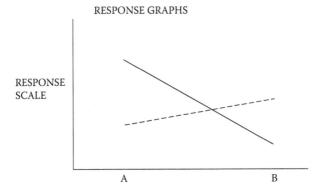

RESPONSE SCALE

A   B

---

* where k is the factor and 2 is the level of each factor.

Formulae

$$SS_A = 2n\left[\sum_{j=1}^{2}\left(\overline{Y}_{\cdot j\cdot} - \overline{Y}...\right)^2\right] = \sum_{j=1}^{2}\frac{Y_{\cdot j\cdot}^{\;2}}{2n} - \frac{\left(Y...\right)^2}{4n}$$

$$SS_B = 2n\left[\sum_{i=1}^{2}\left(\overline{Y}_{i\cdot\cdot} - \overline{Y}...\right)^2\right] = \sum_{i=1}^{2}\frac{Y_{i\cdot\cdot}^{\;2}}{2n} - \frac{\left(Y...\right)^2}{4n}$$

$$SS_{AB} = n\left[\sum_{i=1}^{2}\sum_{j=1}^{2}\left(\overline{Y}_{ij\cdot} - \overline{Y}_{\cdot j\cdot} - \overline{Y}_{i\cdot\cdot} + \overline{Y}...\right)^2\right]$$

$$= \sum_{i=1}^{2}\sum_{j=1}^{2}\frac{Y_{ij\cdot}^{\;2}}{n} - \frac{\left(Y...\right)^2}{4n} - SS_A - SS_B$$

$$SS_{TOTAL} = \sum_{i=1}^{2}\sum_{j=1}^{2}\sum_{n=1}^{2}Y_{ijk}^{\;2} - \frac{\left(Y...\right)^2}{4n}$$

$$SS_{ERROR} = SS_{TOTAL} - SS_A - SS_B - SS_{AB}$$

EXAMPLE

A

		OB	CB	SUM
		37.5	38.2	
		39.2	37.3	
R		40.3	39.1	363.7
		32.8	31.4	
		35.5	32.4	
B				
		33.4	28.2	
		32.1	30.4	
NR		35.2	27.6	301.4
		31.8	29.2	
		28.3	25.2	
		346.1	319.0	665.1

Cell summary

OB

$$\sum_{k=1}^{5} Y_{11k} = 185.3$$

CB

$$\sum_{k=1}^{5} Y_{12k} = 178.4$$

R

$$\sum_{k=1}^{5} Y_{11k}{}^2 = 6903.07$$

$$\sum_{k=1}^{5} Y_{12k}{}^2 = 6415.06$$

$$S_{11}{}^2 = 8.9630$$

$$S_{12}{}^2 = 12.4370$$

$$\overline{Y}_{11.} = 37.06$$

$$\overline{Y}_{12.} = 35.68$$

$$\sum_{k=1}^{5} Y_{21k} = 160.8$$

$$\sum_{k=1}^{5} Y_{12k} = 140.6$$

NR

$$\sum_{k=1}^{5} Y_{21k}{}^2 = 5197.14$$

$$\sum_{k=1}^{5} Y_{22k}{}^2 = 3968.84$$

$$S_{21}{}^2 = 6.4530$$

$$S_{22}{}^2 = 3.7920$$

$$\overline{Y}_{21.} = 32.16$$

$$\overline{Y}_{22.} = 28.12$$

Calculations

$$SS_A = \frac{346.1^2}{10} + \frac{319.0^2}{10} - \frac{655.1^2}{20}$$
$$= 22154.621 - 22117.900 = 36.721$$

$$SS_B = \frac{363.7^2}{10} + \frac{301.4^2}{10} - \frac{655.1^2}{20}$$
$$= 22{,}311.965 - 22117.900 = 194.065$$

$$SS_{AB} = \frac{185.3^2}{5} + \frac{160.8^2}{5} + \frac{178.4^2}{5} + \frac{140.6^2}{5} - \frac{655.1^2}{20} - SS_A - SS_B$$
$$= 22{,}357.530 - 22{,}117.900 - 36.721 - 194.065 = 8.844$$

$$SS_{TOTAL} = 22{,}484.11 - 22{,}117.900 = 366.210$$

$$SS_{ERROR} = 366.210 - 36.721 - 194.065 - 8.844 = 126.580$$

As a check

$$SS_{ERROR} = 4(8.9630 + 12.4370 + 6.4530 + 3.7920) = 126.580$$

$$(126.58 \div 16 = 7.91)$$

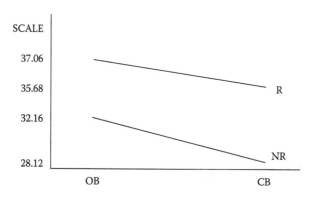

$$F = 36.72 \div 7.91 = 4.64 \text{ STG. @ .05 LEVEL}$$

$$F = 194.06 \div 7.91 = 24.54 \text{ STG. @ .01 LEVEL}$$

$$F = 8.84 \div 7.91 = 1.12 \text{ NOT STG.}$$

Nested Designs

A nested design is employed so that the required estimates of the components of variance are obtained. They take the form of a hierarchy.
   For example

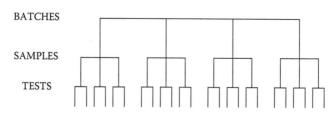

Characteristics

1. No information about interactions.
2. Nested treatment variables can be included in higher dimensional designs (either factorial or randomized).
3. In partial nesting, the nested variable is crossed (in factorial fashion) with some of the treatment dimensions, but is nested within others.
4. Complete nesting is also referred to as a hierarchical design.

Fractional Factorial

$$2^5 = 2 \times 2 \times 2 \times 2 \times 2 \times = 32 \text{ Runs}$$

If 16 runs are employed, then a half-fraction is employed.

$$1/2 \, \xi \, 2^5 = 2^{-1}2^5 = 2^{5-1} = 2^4 = 16 \text{ Runs}$$

Highest resolution

1. Write a full factorial design for the first $k - 1$ variables.
2. Associate the kth variable with plus or minus the interaction column $1\,2\,3\ldots(k-1)$.

One must learn by doing the thing; for though you think you know it, you have no certainty until you try. (Sophocles).

After ANOVA—What?

1. Prior to experimentation—e.g. Orthogonal contrasts
2. Tests of means after experimentation

Range Test: (Newman-Keuls Test)

1. Arrange the k means in order from low to high.
2. Enter the ANOVA table and take the error mean square with its df.
3. Obtain the std. error of the mean for each treatment

$$S_{\bar{Y}.J} = \sqrt{\frac{\text{ERROR MEAN SQUARE}}{\text{NUMBER OF OBSERVATIONS IN } \bar{Y}.J}}$$

where the error mean square is the one used as the denominator in the f test on means $\bar{Y}.J$.

4. Enter a studentized range table of significant ranges at the α level desired using m_2 = df for error mean square and p = 2, 3,. . ., k and list these $k - 1$ ranges.

5. Multiply these ranges by S_{YJ} to form a group of $k - 1$ least significant ranges.

6. Test the observed ranges between means, beginning with largest versus smallest, which is compared with the least significant range for $p = k$; then test largest versus second smallest with the least significant range for $p = k - 1$, and so on.

Scheffe's Test

Since Newman-Keuls test is restricted to comparing pairs of means and it is often desirable to examine other contrasts which represent combinations of treatments, Scheffe's test is preferred.

Scheffe's method used the concept of contrasts; however, these contrasts need not be orthogonal. In fact, any and all conceivable contrasts may be tested for significance.

Scheffe's scheme is more general than the Newman-Keuls and as a consequence is valid for a large set of possible constants. It also requires larger observed differences to be significant than some of the other schemes. The following five steps are the sequence of using the test.

1. Set up all contrasts of interests and compute their numerical value.

2. Determine the significant F statistic for the ANOVA just performed based on α and df $k - 1, n - k$.

3. Compute $A = \sqrt{(k-1)F}$, using the f from step 2.

4. Compute the std. error of each contrast to be tested

$$S_{cm} = \sqrt{(\text{ERROR MEAN SQUARE}) \Sigma_j n_j C_{jm}^2}$$

5. If a contrast c_m is numerically larger than a times s_{cm}, it is declared significant. Or, if $|c_m| > a_{scm}$ reject the H_0 that the true contrast among means is zero.

Duncan Test

This test is derived by considering the set of sample means in rank order of magnitude. Then critical values are selected so that the type i error for all contrasts involving adjacent pairs of means is a. Contrasts between pairs of means which span k rank position are tested with a type i error of $1 - (1 - 2)^{k-1}$. These are referred to as "protection levels."

The Duncan test represents a middle of the road procedure since the level of significance is greater for pair-wise contrasts involving means separated by larger numbers of rank positions. The following five steps are the sequence of using the test.

1. Using the ms for error from the ANOVA table, find the standard error of a mean

$$S_m = \sqrt{mS_{ERROR}/m}$$

2. Find appropriate table and corresponding value, and find the studentized ranges for rank differences from 1 though $p-1$
3. Compute the set of shortest significant ranges by multiplying each studentized ranges by s_m.
4. Rank-order the sample means and systematically compute all pair-wise contrasts.
5. Hypothesis decisions are made by comparing each pair-wise contrast with the appropriate shortest significant range. Any pair-wise contrast which exceeds the appropriate shortest significant range is judged significant.

Dunnett's Test

This test evaluates the differential effects of two or more other treatment conditions in comparison with the control.

The research fixes the level of significance for the set of tests at .05, 01, or any other value.

The test statistic is

$$t = (\bar{Y}_{\cdot j} - \bar{Y}.\rho)S_m$$

where

$$S_m = \sqrt{2mS_{ERROR}/n}$$

Each of the $p-1I$ value is compared with the same tabular critical value.

Selecting an Orthogonal Array

Degrees of freedom are used for the proper selection of an array. We start by adding up all the required df of the factors, the number of levels for each factor and the interactions we wish to investigate.

Examples

Two level factors a b c d e and the interaction a × b, a × c

1. Add up the required df. Each two-level factor has $2 - 1 = 1$ df; each interaction has $1 \times 1 = 1$ df. Total df = (s factors × 1 df) + (2 interactions × 1 df) = 7 df

Hence, 7 df are required to obtain the desired information. The L_8 array is a two-level design with exactly 7 df.

Linear Graphs

Linear graphs represent pictorial equivalents of triangular matrices. They are a graphic tool used to facilitate complicated assignments of factors and interactions to an orthogonal array.

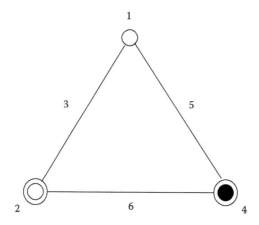

Factors are assigned to points.

Interaction between two factors is assigned to the line segment connecting the corresponding points.

If an interaction is **not** expected, then a factor may be assigned to the corresponding line segment.

Appendix G: When to Use Quality Tools—A Selected List

Activity Network: The activity network is generally used to schedule dependent activities within a plan. It can be used for describing and understanding the activities within a standard work process. The resulting diagram is useful for communicating the plan and risks to other people.

When to use it:

- Use it when planning any project or activity, which is composed of a set of interdependent actions.
- Use it to calculate the earliest date the project can be completed and to find ways of changing this.
- Use it to identify and address risk to completing a project on time.

Affinity Diagram: The affinity diagram provides a visual method of structuring a large number of discrete pieces of information.

When to use it:

- Use it to bring order to fragmented and uncertain information and where there is no clear structure.
- Use it when information is subjective and emotive, to gain consensus while avoiding verbal argument.
- Use it when current opinions, typically about an existing system, obscure potential new solutions.
- Use it, rather than a relations diagram, when the situation calls more for creative organization than for logical organization.

Bar Chart: Bar charts are used to show the differences between related groups of measurements.

When to use it:

- Use it when a set of measurements can be split into discrete and comparable groups, to show the relative change between these groups.
- Use it when there are multiple sets of measurement groups, to show the relationship and change within and between groups.
- Use it, rather than a line graph, to display discrete quantities rather than continuing change.

- Use it, rather than a Pareto chart, when a consistent ordering of bars is wanted. This can ease recognition and comparison of current and previous charts.

Brainstorming: Brainstorming is used to creatively generate new ideas. The creative synergy of a brainstorming session is also useful in helping a team bind together.
  When to use it:

- Use it when new ideas are required, to generate a large list of possibilities.
- Use it when a solution to a problem cannot be logically deduced.
- Use it when information about a problem is confused and spread across several people, to gather the information in one place.

Cause and Effect: The cause–effect diagram is used to identify and structure the causes of a given effect. Use it in preference to a relations diagram where there is one problem and causes are mostly hierarchical (this will be most cases).
  When to use it:

- Use it when investigating a problem, to identify and select key problem causes to investigate or address.
- Use it when the primary symptom (or effect) of a problem is known, but possible causes are not all clear.
- Use it when working in a group, to gain a common understanding of problem causes and their relationship.
- Use it to find other causal relationships, such as potential risks or causes of desired effects.

Check Sheet: The check sheet is used to manually collect data in a reliable, organized way.
  When to use it:

- Use it when data are to be recorded manually, to ensure the data are accurately recorded and are easy to use later, either for direct interpretation or for transcription, for example, into a computer.
- Use it when the recording involves counting, classifying, checking, or locating.
- Use it when it is useful to check each measurement as it is recorded, for example, that it is within normal bounds.
- Use it when it is useful to see the distribution of measures as they are built up.

Control Chart: The control chart is used to identify dynamic and special causes of variation in a repeating process. It is only practical to use it when regular measurements of a process can be made.
  When to use it:

- Use when investigating a process, to determine whether it is in a state of statistical control and thus whether actions are required to bring the process under control.
- Use it to differentiate between special and common causes of variation, identifying the special causes which need to be addressed first.
- Use it to detect statistically significant trends in measurements; for example, to identify when and by how much a change to the process has improved it.
- Use it as an ongoing "health" measure of a process, to help spot problems before they become significant.

Decision Tree: The decision tree is used to select from a number of possible courses of action.
  When to use it:

- Use it when making important or complex decisions, to identify the course of action that will give the best value according to a selected set of rules
- Use it when decision making, to identify the effects of risks.
- Use it when making plans, to identify the effects of actions and possible alternative courses of action.
- Use it when there are chains of decisions, to determine the best decision at each point.
- Use it only when data are available, or can reasonably be determined, on costs and probabilities of different outcomes.

Design of Experiments: "Design of experiments" is use to understand the effects of different factors in a given situation.
  When to use it:

- Use it when investigating a situation where there are many variable factors, one or more of which may be causing problems.
- Use it when variable factors may be interacting to cause problems.
- Use it when testing a solution, to ensure that there are no unexpected side effects.
- Use it when there is not the time or money to try every combination of variables.

FMEA: FMEA is used to identify and prioritize how items fail, and the effects of failure.
  When to use it:

- Use it when designing products or processes, to identify and avoid failure-prone designs.
- Use it when investigating why existing systems have failed, to help identify possible causes and remedies.
- Use it when investigating possible solutions, to help select one with an acceptable risk for the known benefit of implementing it.
- Use it when planning actions, in order to identify risks in the plan and hence identify countermeasures.

Fault Tree: Fault tree analysis is used to show combinations of failures that can cause overall system failure.
  When to use it:

- Use it when the effect of a failure is known, to find how this might be caused by combinations of other failures.
- Use it when designing a solution, to identify ways it may fail, and consequently find ways of making the solution more robust.
- Use to identify risks in a system, and consequently identify risk reduction measures.
- Use it to find failures that can cause the failure of all parts of a "fault-tolerant" system.

Flowchart: The flowchart is used to show the sequential steps within a process.
  When to use it:

- Use it when analyzing or defining a process, to detail the actions and decisions within it.
- Use it when looking for potential problem points in a process.
- Use it when investigating the performance of a process, to help identify where and how it is best measured.
- Use it as a communication or training aid, to explain or agree the detail of the process.

Flow Process Chart: The flow process chart is used to record and illustrate the sequence of actions within a process.
  When to use it:

- Use it when observing a physical process, to record actions as they happen, and thus get an accurate description of the process.

- Use it when analyzing the steps in a process, to help identify and eliminate waste.
- Use it, rather than a flowchart, when the process is mostly sequential, containing few decisions.

Force Field Analysis: The force field diagram is used to weigh up the points for and against a potential action.
   When to use it:

- Use it when decision making is hindered by a number of significant points for and against a decision.
- Use it when there is a lot of argument and indecision over a point, to clarify and agree the balance of disagreement.
- Use it to help identify risks to a planned action and to develop a strategy for counteracting them.
- Use it to help identify the key causes of successful or unsuccessful actions.

Gantt Chart: Gantt charts are used to show the actual time to spend in tasks.
   When to use it:

- Use it when doing any form of planning, to show the actual calendar time spent in each task.
- Use it when scheduling work for individuals, to control the balance between time spent during normal work hours and overtime work.
- Use it when planning work for several people, to ensure that those people who must work together are available at the same time.
- Use it for tracking progress of work against the scheduled activities.
- Use it when describing a regular process, to show who does what, and when.
- Use it to communicate the plan to other people.

Histogram: The histogram is used to show the frequency distribution of a set of measurements.
   When to use it:

- Use it to investigate the distribution of a set of measurements.
- Use it when it is suspected that there are multiple factors affecting a process, to see if this shows up in the distribution.
- Use it to help define reasonable specification limits for a process by investigating the actual distribution.
- Use it when you want to see the actual shape of the distribution, as opposed to calculating single figures like the mean or standard deviation.

***IDEFO (ICOM DEFinitions)*:** IDEFO is used to make a detailed and clear description of a process or system.
When to use it:

- Use it when formally describing a process, to ensure a detailed, clear, and accurate result.
- Use it when the process is complex, and other methods would result in more complex diagrams.
- Use it when mapping a wide variety of processes, as a consistent and scalable process description language.
- Use it when there is time available to work on understanding and producing a complete and correct description of the process.

***Kano Model*:** Kano analysis is a quality measurement tool used to prioritize customer requirements based on their impact to customer satisfaction.
When to use:

- Kano analysis is a tool which can be used to classify and prioritize customer needs. This is useful because customer needs are not all of the same kind, not all have the same importance, and are different for different populations. The results can be used to prioritize your effort in satisfying different customers.

***Kappa Statistic*:** The Kappa is the ratio of the proportion of times the appraisers (see Gage R&R) did agree to the proportion of times the appraisers could agree. If you have a known standard for each rating, you can assess the correctness of all appraisers' ratings compared to the known standard. If Kappa = 1, then there is perfect agreement. If Kappa = 0, then there is no agreement. The higher the value of Kappa, the stronger the agreement. Negative values occur when agreement is weaker than expected by chance, but this rarely happens. Depending on the application, Kappa less than 0.7 indicates that your measurement system needs improvement. Kappa values greater than 0.9 are considered excellent.
When to use it:

- Use it when there is a need for comparison

***Line Graph*:** The line graph is used to show patterns of change in a sequence of measurements.
When to use it:

- Use it when an item is repeatedly measured, to show changes across time.
- Use it when measuring several different items that can be shown on the same scale, to show how they change relative to one another.

- Use it when measuring progress toward a goal, to show the relative improvement.
- Use it, rather than a bar chart, to show continuous change, rather than discrete measurements. It is also better when there are many measurements.
- Use it, rather than a control chart, when not measuring the degree of control of a process.

Matrix Data Analysis Chart: The matrix data analysis chart (MDAC) is used to identify clusters of related items within a larger group.
   When to use it:

- Use it when investigating factors which affect a number of different items, to determine common relationships.
- Use it to determine whether or not logically similar items also have similar factor effects.
- Use it to find groups of logically different items which have similar factor effects.

Matrix Diagram: The matrix diagram is used to identify the relationship between pairs of lists.
   When to use it:

- Use it when comparing two lists to understand the many-to-many relationship between them (it is not useful if there is a simple one-to-one relationship).
- Use it to determine the strength of the relationship between either single pairs of items or a single item and another complete list.
- Use it when the second list is generated as a result of the first list, to determine the success of that generation process. For example, customer requirements versus design specifications.

Nominal Group Technique: The nominal group technique (NGT) is used to collect and prioritize the thoughts of a group on a given topic.
   When to use it:

- Use it when a problem is well understood, but knowledge about it is dispersed among several people.
- Use it when a rapid consensus is required from a team, rather than a more detailed consideration.
- Use it when the team is stuck on an issue; for example, when they disagree about something.

- Use it when the group prefers a structured style of working together.
- Use it, rather than brainstorming, when a limited list of considered opinions is preferred to a long list of wild ideas, or when the group is not sufficiently comfortable together to be open and creative.

Pareto Chart: The Pareto chart is used to show the relative importance of a set of measurements.
    When to use it:

- Use it when selecting the most important things on which to focus, thus differentiating between the "vital few" and the "trivial many."
- Use it after improving a process, to show the relative change in a measured item.
- Use it when sorting a set of measurements, to visually emphasize their relative sizes.
- Use it, rather than a bar chart or pie chart to show the relative priority of a set of numeric measurements.

Prioritization Matrix: The prioritization matrix is used to sort a list of items into an order of importance.
    When to use it:

- Use it to prioritize complex or unclear issues, where there are multiple criteria for deciding importance.
- Use it when there is data available to help score criteria and issues.
- Use it to help select the most important items to work on.
- When used with a group, it will help to gain agreement on priorities and key issues.
- Use it, rather than simple Voting, when the extra effort that is required to find a more confident selection is considered to be worthwhile.

Process Capability: Process capability calculations indicate the ability of a process to meet specification limits.
    When to use it:

- Use it when setting up a process, to ensure it can meet its specification limits.
- Use it when setting specification limits, to ensure they are neither too wide nor too narrow.
- Use it when investigating a process that is not meeting its specification limits.
- Use it only when the process is stable and has a normal distribution.

Process Decision Program Chart: The process decision program chart (commonly just referred to as PDPC) is used to identify potential problems and countermeasures in a plan.
When to use it:

- Use it when making plans, to help identify potential risks to their successful completion.
- When risks are identified, use it to help identify and select from a set of possible countermeasures.
- Also use it to help plan for ways of avoiding and eliminating identified risks.
- It is of best value when risks are nonobvious, such as in unfamiliar situations or in complex plans, and when the consequences of failure are serious.

Relations Diagram: The relations diagram is used to clarify and understand complex relationships.
When to use it:

- Use it when analyzing complex situations where there are multiple interrelated issues.
- Use it where the current problem is perceived as being a symptom of a more important underlying problem.
- It is also useful in building consensus within groups.
- It is commonly used to map cause–effect relationships, but also can be used to map any other type of relationship.
- Use it, rather than an affinity diagram, when there are logical, rather than subjective, relationships.
- Use it, rather than a cause–effect diagram, when causes are non-hierarchic or when there are complex issues.

Scatter Diagram: The scatter diagram is used to show the type and degree of any causal relationship between two factors.
When to use it:

- Use it when it is suspected that the variation of two items is connected in some way, to show any actual correlation between the two.
- Use it when it is suspected that one item may be causing another, to build evidence for the connection between the two.
- Use it only when both items being measured can be measured together, in pairs.

Simulation in Lean: Simulation is a practical tool that changes complex parameters into mathematical, logical models which are possible to analyze under different scenarios. Models can be classified from various aspects. They can be categorized as being mathematical or physical, static (Monte Carlo) or dynamic (shows a system that always changes), deterministic (includes no random variables) or stochastic (contain at least one random variable), and also discrete or continuous models. But the simulation beside its benefits has also some disadvantages.

When to use:

- Assessing hospital operations in different wards.
- Estimating the number of needed doctors and nurses in emergency ward.
- Reducing the length of time that patient has to stay in emergency ward.
- Predicating the required number of beds according to the patient interval.

Advantages:

1. The most important benefit of simulation is to get results at less cost.
2. Assumptions about specific phenomena can be checked for feasibility.
3. The simulation answers "what if" questions.
4. Analysis of bottlenecks and finding the roots is easier.
5. The system can be tested without allocating any resources.

Disadvantages:

1. Working with simulation modeling requires experience and specific training.
2. Understanding the simulation results is difficult because they are based on random inputs.

String Diagram: The string diagram is used to investigate the physical movement in a process.

When to use it:

- Use it when analyzing a manual or physical process that involves significant physical movement, in order to make movements easier and quicker. Movements may be of people, materials, or machines.
- Use it when designing the layout of a work area, to identify the optimum positioning of machines and furniture

Surveys: Surveys are used to gather information from people.
  When to use it:

- Use it when information that is required is held by an identifiable and dispersed group of people.
- Use it to help decision making, by turning disparate qualitative data into useful quantitative information.
- Use it only when the time and effort are available to complete the survey.
- Use it, rather than brainstorming or NGT to gather real data about what a diverse group of people think (rather than an opinion of what they think).

Tables: Tables are used to organize and relate multiple pieces of information.
  When to use it:

- Use it when gathering information, to help prompt for a complete set of data.
- Use it when information is disorganized, to help collate and understand it.
- Use it for summarizing information to make it easier to present and understand.
- Use specific table tools as frameworks for particular tasks, either to organize existing information or to prompt for specific categories of information.

Tree Diagrams: The tree diagram is used to break down a topic into successive levels of detail.
  When to use it:

- Use it when planning, to break down a task into manageable and assignable units.
- Use it when investigating a problem, to discover the detailed component parts of any complex topic.
- Use it only when the problem can be broken down in a hierarchical manner.
- Use it, rather than a relations diagram, to break down a problem when the problem is hierarchical in nature.

Value Analysis: Value analysis is used to determine and improve the value of a product or process.
  When to use it:

- Use it when analyzing a product or process, to determine the real value of each component.

- Use it when looking for cost savings, to determine components that may be optimized.
- Use it only when the item to be analyzed can be broken down into subcomponents and realistic costs and values allocated to these.

Voting: Voting is used to prioritize and select from a list of items.
When to use it:

- Use it as a quick tool when a group must select one or more items from a list, for example as generated by a brainstorming session.
- Use it when it is important that the group accept the result as fair.
- Use it when the knowledge to enable selection is spread within the group.
- Use it when the opinion of all group members is equally valued.
- It can also be used to "test the water," to determine opinions without committing to a final selection.
- Use it, rather than a prioritization matrix, when the added accuracy of the prioritization matrix is not worth the extra effort.

Glossary

The following glossary has been prepared so that a quality professional becomes familiar with some definitions and concepts used in the statistical world for improvement. The list is not exhaustive by any means. However, we believe it is functional and helpful.

3K: Kiken (dangerous), Kitanai (dirty), Kitsui (stressful)—general workplace hazards

3M: Muda (waste), Mura (irregular, inconsistent) and Muri (unreasonable strain)

3P: Production, Preparation, Process

3R: Recording, Recalling, Reconstructing—when generating new ideas

3S: Stabilize, Synchronize, Standardize—steps in lean product development

4S: Sort (arrangement), Seiton (organization), Seiso (cleanliness), Seiketsu (act of cleaning).

5M: Manpower, Machine, Method, Material and Measurement—sources of variation

5P: Plant, Production, People, Policies, Procedures—sources of variation (for fishbone)

5R: Responsiveness, Reliability, Rhythm, Responsibility, Relevance

5S: Seiri (sort), Seiton (straighten), Seiso (shine), Seiketsu (standardize) and Shitsuke (sustain)

6S: Seiri (sort), Seiton (straighten), Seiso (shine), Seiketsu (standardize) and Shitsuke (sustain), Safety

5W2H: Who, What, When, Where, Why & How and How Many (root cause analysis)

7P: Proper prior planning prevents pitifully poor performance

7 Wastes (sins): Over-production, Transport, Waiting, Inventory, Defects, Over-processing, Unnecessary Movement

8D: Eight discipline problem solving method and report. (A Ford Motor approach to problem solving that has become the default analysis in the automotive industry with very minor modifications. It is based on nine steps, which are: D0 = prepare for the 8D process, D1 = establish the team, D2 = describe the problem, D3 = develop interim containment action (ICA), D4 = define and verify root cause and escape point, D5 = choose and verify permanent corrective actions (PCAs), D6 = implement and validate permanent corrective actions, D7 = prevent recourance, D8 = congratulate the team).

A priori **test:** See *Planned comparison.*

Absolute fit measure: Measure of overall *goodness-of-fit* for both the *structural* and *measurement models* collectively. This type of measure does not make any comparison to a specified *null model* (*incremental fit measure*) or adjust for the number of parameters in the estimated model (*parsimonious fit measure*).

Accelerated test methods: A test strategy that shortens the time to failure of the product under test by inducing stresses that are more severe than operationally encountered stresses (e.g., more severe than the 90th percentile loads) in a way that does not change the underlying failure mechanisms/modes that customers may potentially encounter. The technique of accelerated testing involves: selection of a mathematical model, selection of accelerated stress parameters and acceleration levels, generation of test procedures, and analysis of the test data.

Acceptable quality level (AQL): Used in acceptance sampling: maximum percentage or proportion of nonconforming units that can be considered acceptable as a process average. When a continuing series of lots is considered, a quality level that, for the purposes of sampling inspection, is the limit of a satisfactory process average.

Acceptance sampling: Sampling inspection in which decisions concerning acceptance or rejection of materials or services are made. Also includes procedures used in determining the acceptability of the items in question based on the true value of the quantity being measured. In essence it is a specific plan that indicates whether to accept or not accept the particular lot of interest. There are two types: attributes sampling and variables sampling. In attributes sampling, the presence or absence of a characteristic is noted in each of the units inspected. In variables sampling, the numerical magnitude of a characteristic is measured and recorded for each inspected unit; this involves reference to a continuous scale of some kind.

Acceptance sampling plan: A specific plan that indicates the sampling sizes and the associated acceptance or nonacceptance criteria to be used. In attributes sampling, for example, there are single, double, multiple, sequential, chain, and skip-lot sampling plans. In variables sampling, there are single, double, and sequential sampling plans.

Accountability/responsibility matrix: A structure that relates the *project* organization Structure to the *work breakdown structure*; assures that each element of the project *scope* of work is assigned to a responsible individual.

Accuracy (of measurement): Difference between the average result of a measurement with a particular instrument and the true value of the quantity being measured.

Accuracy: A characteristic of measurement that addresses how close an observed value is to the true value. It answers the question, "Is it right?".

Action plan: The detail plan to implement the actions needed to achieve strategic goals and objectives (similar, but not as comprehensive as a project plan).

Activation function: Mathematical function within the *node* that translates the summated score of the weighted input values into a single output value. Although the function can be any type, the most common form is a *sigmoid function*.

Adaptive conjoint: Methodology for conducting a conjoint analysis that relies on information from the respondents (e.g., importance of attributes) to adapt the conjoint design to make the task even simpler. Examples are the *self-explicated* and *adaptive* or *hybrid models*.

Adaptive model: Technique for simplifying conjoint analysis by combining the *self-explicated* and traditional conjoint models.

Additive model: Model based on the additive *composition rule*, which assumes that individuals just "add up" the *part-worths* to calculate an overall or "total worth" score indicating *utility* or preference. It is also known as a *main effects model*. It is the simplest conjoint model in terms of the number of evaluations and the estimation procedure required.

Adjusted coefficient of determination (adjusted R^2): Modified measure of the *coefficient of determination* that takes into account the number of independent variables included in the regression equation and the sample size. Although the addition of independent variables will always cause the coefficient of determination to rise, the adjusted coefficient of determination may fall if the added independent variables have little explanatory power and/or if the *degrees of freedom* become too small. This statistic is quite useful for comparison between equations with different numbers of independent variables, differing sample sizes, or both.

Adjusted SS: Adjustment to the sum of squares used in general linear models (GLM) to account for designs without orthogonality. A Type II adjustment to the SS of a term involves adjusting for all other terms in the model that do not contain the term in question. A Type III adjustment involves adjusting for all other terms in the model, including those containing the term in question. Only Type II SS are suitable for models with fixed cross-factors, and only Type III SS are suitable for models with random cross-factors. Designs that are both orthogonal and balanced have adjusted SS equal to sequential SS.

Affinity chart (diagram): Brainstorming tool used to gather large quantities of information from many people; ideas usually are put on sticky notes, then categorized into similar columns; columns are named giving an overall grouping of ideas. In other words, it is a management and planning tool used to organize ideas into natural groupings in a way that stimulates new, creative ideas. Also known as the "KJ" method.

Agglomerative methods: *Hierarchical procedure* that begins with each *object* or observation in a separate cluster. In each subsequent step, the two object clusters that are most similar are combined to build a new aggregate cluster. The process is repeated until all objects are finally combined into a single cluster.

Aggregate analysis: Approach to MDS in which a *perceptual map* is generated for a group of respondents' evaluations of *objects*. This composite perceptual map may be created by a computer program or by the researcher to find a few "average" or representative subjects.

Aggregated data: Data created through a process of combination or summarization. They can then be stored in this summarized form in a *data warehouse*.

Agile approach: See *lean approach*.

Algorithm: Set of rules or procedures; similar to an equation.

All-available approach: *Imputation method* for missing data that computes values based on all available valid observations.

All-possible-subsets regression: Method of selecting the variables for inclusion in the regression model that considers all possible combinations of the independent variables. For example, if the researcher has specified four potential independent variables, this technique would estimate all possible regression models with one, two, three, and four variables. The technique would then identify the model(s) with the best predictive accuracy.

Alpha (α): Significance level associated with the statistical testing of the differences between two or more groups. Typically, small values, such as 0.05 or 0.01, are specified to minimize the possibility of making a *Type I error*. See *Type I error*.

Alternative analysis: Breaking down a complex *scope* situation for the purpose of generating and evaluating different solutions and approaches.

Alternatives: Review of the means available and the impact of tradeoffs to attain the *objectives*.

AMADEOS: Advanced Multi-Attribute Design Evaluation & Optimization Software.

Amount at stake: The extent of adverse consequences that could occur to the *project*.

Analysis: The study and examination of something complex and separation into its more simple components. Analysis typically includes discovering not only what are the parts of the thing being studied but also how they fit together and why they are arranged in this particular way. A study of schedule variances for cause, impact, corrective action, and results.

Analysis of covariance (ANCOVA): A general linear model containing one or more covariates, usually in addition to one or more categorical factors. Each covariate X is tested for a linear trend with the continuous response Y. Transformations of response or covariate may be necessary to achieve linearity prior to analysis. A covariate factor X varies at least on an ordinal scale, and usually on a continuous scale (such as time, distance, etc.) and is, therefore, a covariate of the response Y.

Analysis of means (ANOM): A statistical procedure for troubleshooting industrial processes and analyzing the results of experimental designs with factors at fixed levels. It provides a graphical display of data. Ellis R. Ott developed the procedure in 1967 because he observed that nonstatisticians had difficulty understanding analysis of variance. Analysis of means is easier for quality practitioners to use because it is an extension of the control chart. In 1973, Edward G. Schilling further extended the concept, enabling analysis of means to be used with normormal distributions and attributes data where the normal approximation to the binomial distribution does not apply. This is referred to as analysis of means for treatment effects.

Analysis of results: The effect of each factor on the response of the system is determined. Using simple statistical techniques, the largest effects are isolated and a prediction equation is formulated to predict the behavior of the system more accurately.

Analysis of variance (ANOVA): Statistical technique used to determine whether samples from two or more groups come from populations with equal means. Analysis of variance employs one dependent measure, whereas multivariate analysis of variance compares samples based on two or more dependent variables. Furthermore, it is an analysis of the relative contributions of explained and unexplained sources of variance in a continuous response variable. In its broadest sense ANOVA includes explanatory factors that vary on continuous as well as categorical scales. In its narrower sense, ANOVA refers to balanced designs with categorical factors, while general linear models (GLM) encompasses also unbalanced designs and covariates. Parametric ANOVA and GLM partition the total variance in the response by measuring sums of squared deviations from modelled values. Significant effects are tested with the F-statistic, which assumes random sampling of independent replicates, homogeneous within-sample variances, and a normal distribution of the residual error variation around sample means.

Analysis sample: Group used in estimating the discriminant function(s) or the logistic regression model. When constructing *classification matrices*, the original sample is divided randomly into two groups, one for model estimation (the analysis sample) and the other for validation (the *holdout sample*).

Analytical data: Operational data that have been integrated and combined with existing analytical data through a *data warehousing* operation to form a historical record of a business event or object (e.g., customer, firm, or product).

Analytical thinking: Breaking down a problem (the whole) or situation into discrete parts to understand how each part contributes to the whole.

Analyze: DMAIC phase **A** where process detail is scrutinized for improvement opportunities. Note that (1) data are investigated and verified to prove suspected root causes and substantiate the problem statement (see also Cause and effect), and (2) process analysis includes reviewing process maps for value-added/non-value-added activities. See also Process map, Value-adding activities, Non-value-adding activities.

Andon board: A visual device (usually lights) displaying status alerts that can easily be seen by those who should respond.

Anti-image correlation matrix: Matrix of the partial correlations among variables after factor analysis, representing the degree to which the factors "explain" each other in the results. The diagonal contains the *measures of sampling adequacy* for each variable, and the off-diagonal values are partial correlations among variables.

AOQ (average outgoing quality): The expected average outgoing quality following the use of an acceptance sampling plan for a given level of quality. Any lots not meeting this level of quality must be inspected and nonconforming units removed and replaced. These lots are then included with the acceptable lots.

AOQL: Average outgoing quality limit.

Appraisal costs: Costs incurred to determine the degree of conformance to quality requirements.

Approve: To accept as satisfactory. Approval implies that the thing approved has the endorsement of the approving agency; however, the approval may still require confirmation by somebody else. In management use, the important distinction is between *approve* and *authorize.* Persons who approve something are willing to accept it as satisfactory for their purposes, but this decision is not final. Approval may be by several persons. The person who authorizes has final organization authority. This authorization is final approval.

Approved changes: *Changes* that have been *approved* by higher authority.

APQP: Advanced product quality planning.

AQL: Acceptable quality level.

Arrow diagram: A management and planning tool used to develop the best possible schedule and appropriate controls to accomplish the schedule; the critical path method (CPM) and the program evaluation review technique (PERT) make use of arrow diagrams.

Artificial intelligence (AI): Generalized area of computer science dealing with the creation of computer programs that attempt to emulate the learning properties of the human brain. AI has evolved into a number of more specialized areas, including *neural networks* and *genetic algorithms.*

Assignable causes (of variation): Significant, identifiable changes in the relationships of materials, methods, machines, measurement, mother nature, and people (5M & P).

Associative thinking: Joining parts into a whole.

Association rules: Rules based on the correlation between attributes or *dimensions* of *data elements.*

Assumptions for ANOVA: These are the necessary preconditions for fitting a given type of model to data. No form of generalisation from particular data is possible without assumptions. They provide the context for, and the means of evaluating, scientific statements purporting to truly explain reality. As with any statistical test, ANOVA assumes unbiased sampling from the population of interest. Its other assumptions concern the error variation against which effects are tested by the ANOVA model: (1) that the random variation around fitted values is the same for all sample means of a factor, or across the range of a covariate; (2) that the residuals contributing to this variation are free to vary independently of each other; and (3) that the residual variation approximates to a normal distribution. Underlying assumptions should be tested where possible, and otherwise acknowledged as not testable for a given reason of design or data deficiency.

Assurance: To examine with the intent to verify.

Attribute data: Go/no-go information. The control charts based on attribute data include fraction defective chart, number of affected units chart, count chart, count per-unit chart, quality score chart, and demerit chart.

Attributes data (quality): Data coming basically from GO/ NO-GO, pass/ fail determinations of whether units conform to standards. Also includes noting presence or absence of a quality characteristic. May or may not include weighting by seriousness of defects etc.

Attributes data: Qualitative data that can be counted for recording and analysis. Examples include characteristics such as the presence of a required label, the installation of all required fasteners, and the number of errors on an expense report. Other examples are characteristics that are inherently measurable (i.e., could be treated as variables data), but where the results are recorded in a simple yes/ no fashion, such as acceptability of a shaft diameter when measured on a go/no-go gauge, or the presence of engineering changes on a drawing. Warranty data (R/1000 and cost per unit) are attribute data, although with the large divisors involved, these measures approach being variables data (see also Variables data).

Attribute requirement list (ARL): A list of requirements for each product-level attribute.

Audit program: The organizational structure, commitment, and documented methods used to plan and perform audits.

Audit team: The group of individuals conducting an audit under the direction of a team leader, relevant to a particular product, process, service, contract, or project.

Audit: a planned, independent, and documented assessment to determine whether agreed-upon requirements are being met.

Auditee: The individual or organization being audited.

Auditor: An individual or organization carrying out an audit.

Audits: A planned and documented *activity* performed by qualified personnel to determine by investigation, examination, or evaluation of *objective evidence*, the adequacy and compliance with established *procedures*, or the applicable documents, and the effectiveness of implementation.

Authoritarian: Lets individuals know what is expected of them, gives specific guidance as to what should be done, makes his part of the group understood, schedules work to be done, and asks group members to follow standard rules and regulations.

Authorize: To give final approval. A person who can authorize something is vested with authority to give final endorsement, which requires no further approval.

Authorized work: An effort that has been approved by higher authority and may or may not be defined or finalized.

Autocratic management: Autocratic managers are concerned with developing an efficient workplace and have little concern for people (theory X assumptions about people). They typically make decisions without input from subordinates, relying on their positional power.

Autonomation (Jidoka): Use of specially equipped automated machines capable of detecting a defect in a single part, stopping the process, and signaling for assistance.

Availability: The ability of a product to be in a state to perform its designated function under stated conditions at a given time. Availability can be expressed by the ratio: [Uptime]/[Uptime + Downtime].

Average: See Mean.

Average chart: A control chart in which the subgroup average, Xbar, is used to evaluate the stability of the process level.

Average outgoing quality (AOQ): The expected average quality level of outgoing product for a given value of incoming product quality.

Average outgoing quality limit (AOQL): The maximum average outgoing quality over all possible levels of incoming quality for a given acceptance sampling plan and disposal specification.

Average linkage: *Algorithm* used in *agglomerative methods* that represents *similarity* as the average distance from all objects in one cluster to all objects in another. This approach tends to combine clusters with small variances.

Axiomatic design: An approach to the design of a product or process based on the independence and information design axioms that provide principles for the development of a robust design.

Background variation: Sources of variation that are always present and are part of the natural (random) variation inherent in the process itself (e.g., lighting, raw material variation). Their origin can usually be traced to elements of the system that can be eliminated only by changing the process.

Backpropagation: Most common *learning* process in *neural networks*, in which errors in estimating the output *nodes* are "fed back" through the system and used as indicators as to how to recalibrate the weights for each node.

Backward elimination: Method of selecting variables for inclusion in the regression model that starts by including all independent variables in the model and then eliminating those variables not making a significant contribution to prediction.

Backward pass: Calculation of late finish times (dates) for all uncompleted network activities. Determined by working backwards through each activity.

Balance: A balanced design has the same number of replicate observations in each sample. Balance is a desirable attribute particularly of cross factored models, where loss of balance generally (though not inevitably) leads to loss of orthogonality. The consequent complications to the partitioning of sources of variance in the response are accommodated by general linear models.

Balanced design: Stimuli *design* in which each level within a *factor* appears in equal number of times.

Balanced incomplete block: It is a reduced version of the balanced complete block design. The design is incomplete if each of the n blocks tests only c of the a levels of treatment A, where $c < a$. It is balanced provided that each treatment level is tested the same number of times, given by $r = nc/a$, and each pair of treatment levels appears in the same number of blocks, given by $\lambda = nc(c - 1)/[a(a - 1)]$. For example: $a = 4, n = 6, c = 2$, so $r = 3$ and $\lambda = 1$.

Balanced scorecard: Categorizes ongoing measures into four significant areas: finance, process, people, and innovation. Used as a presentation tool to update sponsors, senior management, and others on the progress of a business or process; also useful for process owners. Translates an organization's mission and strategy into a comprehensive set of performance measures to provide a basis for strategic measurement and management, utilizing four balanced views: financial, customers, internal business processes, and learning and growth.

Baseline concept: Management's project management plan for a *project* fixed prior to commencement.

Baseline measures: Data signifying the level of process performance as it is/was operating at the initiation of an improvement project (prior to solutions).

Baseline: Management plan and/or *scope* document fixed at a specific point in time in the *Project life cycle*.

Bartlett test of sphericity: Statistical test for the overall significance of all correlations within a correlation matrix.

Batch processing: Running large batches of a single product through the process at one time, resulting in queues awaiting next steps in the process.

Bathtub curve: Also called "life-history curve." The plot of instantaneous failure rate versus time is known as a Hazard Curve. It is more often called a Bathtub Curve because of its shape. It consists of three distinct patterns of failures: failures with decreasing rates (e.g., infant mortality), failures with constant rates (e.g., accidental, sudden overload, impact), and failures with increasing rates (e.g., wear-out, fatigue).

Bayes theorem: A theorem of statistics relating conditional probabilities.

Bayesian statistics: A method that provides a formal means of including prior information within a statistical analysis. It is a way of expressing and updating a decision-oriented learning process. A software package called ReDCAS is available to perform Bayesian analysis.

Bell-shaped curve: A curve or distribution showing a central peak and tapering off smoothly and symmetrically to "tails" on either side. A normal (Gaussian) curve is an example.

Benchmark: A fixed point of reference; standard for comparison; an outstanding example, appropriate for use as a model.

Benchmarking: An improvement process in which a company measures its performance against that of best-in-class companies (or others who are good performers), determines how those companies achieved their performance levels, and uses the information to improve its own performance. The areas that can be benchmarked include strategies, operations, processes, and procedures.

Benefit–cost analysis: Collection of the dollar value of benefits derived from an initiative and the associated costs incurred and computing the ratio of benefits to cost.

Benefits administration: The formal system by which the organization manages its nonfinancial commitment to its employees. Includes such benefits as vacation, leave time, and retirement.

Best and final contract offer: Final offer by the supplier to perform the work after incorporating negotiated and agreed *changes* in the procurement documents.

Beta (β): See *Type II error.*

Beta (β) risk: The maximum probability *of* saying a process or lot is acceptable when, in fact, it should be rejected (see Consumer's risk).

Beta coefficient: Standardized regression coefficient (see Standardization) that allows for a direct comparison between coefficients as to their relative explanatory power of the dependent variable. Whereas *regression coefficients* are expressed in terms of the units of the associated variable, thereby making comparisons inappropriate, beta coefficients use standardized data and can be directly compared. In other words, a regression coefficient is explaining the data for the experiment at hand, whereas the beta coefficient can project the results of the relationship to populations because of its standardized value.

Bias (in measurement): A characteristic of measurement that refers to a systematic difference. That systematic difference is an error that leads to a difference between the average result of a population of measurements and the true, accepted value of the quantity being measured.

Bias: A characteristic of measurement that refers to a systematic difference.

Bias node: Additional *node* to a *hidden layer*, which has a constant value. This node functions in a manner similar to the constant term in multiple regression.

Big Q, little q: A term used to contrast the difference between managing for quality in all business processes and products (big Q) and managing for quality in a limited capacity, traditionally in only factory products and processes (little q).

Bimodal distribution: A frequency distribution that has two peaks. Usually an indication of samples from two processes incorrectly analyzed as a single process.

Binomial distribution (probability distribution): Given that a trial can have only two possible outcomes (yes/no, pass/fail, heads/tails), of which one outcome has probability p and the other probability q ($p + q = 1$), the probability that the outcome represented by p occurs r times in n trials is given by the binomial distribution.

Biserial correlation: Correlation measure used to replace the product–moment correlation when a metrically measured variable is associated with a nonmetric binary (0,1) measure. Also see *Polyserial correlation*.

Bivariate partial correlation: Simple (two-variable) correlation between two sets of residuals (unexplained variances) that remain after the association of other independent variables is removed.

Black belt: A team leader, trained in the Six Sigma (DMAIC and DCOV) process and facilitation skills, responsible for guiding an improvement project to completion. They are technically oriented individuals held in high regard by their peers. They should be actively involved in the organizational change and development process. Candidates may come from a wide range of disciplines and need not be formally trained statisticians or engineers. Six Sigma technical leaders work to extract actionable knowledge from an organization's information warehouse. Successful candidates should understand one or more operating systems, spreadsheets, database managers, presentation programs, and word processors. As part of their training they will be required to become proficient in the use of one or more advanced statistical analysis software packages.

Blemish: An imperfection that is severe enough to be noticed but should not cause any real impairment with respect to intended normal or reasonably foreseeable use (see also Defect, Imperfection, and Nonconformity).

Block diagram: A diagram that shows the operation, interrelationships, and interdependencies of components in a system. Boxes, or blocks (hence the name), represent the components; connecting lines between the blocks represent interfaces. There are two types of block diagrams: a functional block diagram, which shows a system's subsystems and lower-level products, their interrelationships, and interfaces with other systems; and a reliability block diagram, which is similar to the functional block diagram except that it is modified to emphasize those aspects influencing reliability. Also known as boundary diagram.

Block: A level of a random factor designated to sample unmeasured spatial or temporal variation in the environment. Designs for randomized complete blocks have every level of every factor (or factor combination if several) represented in each block. The Latin square is a variant on this design for two blocking factors.

Blocked designs: Analysis of one or more categorical factors with levels, or combinations of levels, randomly assigned in blocked sampling units of plots within blocks, and replicated only across blocks (including orthogonal contrasts, and balanced incomplete block, Latin squares, and Youden square variants on the one-factor complete block design)

Blocking factor: Characteristic of respondents in the ANOVA or MANOVA that is used to reduce within-group variability. This characteristic becomes an additional *treatment* in the analysis. In doing so, additional groups are formed that are more homogeneous. As an example, assume that customers were asked about buying intentions for a product and the independent measure used was age. Examination of the data found that substantial variation was due to gender. Then gender could be added as a further treatment so that each age category was split into male and female groups with greater within-group homogeneity.

Bogey testing: Conducting a test only to a specified time, mileage, or cycles.

Bonferroni inequality: Approach for adjusting the selected *alpha* level to control for the overall *Type I error* rate. The procedure involves (1) computing the adjusted rate as α divided by the number of statistical tests to be performed, and (2) using the adjusted rate as the critical value in each separate test.

Bootstrapping: Form of resampling in which the original data are repeatedly sampled with replacement for model estimation. Parameter estimates and standard errors are no longer calculated with statistical assumptions, but instead are based on empirical observations.

B_q life: The life at which "q" percent of the items in a population are expected to fail. For example, "B_{10} = 70,000 miles," means that 10% of the items are expected to fail by 70,000 miles (and that 90% had a life more than 70,000 miles).

Box test: Statistical test for the equality of the variance/covariance matrices of the dependent variables across the groups. It is very sensitive, especially to the presence of nonnormal variables. A significance level of 0.01 or less is used as an adjustment for the sensitivity of the statistic.

Box plot: Method of representing the distribution of a variable. A box represents the major portion of the distribution, and the extensions—called whiskers—reach to the extreme points of the distribution. Very useful in making comparisons of one or more variables across groups.

Box's M: Statistical test for the equality of the covariance matrices of the independent variables across the groups of the dependent variable. If the statistical significance is greater than the critical level (e.g., .01), then the equality of the covariance matrices is supported. If the test shows statistical significance, then the groups are deemed different and the assumption is violated.

BPR (business process reengineering): See Reengineering.

Brainstorming: A problem-solving tool that teams use to generate as many ideas as possible and as fast as possible related to a particular subject. Team members begin by offering all their ideas; the ideas are not discussed or reviewed until after the brainstorming session.

Breakthrough: A method of solving chronic problems that results from the effective execution of a strategy designed to reach the next level of quality. Such change often requires a paradigm shift within the organization.

Bridging design: Stimuli *design* for a large number of factors (attributes) in which the attributes are broken into a number of smaller groups. Each attribute group has some attributes contained in other groups, so the results from each group can be combined, or bridged.

Business processes: Processes that focus on what the organization does as a business and how it goes about doing it. A business has functional processes (generating output within a single department) and cross-functional processes (generating output across several functions or departments).

Calibration (of instrument): Adjusting an instrument using a reference standard to reduce the difference between the average reading of the instrument and the "true" value of the standard being measured, that is, to reduce measurement bias.

Calibration: The comparison of a measurement instrument or system of unverified accuracy to a measurement instrument or system of a known accuracy to detect any variation from the true value.

Camp–Meidell conditions: For frequency distribution and histograms: A distribution is said to meet Camp–Meidell conditions if its mean and mode are equal and the frequency declines continuously on either side of the mode.

Canonical correlation: Measure of the strength of the overall relationships between the linear composites *(canonical variates)* for the independent and dependent variables. In effect, it represents the bivariate correlation between the two canonical variates.

Canonical cross-loadings: Correlation of each observed independent or dependent variable with the opposite *canonical variate*. For example, the independent variables are correlated with the dependent canonical variate. They can be interpreted like *canonical loadings*, but with the opposite canonical variate.

Canonical function: Relationship (correlational) between two linear composites *(canonical variates)*. Each canonical function has two canonical variates, one for the set of dependent variables and one for the set of independent variables. The strength of the relationship is given by the *canonical correlation*.

Canonical loadings: Measure of the simple linear correlation between the independent variables and their respective *canonical variates*. These can be interpreted like factor loadings and are also known as canonical structure correlations.

Canonical roots: Squared *canonical correlations*, which provide an estimate of the amount of shared variance between the respective optimally weighted *canonical variates* of dependent and independent variables. Also known as *eigenvalues*.

Canonical variates: Linear combinations that represent the weighted sum of two or more variables and can be defined for either dependent or independent variables. Also referred to as *linear composites*, linear compounds, and linear combinations.

Capability (of process): The uniformity of product that a process is capable of producing. Can be expressed numerically using C_p, C_r, C_{pk}, and $Z_{max}/3$ when the data are normally distributed.

Capability ratio (C_p): Is equal to the specification tolerance width divided by the process capability.

Cascading training: Training implemented in an organization from the top down, where each level acts as trainers to those below.

Cash-flow analysis: The activity of establishing cash flow (dollars in and out of the *project*) by month and the accumulated total *cash flow* for the project for the measurement of actual versus the *budget* costs. This is necessary to allow for funding of the project at the lowest carrying charges and is a method of measuring project progress.

Catchball: A term used to describe the interactive process of developing and deploying policies and plans with Hoshin planning.

Categorical variable: Variable that uses values that serve merely as a label or means of identification. It is also referred to as a nonmetric, nominal, binary, qualitative, or taxonomic variable. The number on a football jersey is an example.

Causal part: Part determined by the dealer to be the part that caused the product repair and, thus, the warranty claim against the company. Part causing the repair or replacement of other parts. The condition code appears next to the causal part number. The Causal Part is charged with the entire repair cost.

Causal relationship: Dependence relationship of two or more variables in which the researcher clearly specifies that one or more variables "cause" or create an outcome represented by at least one other variable. Must meet the requirements for *causation*.

Causation: Principle by which cause and effect are established between two variables. It requires that there be a sufficient degree of association (correlation) between the two variables, that one variable occurs before the other (that one variable is clearly the outcome of the other), and that there be no other reasonable causes for the outcome. Although in its strictest terms causation is rarely found, in practice strong theoretical support can make empirical estimation of causation possible.

Cause-and-effect analysis: A visually effective way of identifying and recording the possible causes of a problem and the relationships between them as they are suggested by the team. A pictorial diagram showing all the cause (process inputs) and effect (resulting problem being investigated) relationships among the factors that affect the process. In essence, it is a tool for analyzing process variables. It is also referred to as the Ishikawa diagram, after the name of its developer–Kaoru Ishikawa, or fishbone diagram and or feather diagram after its appearance. The complete diagram resembles a fish skeleton (causes are the bones of the fish and the effect are shown as the head of the fish). The diagram illustrates the main causes and subcauses leading to an effect (symptom). The cause-and-effect diagram is one of the seven basic tools of quality.

c-Chart: For attributes data: A control chart of the number of defects found in a subgroup of fixed size. The c-chart is used where each unit typically has a number of defects.

CDF: Cumulative distribution function

CDS: Component design specification

Cell (*of frequency distribution and/or* histogram): For a sample based on a continuous variable, a cell is an interval into which individual data points are grouped. The full range of the variable is usually broken into intervals of equal size and the number of points in each cell totaled. These intervals (cells) make up a frequency distribution or histogram. This greatly reduces the amount of information that must be dealt with, as opposed to treating each element (data point) individually.

Cell: A layout of workstations and/or various machines for different operations (usually in a U shape) in which multitasking operators proceed, with a part, from machine to machine, to perform a series of sequential steps to produce a whole product or major subassembly.

Censored data: Observations that are incomplete in a systematic and known way. One example occurs in the study of causes of a particular event in which some characteristics are not taken into account. Censored data are an example of *ignorable missing data*.

Center line: For control charts: the horizontal line marking the center of the chart, usually indicating the grand average of the quantity being charted.

Central limit theorem: If samples of a population with size *n* are drawn, and the values of Xbar are calculated for each sample group, and the distribution of Xbar is found, the distribution's shape is found to approach a normal distribution for sufficiently large *n*. This theorem allows one to use the assumption of a normal distribution when dealing with Xbar. "Sufficiently large" depends on the population's distribution and what range of Xbar is being considered; for practical purposes, the easiest approach may be to take a number of samples of a desired size and see whether their means are normally distributed. If not, the sample size should be increased.

Central tendency: A measure of the point about which a group of values is clustered; some measures of central tendency are mean, mode, and median. Another way of saying it is the propensity of data collected on a process to concentrate around a value situated somewhere midway between the lowest and highest value.

Centroid method: *Agglomerative algorithm* in which *similarity* between clusters is measured as the distance between *cluster centroids*. When two clusters are combined, a new centroid is computed. Thus, cluster centroids migrate, or move, as the clusters are combined.

Centroid: Mean value for the discriminant *Z* scores of all objects within a particular category or group. For example, a two-group discriminant analysis has two centroids, one for the objects in each of the two groups.

Chaku-chaku (Japanese): Meaning "load-load" in a cell layout where a part is taken from one machine and loaded into the next.

Champion: An individual who has accountability and responsibility for many processes or who is involved in making strategic-level decisions for the organization. The champion ensures ongoing dedication of project resources and monitors strategic alignment (also referred to as a sponsor).

Change agent: The person who takes the lead in transforming a company into a quality organization by providing guidance during the planning phase, facilitating implementation, and supporting those who pioneer the changes.

Change in scope: A change in *objectives*, work plan, or *schedule* that results in a material difference from the terms of an approval to proceed previously granted by higher authority. Under certain conditions (normally so stated in the approval instrument), change in resources application may constitute a change in *scope*.

Change order/purchase order amendment: Written order directing the contractor to make *changes* according to the provisions of the *contract documents*.

Change: An increase or decrease in any of the *project* characteristics.

Changed conditions (contract): A change in the *contract* environment, physical or otherwise, compared to that contemplated at the time of *bid*.

Changeover: Changing a machine or process from one type of product or service to another.

Characteristic: A dimension or parameter of a part that can be measured and monitored for control and capability. A property that helps to identify or to differentiate between entities and that can be described or measured to determine conformance or nonconformance to requirements.

Charter: A documented statement officially initiating the formation of a committee, team, project, or other effort in which a clearly stated purpose and approval is conferred. It is the specific team document defining the context, specifics, and plans of an improvement project; includes business case, problem and goal statements, constraints and assumptions, roles, preliminary plan, and scope. Periodic reviews with the sponsor ensure alignment with business strategies; review, revise, refine periodically throughout the DMAIC process based on data.

Check list and recording sheet: A form used to ensure the factor-level settings are correct for each run of the experiment. The form can also be used to record the results of each experimental run and to note any unplanned events that may distort or bias the results.

Check list: A tool used to ensure that all important steps or actions in an operation have been taken; checklists contain items that are important or relevant to an issue or situation; a form used to ensure the factor-level settings are correct for each run of the experiment. The form can also be used to record the results of each experimental run and to note any unplanned events that may distort or bias the results. Checklists are often confused with check sheets and data sheets (see also Check sheet).

Check sheet: A simple data-recording device. The check sheet is custom designed for the particular use, allowing ease in collecting and interpreting the results and to reduce the likelihood of errors in recording data. The check sheet is one of the seven basic tools of quality. Check sheets are often confused with data sheets and checklists (see also Checklist). A sheet for the recording of data on a process or its product. (These may be forms, tables, or worksheets facilitating data collection and compilation; allows for collection of stratified data. See also Stratification.) The check sheet is designed to remind the user to record each piece of information required for a particular study and to reduce the likelihood of errors in recording data. The data from the check sheet can be typed into a computer for analysis when the data collection is complete.

Chi-square: Method of standardizing data in a *contingency table* by comparing the actual cell frequencies to an expected cell frequency. The expected cell frequency is based on the marginal probabilities of its row and column (probability of a row and column among all rows and columns). It is also used as a test for goodness of fit: a measure of how well a set of data fits a proposed distribution, such as the normal distribution. The data are placed into classes and the observed frequency (O) is compared to the expected frequency (E) for each class of the proposed distribution. The result for each class is added to obtain a chi-square value. This is compared to a critical chi-square value from a standard table for a given (alpha) risk and degrees of freedom. If the calculated value is smaller than the critical value, we can conclude that the data follow the proposed distribution at the chosen level of significance.

Choice set: Set of full-profile stimuli constructed through experimental design principles and used in the *choice-based approach.*

Choice simulator: Procedure that allows the researcher to assess many "what-if" scenarios, including the preference for possible product or service configurations or the competitive interactions among stimuli assumed to constitute a market. Once the conjoint *part-worths* have been estimated for each respondent, the choice simulator analyzes a set of *full-profile stimuli* and predicts both individual and aggregate choices for each stimulus in the set.

Choice-based conjoint approach: Alternative form of collecting responses and estimating the conjoint model. The primary difference is that respondents select a single *full-profile stimulus* from a set of stimuli (known as a *choice set*) instead of rating or ranking each stimulus separately.

Chronic condition: Long-standing adverse condition that requires resolution by changing the status quo. For example, actions such as revising an unrealistic manufacturing process or addressing customer defections can change the status quo and remedy the situation.

Chronic problem: A long-standing adverse situation that can be remedied by changing the status quo. For example, actions such as revising an unrealistic manufacturing process or addressing customer defections can change the status quo and remedy the situation.

City-block approach: Method of calculating distances based on the sum of the absolute differences of the coordinates for the objects. This method assumes the variables are uncorrelated and unit scales are compatible.

Classification function: Method of classification in which a linear function is defined for each group. Classification is performed by calculating a score for each observation on each group's classification function and then assigning the observation to the group with the highest score. This differs from the calculation of the *discriminant z-score*, which is calculated for each *discriminant function*.

Classification matrix: Matrix assessing the predictive ability of the discriminant function(s) or logistic regression. It is also called a confusion, assignment, or prediction matrix. Created by cross-tabulating actual group membership with predicted group membership, this matrix consists of numbers on the diagonal representing correct classifications, and off-diagonal numbers representing incorrect classifications.

Cluster: For control charts: a group of points with similar properties. Usually an indication of short duration, assignable causes.

Cluster analysis: Multivariate technique with the objective of grouping respondents or cases with similar profiles on a defined set of characteristics. Similar to Q *factor analysis.*

Cluster centroid: Average value of the objects contained in the cluster on all the variables in the *cluster variate.*

Cluster seeds: Initial *centroids* or starting points for clusters. These values are selected to initiate *nonhierarchical clustering* procedures, in which clusters are built around these prespecified points.

Cluster variate: Set of variables or characteristics representing the *objects* to be clustered and used to calculate the *similarity* between objects.

Coaching: A continuous improvement technique by which people receive one-to-one learning through demonstration and practice and that is characterized by immediate feedback and correction.

Coded plan matrix: The levels of each factor within the plan matrix are represented by a code. The codes can be '1' and '2' or '−' and '+'. The use of '−' and '+' is preferred as these simplify the use of the matrix when calculating the effect of each factor. Taguchi, on the other hand, prefers the "1" and "2" designation.

Coefficient of determination (R^2): Measure of the proportion of the variance of the dependent variable about its mean that is explained by the independent, or predictor, variables. The coefficient can vary between 0 and 1. If the regression model is properly applied and estimated, the researcher can assume that the higher the value of R^2, the greater the explanatory power of the regression equation, and therefore, the better the prediction of the dependent variable.

Collinearity: Expression of the relationship between two (collinearity) or more (multicollinearity) independent variables. Two independent variables are said to exhibit complete collinearity if their correlation coefficient is 1, and complete lack of collinearity if their correlation coefficient is 0. *Multicollinearity* occurs when any single independent variable is highly correlated with a set of other independent variables. An extreme case of collinearity/multicollinearity is *singularity*, in which an independent variable is perfectly predicted (i.e., correlation of 1.0) by another independent variable (or more than one).

Commissioning: *Activities* performed for the purpose of substantiating the capability of the *project* to function as designed.

Commitment: An agreement to consign, or reserve the necessary *resources* to fulfill a requirement, until *expenditure* occurs. A commitment is an *event*.

Common cause: A common cause is any input, or work task, which is defined by the process, and thus, is always present. Problems from common causes are referred to as "chronic pain." See also Control Charts, Run Chart or Time Plot, Special Cause, Variation. Those sources of variability in a process that are truly random, that is, inherent in the process itself. An example is the slight vibration emitted by a cutting machine that affects the precision of the part it is cutting. Common causes are predictable in that they will stay on the job until eliminated or reduced. On control charts, an unchanging level of common causes manifest themselves by measurements that randomly hover about a fixed average and within control limits (almost 100% of the time). A process that has only common causes of variation is said to be stable or predictable or in control. Also called "chance causes." See also Special causes.

Common cause shift: A common cause shift occurs when some element or elements of the process is permanently changed, resulting in a new (better or worse) level of performance. With warranty data, the common cause shift can be due to any number of design or manufacturing process changes. It is also important to understand that nonprofit common causes, such as a change in warranty policy administration, could also be responsible for a common cause shift.

Common causes of variation: Causes that are inherent in any process all the time.

Common factor analysis: Factor model in which the factors are based on a reduced correlation matrix. That is, *communalities* are inserted in the diagonal of the *correlation* matrix, and the extracted factors are based only on the *common variance*, with *specific* and *error variances* excluded.

Common variance: Variance shared with other variables in the factor analysis.

Communality: Total amount of variance an original variable shares with all other variables included in the analysis.

Company-wide quality control (CWQC): Similar to Total Quality Management.

Comparison group: The category of a nonmetric variable that receives all zeros when *indicator coding* is used or all minus ones (–1s) when *effects coding* is used in creating *dummy variables.*

Competence: Refers to a person's ability to team and perform a particular activity. Competence generally consists of skill, knowledge, experience, and attitude components.

Competency-based training: A training methodology that focuses on building mastery of a predetermined segment or module before moving on to the next.

Competitive analysis: The gathering of intelligence relative to competitors in order to identify opportunities or potential threats to current and future strategy.

Competing models strategy: Strategy that compares the proposed model with a number of alternative models in an attempt to demonstrate that no better-fitting model exists. This is particularly relevant in structural equation modeling because a model can be shown only to have acceptable fit, but acceptable fit alone does not guarantee that another model will not fit better or equally well.

Complete case approach: Approach for handling missing data that compute values based on data from only complete cases, that is, cases with no *missing data.*

Complete linkage: *Agglomerative algorithm* in which *interobject similarity* is based on the maximum distance between *objects* in two clusters (the distance between the most dissimilar members of each cluster). At each stage of the agglomeration, the two clusters with the smallest maximum distance (most similar) are combined.

Component analysis: Factor model in which the factors are based on the total variance. With component analysis, unities (1s) are used in the diagonal of the *correlation matrix*; this procedure computationally implies that all the variance is *common* or shared.

Composite measure: See *Summated scale.*

Composition rule: Rule used in combining attributes to produce a judgment of relative value or *utility* for a product or service. For illustration, let us suppose a person is asked to evaluate four objects. The person is assumed to evaluate the attributes of the four objects and to create some overall relative value for each. The rule may be as simple as creating a mental weight for each perceived attribute and adding the weights for an overall score *(additive model),* or it may be a more complex procedure involving *interaction effects.*

Compositional method: Alternative approach to the more traditional *decompositional* methods of perceptual mapping that derive overall *similarity* or *preference* evaluations from evaluations of separate attributes by each respondent. These separate attribute evaluations are combined (composed) for an overall evaluation. The most common examples of compositional methods are the techniques of factor analysis and discriminant analysis.

Compositional model: Class of multivariate models that base the dependence relationship on observations from the respondent regarding both the dependent and the independent variables. Such models calculate or "compose" the dependent variable from the respondent-supplied values for the independent variables. Principal among such methods are regression analysis and discriminant analysis. These models are in direct contrast to *decompositional models.*

Communicating with groups: The means by which the *Project Manager* conducts meetings, presentations, negotiations, and other activities necessary to convey the project's needs and concerns to the *project team* and other groups.

Communicating with individuals: Involves all activities by which the *Project Manager* transfers information or ideas to individuals working on the *project.*

Communications management (framework): The proper organization and *control* of information transmitted by whatever means to satisfy the needs of the project. It includes the processes of transmitting, filtering, receiving, and interpreting or understanding information using appropriate skills according to the application in the project environment. It is at once the master and the servant of a project in that it provides the means for interaction between the many disciplines, functions, and activities, both internal and external to the project, and which together result in the successful completion of that project. Conducting or supervising the exchange of information.

Company-wide quality control (CWQC): Similar to Total Quality Management.

Compensation and evaluation: The measurement of an individual's performance and the financial payment provided to employees as a reward for their performance and as a motivator for future performance.

Competence: Refers to a person's ability to team and perform a particular activity. Competence generally consists of skill, knowledge, experience, and attitude components.

Competency-based training: A training methodology that focuses on building mastery of a predetermined segment or module before moving on to the next.

Competitive analysis: The gathering of intelligence relative to competitors in order to identify opportunities or potential threats to current and future strategy.

Conceptual definition: Specification of the theoretical basis for a concept that is represented by a factor.

Concerns: Number of defects (nonconformities) found on a group of samples in question.

Concurrent engineering: A process in which an organization designs a product or service using input and evaluations from business units and functions early in the process, anticipating problems, and balancing the needs of all parties. The emphasis is on upstream prevention versus downstream correction.

Condition index: Measure of the relative amount of variance associated with an *eigenvalue* so that a large condition index indicates a high degree of *collinearity*.

Confidence: The proportion of times two or more events occur jointly. For example, if product A is purchased 60% of the time a customer also purchases product B, we have a confidence of 60% in saying products A and B are purchased jointly.

Confidence interval: Range within which a parameter of a population (e.g., mean, standard deviation, etc.) may be expected to fall, on the basis of measurement, with some specified confidence level.

Confidence level: The probability set at the beginning of a hypothesis test that the variable will fall within the confidence interval. A confidence level of 0.95 is commonly used.

Confidence limits: The upper and lower boundaries of a confidence interval.

Confirmatory analysis: Use of a multivariate technique to test (confirm) a prespecified relationship. For example, suppose we hypothesize that only two variables should be predictors of a dependent variable. If we empirically test for the significance of these two predictors and the nonsignificance of all others, this test is a confirmatory analysis. It is the opposite of *exploratory analysis*.

Confirmatory modeling strategy: Strategy that statistically assesses a single model for its fit to the observed data. This approach is actually less rigorous than the *competing models strategy* because it does not consider alternative models that might fit better or equally well than the proposed model.

Conformance (of product): Adherence to some standard of the product's properties. The term is often used in attributes studies of product quality; that is, a given unit of the product is either in conformance to the standard or it is not.

Confusion data: Procedure to obtain respondents' perceptions of *similarities data*. Respondents indicate the similarities between pairs of stimuli. The pairing (or "confusing") of one stimulus with another is taken to indicate similarity. Also known as *subjective clustering*.

Concerns: Number of defects (nonconformities) found on a group of samples in question.

Conciliatory: A project manager who is friendly and agreeable, one that attempts to assemble and unite all project parties involved to provide a compatible working team.

Concurrent engineering: A process in which an organization designs a product or service using input and evaluations from business units and functions early in the process, anticipating problems, and balancing the needs of all parties. The emphasis is on upstream prevention versus downstream correction.

Conducting the experiment: The experiment, once planned and designed, is performed by testing the system under investigation according to the combination of factor-level settings determined by each experimental run.

Confidence interval: An interval estimate of a parameter which contains the true value of the parameter with given probability. The width of the interval is affected by the degree of confidence, sample size, and variability. It is usually a probability set at the beginning of a hypothesis test that the variable will fall within the confidence interval. A confidence level of 0.95 is commonly used.

Confidence limits: The upper and lower boundaries of a confidence interval.

Configuration (baseline) control: A system of procedures that monitors emerging *project scope* against the *scope baseline*. Requires documentation and management approval on any *change* to the baseline.

Confirmation run: An experiment performed under the optimal conditions predicted by the analysis of the results from the designed experiment. This test verifies the quantitative knowledge of the system (which was determined through the analysis).

Conflict management: The process by which the *Project Manager* uses appropriate managerial *techniques* to deal with the inevitable disagreements, both technical and personal in nature, which develop among those working toward *project* accomplishment.

Conflict resolution: A process for resolving disagreements in a manner acceptable to all parties. To seek a solution to a problem, five methods in particular have been proven through confrontation, compromise, smoothing, forcing, and withdrawal

Conformance (of product): Adherence to some standard of the product's properties. The term is often used in attributes studies of product quality; that is, a given unit of the product is either in conformance to the standard or it is not.

Conformance: An affirmative indication or judgment that a product or service has met the requirements of a relevant specification, contract, or regulation.

Confounding: When interaction effects cannot be separated from the main effects, the main effects are said to be confounded with interactions.

Confrontation: Where two parties work together toward a solution of the problem.

Conjoint variate: Combination of variables (known as *factors*) specified by the researcher that constitute the total worth or *utility* of the stimuli. The researcher also specifies all the possible values for each factor, with these values known as *levels*.

Consensus: Finding a proposal acceptable enough that all team members can support the decision and no member opposes it.

Constancy of purpose: Occurs when goals and objectives are properly aligned to the organizational vision and mission.

Constant-cause system: A system or process in which the variations are random and are constant in time.

Constraint management: Pertains to identifying a constraint and working to remove or diminish the constraint, while dealing with resistance to change.

Constraint: A constraint may range from the intangible (e.g., beliefs, culture) to the tangible (e.g., posted rule prohibiting smoking, build-up of work-in process awaiting the availability of a machine or operator).

Constraints: Applicable restrictions that will affect the *scope*. Any factor that affects when an activity can be scheduled (see Restraint).

Construct: Concept that the researcher can define in conceptual terms but cannot be directly measured (e.g., the respondent cannot articulate a single response that will totally and perfectly provide a measure of the concept) or measured without error (see *Measurement error*). Constructs are the basis for forming *causal relationships*, as they are the "purest" possible representation of a concept. A construct can be defined in varying degrees of specificity, ranging from quite narrow concepts, such as total household income, to more complex or abstract concepts, such as intelligence or emotions. No matter what its level of specificity, however, a construct cannot be measured directly and perfectly but must be approximately measured by *indicators*.

Consultative: A decision-making approach in which a person talks to others and considers his or her input before making a decision.

Consumer market customers: End-users of a product or service (customers and noncustomers alike).

Consumer's risk: The maximum probability of saying a process or lot is acceptable when, in fact, it should be rejected. In a sampling plan, it refers to the probability of acceptance of a lot, the quality of which has a designated numerical value representing a level that is seldom desirable. Usually the designated value will be the lot tolerance percent defective (LPTD). Also called beta risk or Type II error.

Consumption: Conversion of a liquid resource to a less recoverable state, that is, expenditures of time, human resources, dollars to produce something of value, or the incorporation of inventoried materials into fixed assets.

Content validity: Assessment of the degree of correspondence between the items selected to constitute a *summated scale* and its *conceptual definition*.

Contingencies: Specific provision for unforseeable elements of *cost* within the defined *project scope*; particularly important where previous experience relating estimates and actual costs has shown that unforeseeable events that will increase costs are likely to occur. If an allowance for escalation is included in the contingency it should be a separate item, determined to fit expected escalation conditions for the project.

Contingency allowances: Specific provision for unforeseen elements of cost within the defined *project scope*; particularly important where previous experience relating *estimates* and actual *costs* has shown that unforeseen *events* that will increase costs are likely to occur. If an allowance for escalation is included in the contingency it should be as a separate item, determined to fit expected escalation conditions of the project.

Contingency plan: A plan that identifies key assumptions beyond the *Project Manager's* control and their probability of occurrence. The plan identifies alternative strategies for achieving *project* success.

Contingency planning: The establishment of management plans to be invoked in the event of specified *risk events*. Examples include the provision and prudent management of a *contingency* allowance in the *budget*, the preparation of alternative *schedule activity* sequences or "work-arounds," emergency responses to reduce damages arising out of risk incidents, and the evaluation of liabilities in the *event* of complete *project* shutdown.

Contingency table: Cross-tabulation of two nonmetric or categorical variables in which the entries are the frequencies of responses that fall into each "cell" of the matrix. For example, if three brands were rated on four attributes, the brand-by-attribute contingency table would be a three-row-by-four-column table. The entries would be the number of times a brand was rated as having an attribute.

Continuous: A characteristic of variables data, such as the waiting time for a computer program to be executed. Continuous data can theoretically take on an infinite number of values in a given interval. In practice, there are always a finite number of values, because of the discreetness of the measurement system. Thus, for example, waiting time may only be measured to the nearest second.

Continuous data: The results of measuring a continuous variable. Any variable measured on a continuum or scale that can be infinitely divided; in fact, the measurement depends on the system used; primary types include time, dollars, size, weight, temperature, and speed; also referred to as "variable data" (see also Attribute data).

Continuous probability distribution: A graph or formula representing the probability of a particular numeric value of continuous (variable) data, based on a particular type of process that produces the data.

Continuous process improvement: Includes the actions taken throughout an organization to increase the effectiveness and efficiency of activities and processes in order to provide added benefits to the customer and organization. It is considered a subset of total quality management and operates according to the premise that organizations can always make improvements. Continuous improvement can also be equated with reducing process variation.

Continuous variable: A variable that can assume any of a range of values; an example would be the measured size of a part.

Contrast matrix: When a coded plan matrix using the codes '–' and '+' is modified to show '–1' and '+1', it becomes known as a contrast matrix. Each column of the contrast matrix is known as a contrast and can be used to calculate the effect of the factor or the interactions(s) attributable to that column.

Contrast set: The number of alternative sets of orthogonal contrasts refers to the number of different statistical models available to describe the orthogonal partitioning of variation between i levels of a categorical factor.

Contrast: Procedure for investigating specific group differences of interest in conjunction with ANOVA and MANOVA, for example, comparing group mean differences for a specified pair of groups. Another way of thinking about a contrast is to think of it as a weighted average of the experiment's results (response data), where all the weights sum to zero. Each column of a contrast matrix represents the contrast for a factor or interaction.

Control (general): A statistical concept indicating that a process operating within an expected range of variation is being influenced mainly by "common cause" factors; processes operating in this state are referred to as "in control." See also Control Charts, Process Capability, Variation. In other words, the exercise of *corrective action* as necessary to yield a required outcome consequent upon *monitoring* performance.

Control (*of* process): A process is said to be in a state of statistical control if the process exhibits only random variations (as opposed to systematic variations and/or variations with known sources). When monitoring control with control charts, a state of control is exhibited when all points remain between set control limits.

Control chart: A plot of some parameter of process performance, usually determined by regular sampling of the product, as a function (usually) of time or unit number or other chronological variable. The control limits are also plotted for comparison. The parameter plotted may be the mean value of a particular measurement for a product sample of specified size (Xbar chart), the range of values in the sample (R-chart), the percent of defective units in the sample (p-chart), and so on. Another way to look at the control chart is to view it as a basic tool that consists of a graphical representation of data with upper and lower control limits on which values of some statistical measure for a series of samples or subgroups are plotted. It frequently shows a central line to help detect a trend of plotted values toward either control limit. It is used to monitor and analyze variation from a process to see whether the process is in statistical control.

Control factors: A term frequently used in experimental/robust design to designate those design or process variables that can be freely specified by the engineer and may be set at various levels.

Control group: An experimental group that is not given the treatment under study. The experimental group that is given the treatment is compared to the control group to ensure any changes are due to the treatment applied.

Control limits: The limits within which the product of a process is expected (or required) to remain. If the process leaves the limits, it is said to be out of control. This is a signal that action should be taken to identify the cause and eliminate it if possible. Note: Control limits are not the same as tolerance limits. Control limits always indicate the "voice of the process," the "behavior of the process," and they are always calculated. Tolerance limits, on the other hand, are always given by the customer.

Control plan: A document that may include the characteristics for quality of a product or service, measurements, and methods of control.

Control system: A mechanism that reacts to the current *project* status in order to ensure accomplishment of project *objectives*.

Control: (1) DMAIC phase **C**; once solutions have been implemented, ongoing measures track and verify the stability of the improvement and the predictability of the process. Often includes process management techniques and systems, including process ownership, cockpit charts, and/or process management charts, and so on. *See also* Cockpit charts, Process management. (2) A statistical concept indicating that a process operating within an expected range of variation is being influenced mainly by "common cause" factors; processes operating in this state are referred to as "in control." *See also* Control charts, Process capability, Variation.

Convergent thinking: Convergent thinking means narrowing down to one answer.

Cook's distance (D_i): Summary measure of the influence of a single case (observation) based on the total changes in all other residuals when the case is deleted from the estimation process. Large values (usually greater than 1) indicate substantial influence by the case in affecting the estimated regression coefficients.

COPQ: Cost of poor quality

Core competency: Pertains to the unique features and characteristics of an organization's overall capability.

Corporate business life cycle: A life cycle that encompasses phases of policy planning and identification of needs before a *Project Life Cycle*, as well as product in service and disposal after the project life cycle.

Corporate culture: While the word "corporate" typically appears, the culture referred to may be that of any type of organization, large or small. The term relates to the collective beliefs, values, attitudes, manners, customs, behaviors, and artifacts unique to an organization.

Corporate project strategy: The overall direction set by the corporation, of which the *project* is a part, and the relationship of specific procurement actions to these corporate directions.

Corrective action: (cost management). The development of changes in plan and approach to improve the performance of the *project* (communications management). Measures taken to rectify conditions adverse to specified *quality*, and where necessary, to preclude repetition (general): Action taken to eliminate the root cause(s) and symptom(s) of an existing deviation or nonconformity to prevent recurrence.

Correlation: Refers to the measure of the relationship between two sets of numbers or variables.

Correlation coefficient *(r)*: Coefficient that indicates the strength of the association between any two metric variables. The sign (+ or −) indicates the direction of the relationship. The value can range from −1 to +1, with +1 indicating a perfect positive relationship, 0 indicating no relationship, and −1 indicating a perfect negative or reverse relationship (as one variable grows larger, the other variable grows smaller).

Correlation matrix: Table showing the intercorrelations among all variables.

Correspondence analysis: *Compositional approach* to perceptual mapping that relates categories of a *contingency table.* Most applications involve a set of *objects* and attributes, with the results portraying both objects and attributes in a common *perceptual map.*

Cost of poor quality (COPQ): The costs associated with providing poor-quality products or services. Dollar measures depicting the impact of problems (internal and external failures) in the process as it exists; it includes labor and material costs for handoffs, rework, inspection, and other non-value-added activity. In other words, the cost incurred from generating a defect, or cost that would be avoided if, rather than a "Defect," the "Opportunity" yielded a "Success."

Cost of quality (COQ): Costs incurred in assuring quality of a product or service. There are four categories of quality costs: internal failure costs (costs associated with defects found before delivery of the product or service), external failure costs (costs associated with defects found during or after product or service delivery), appraisal costs (costs incurred to determine the degree of conformance to quality requirements), and prevention costs (costs incurred to keep failure and appraisal costs to a minimum).

Cost performance measurement baseline: The formulation of *budget costs*, measurable goals (particularly time and quantities), for the purposes of comparisons, analysis, and *forecasting* future costs.

Count chart: A control chart for evaluating the stability of a process in terms of the count of events of a given classification occurring in a sample.

Count-per-unit chart: A control chart for evaluating the stability of a process in terms of the average count of events of a given classification per unit occurring in a sample.

Covariate: A factor X that varies at least on an ordinal scale, and usually on a continuous scale (such as time, distance, etc.) and is, therefore, a covariate of the response Y. Analysis of covariance assumes that the response has a linear relation to the covariate, and transformations of response or covariate may be necessary to achieve this prior to analysis.

Covariates, or covariate analysis: Use of regression-like procedures to remove extraneous (nuisance) variation in the dependent variables due to one or more uncontrolled metric independent variables (covariates). The covariates are assumed to be linearly related to the dependent variables. After adjusting for the influence of covariates, a standard ANOVA or MANOVA is carried out. This adjustment process (known as ANCOVA or MANCOVA) usually allows for more sensitive tests of treatment effects.

COVRATIO: Measure of the influence of a single observation on the entire set of estimated regression coefficients. A value close to 1 indicates little influence. If the COVRATIO value minus 1 is greater than $\pm 3p/n$ (where p is the number of independent variables +1, and n is the sample size), the observation is deemed to be influential based on this measure.

C_P—For process capability studies: C_p is a capability index defined by the formula: $C_p = \text{Tolerance}/6\sigma$. C_p shows the process capability potential but does not consider how centered the process is. C_p may range in value from 0 to infinity, with a large value indicating greater potential capability. A value of 1.67 or greater is considered an acceptable value of a C_p. Specifically in the Six Sigma methodology its value is expected to be greater than 2.0.

C_{pk}: A widely used process capability index. It is expressed as a ratio of the smallest answer of [USL-Xbar]/3 sigma or [Xbar − LSL]/3 sigma. If the C_{pk} has a value equal to Cp the process is centered on the nominal; if C_{pk} is negative, the process mean is outside the specification limits; if C_{pk} is between 0 and 1, then some of the Six Sigma spread falls outside the tolerance limits. If C_{pk} is larger than 1, the Six Sigma spread is completely within the tolerance limits. A value of 1.33 or greater is usually desired. Also known as $Z_{min}/3$.

CPR (cost per repair): Computed as the CPU divided by the R/1000 (CPU*1000/R/1000). Different than Average Cost Per Repair (ACPR).

CPU (cost per unit): Total cost of repairs divided by the number of products, usually expressed at a given Time-In-Service (TIS). The repairs and products included in the calculation are selected according to the rules defined by a TIS logic. See also TIS and Logic.

CQI: Continuous quality improvement

C_r: For process capability studies: The inverse of C_p, C_r can range from 0 to infinity in value, with a smaller value indicating a more capable process.

Crawford slip method: Refers to a method of gathering and presenting anonymous data from a group.

Criteria matrix: Decision-making tool used when potential choices must be weighed against several key factors (e.g., cost, ease to implement, impact on customer). Encourages use of facts, data, and clear business objectives in decision making.

Criteria: A statement that provides *objectives*, guidelines, *procedures*, and standards that are to be used to execute the development, design, and/or implementation portions of a PROJECT.

Criterion: A standard, rule, or test upon which a judgment or decision can be based.

Criterion validity: Ability of clusters to show the expected differences on a variable not used to form the clusters. For example, if clusters were formed on performance ratings, marketing thought would suggest that clusters with higher performance ratings should also have higher satisfaction scores. If this was found to be so in empirical testing, then criterion validity is supported.

Criterion variable (Y): See *Dependent variable.*

Critical F: The value of the *F*-statistic at the threshold probability α of mistakenly rejecting a true null hypothesis (the critical Type I error). The *F*-statistic is the test statistic used in ANOVA and GLM, named in honor of R. A. Fisher, who first described the distribution and developed the method of analysis of variance in the 1920s. The continuous *F*-distribution for a given set of test and error degrees of freedom is used to determine the probability of obtaining at least as large a value of the observed *F*-ratio of explained to unexplained variation, given a true null hypothesis. The associated *p*-value reports the significance of the test effect on the response in terms of the probability of mistakenly rejecting a true null hypothesis, which is deemed acceptably small if $p < \alpha$, where α often takes a value of 0.05. The *p*-value is the area under the *F*-distribution to the right of the corresponding *F*-value:

$$P = \int_{F_Q}^{\infty} \left(\frac{\left(p^p.q^q.F^{p-2}.(q+p.F)^{-(p+q)}\right)^{1/2}}{B\left(\frac{p}{2},\frac{q}{2}\right)} \right) dF$$

where p and q are the model test and error degrees of freedom, respectively, and the beta function $B(x, y) = \Gamma(x) \cdot \Gamma(y)/\Gamma(x + y)$, If the integral limit F_Q exceeds the critical F-value $F_{[\alpha]}$, then $P < \alpha$. Figure G.1 shows two examples of the position of the critical F-value along its F-distribution. The distribution has a strong positive skew for small test degrees of freedom, showing in Figure G.1

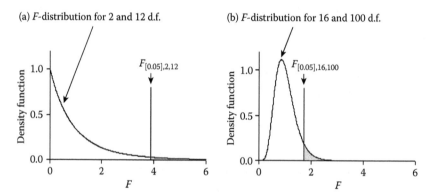

(a) F-distribution for 2 and 12 d.f.   (b) F-distribution for 16 and 100 d.f.

FIGURE G.1
Two examples of the position of the critical F-value along its F-distribution.

In the absence of a treatment effect, the observed $F = \text{TMS/EMS}$ follows the F-distribution, here showing with $\alpha = 0.05$ given by the red-shaded area under its right-hand tail above $F_{[\alpha]}$. (a) $F_{[0.05],2,12} = 3.89$; (b) $F_{[0.05],16,100} = 1.75$. (Source: http://www.soton.ac.uk/~cpd/anovas/datasets/Critical%20F.htm)

Critical value: Value of a test statistic (t-test, F-test) that denotes a specified significance level. For example, 1.96 denotes a 0.05 significance level for the t-test with large sample sizes.

Cronbach's alpha: Measure of *reliability* that ranges from 0 to 1, with values of 0.60 to 0.70 deemed the lower limit of acceptability.

Cross-factor: One factor is crossed with another when each of its levels is tested in each level of the other factor. For example, in a test of influences on crop yield, a watering regime factor (W: wet and dry) is crossed with a sowing density factor (D: high and low) when the response to the wet regime is tested at both high and low sowing density, and so is the response to the dry regime. If each of the four combinations of levels has replicate observations, then a cross-factored analysis can test for an interaction between the two treatment factors in their effect on the response. In a statistics package the fully replicated model is called by writing:

$$W + D + W*D$$

where "W*D" refers to the interaction between W and D.
Or equally:

$$W|D$$

where "|" is the command for main effects and their interaction.

An *interaction* between factors refers to the influence of one factor on a response depending on the level of another factor. For example, students may respond to different tutorial systems (T) according to their gender (G) and age (A), tested with model $Y = T|G|A + \varepsilon$. If older boys respond better to one system and younger girls to another, this may be indicated by a significant interaction effect T*G*A on the response. If one factor is a covariate, the interaction is illustrated by different regression slopes at each level of the categorical factor. Two covariates show a significant interaction in a curved plane for their combined effect on the response.

An interaction term must always be entered in the model after its constituent main effects. The significance of its effect is reported before them, however, because its impact influences interpretation of the main effects. A treatment main effect can be interpreted independently of its interactions with other factors only if the others are random factors. Thus, in the teaching example above, it would be misleading to interpret a significant main effect *T* as meaning that one tutorial system generally works better than another, if the student response to system depends on age and/or gender. Equally in the agricultural example, it would be unwise to report no general effect of sowing density from a nonsignificant main effect D, if the interaction with watering regime is significant. In this case, density will have had opposite effects on yield depending on watering, with the result that it appears to have no effect on pooling the levels of watering. A fully replicated design with two cross-factors can have eight alternative outcomes in terms of the significance of its main effects and interactions.

Cross-functional team: A group consisting of members from more than one department that is organized to accomplish a project.

CSR: Customer service representative.

CTQ: Critical to quality characteristic—a characteristic of a product, service, or information that is important to the customer. CTQs must be measurable in either a "quantitative" manner (i.e., 3.00 mg, etc.) or qualitative manner (correct/incorrect, etc.)

Cross-tabulation table: See *Contingency table.*

Cross-validation: Procedure of dividing the sample into two parts: the *analysis sample* used in estimation of the discriminant function(s) or logistic regression model, and the *holdout sample* used to validate the results. Cross-validation avoids the "overfitting" of the discriminant function or logistic regression by allowing its validation on a totally separate sample.

Culture (general): The integrated pattern of human knowledge, belief, and behavior that depends upon an individual's capacity for learning and transmitting knowledge to succeeding generations.

Culture (organizational): A system of values, beliefs, and behaviors inherent in an organization (also see Corporate culture).

Culture: A system of values, beliefs, and behaviors inherent in an organization (also see Corporate culture).

Cumulative sum chart (CuSum): A cumulative sum chart plots the cumulative deviation of each subgroup's average from the nominal value. The ordinate of each plotted point represents the algebraic sum of the previous ordinate and the most recent deviations from the target. That means, if the process consistently produces parts near the nominal, the CuSum chart shows a line that is essentially horizontal. If the process begins to shift, the line will show an upward or downward trend. The CuSum chart is sensitive to small shifts in process level.

Customer council: A group usually composed of representatives from an organization's largest customers who meet to discuss common issues.

Customer delight: The result achieved when customer requirements are exceeded in ways the customer finds valuable.

Customer relationship management (CRM): Refers to an organization's knowledge of their customers' unique requirements and expectations, and using the information to develop a closer and more profitable link to business processes and strategies.

Customer requirements: Defines the needs and expectations of the customer; translated into measurable terms and used in the process to ensure compliance with the customers' needs.

Customer satisfaction: The result of delivering a product or service that meets customer requirements, needs, and expectations. Also a measure of how satisfied owners are with their product or service or specific aspects or systems of those products or services as determined by surveys.

Customer segmentation: Refers to the process of differentiating customers based on one or more dimensions for the purpose of developing a marketing strategy to address specific segments.

Customer service: The activities of dealing with customer questions; also sometimes the department that takes customer orders or provides postdelivery services.

Customer value: The market-perceived quality adjusted for the relative price of a product.

Customer: Anyone who receives a product, service, or information from an operation or process. The term is frequently used to describe "External" customers—those who purchase the manufactured products or services that are the basis for the existence of the business. However, "Internal" customers, also important, are "customers" who receive the intermediate or internal products or services from internal "Suppliers" (see external customer and internal customer). In other words, a recipient of a product or service who is provided by a supplier.

Customer/client personnel: Those individuals working for the organization who will assume responsibility for the product produced by the project when the project is complete.

Customer–supplier partnership: A long-term relationship between a buyer and supplier characterized by teamwork and mutual confidence. The supplier is considered an extension of the buyer's organization. The partnership is based on several commitments. The buyer provides long-term contracts and uses fewer suppliers. The supplier implements quality assurance processes so that incoming inspection can be minimized. The supplier also helps the buyer reduce costs and improve product and process designs.

Cutting score: Criterion (score) against which each individual's discriminant z-score is compared to determine predicted group membership. When the analysis involves two groups, group prediction is determined by computing a single cutting score. Entities with z scores below this score are assigned to one group, whereas those with scores above it are classified in the other group. For three or more groups, multiple discriminant functions are used, with a different cutting score for each function.

Cycle: A recurring pattern. Refers to the time that it takes to complete a process from beginning to end.

Cycle-time reduction: To reduce the time that it takes, from start to finish, to complete a particular process.

Cycle time: All time used in a process; includes actual work time and wait time.

Data: Facts presented in descriptive, numeric, or graphic form.

Data cleaning: Removal of random and systematic errors from *data elements* through filtering, merging, and translation.

Data element: Level of data stored in a database. Separate data elements may represent different levels of aggregation of summarization, ranging from *primitive data* to *aggregate data*.

Data mining: Extracting valid *and* previously unknown actionable information from large databases and applying it to business models.

Data scrubbing: See *Data cleaning*.

Data transformations: A variable may have an undesirable characteristic, such as nonnormality, which detracts from its use in a multivariate technique. A transformation, such as taking the logarithm or square root of the variable, creates a transformed variable that is more suited to portraying the relationship. Transformations may be applied to either the dependent or independent variables, or both. The need and specific type of transformation may be based on theoretical reasons (e.g., transforming a known nonlinear relationship) or empirical reasons (e.g., problems identified through graphical or statistical means).

Data visualization: Techniques for portraying data in graphical format, typically used *in an* exploratory *manner* to identify basic relationships.

Data warehousing: Assimilation and integration of data from internal and external sources in a format suitable for *online analytical processing* and *decision support system* applications.

DCOV: The design for Six Sigma model: Define, characterize, optimize, and verify.

Decision matrix: A matrix used by teams to evaluate problems or possible solutions. For example, after a matrix is drawn to evaluate possible solutions, the team lists them in the far-left vertical column. Next, the team selects criteria to rate the possible solutions, writing them across the top row. Then, each possible solution is rated on a predetermined scale (such as 1–5) for each criterion and the rating recorded in the corresponding grid. Finally, the ratings of all the criteria for each possible solution are added to determine its total score. The total score is then used to help decide which solution deserves the most attention.

Decision support systems (DSS): Interactive systems developed to provide users access to data for (1) *ad hoc* or ill-defined queries, (2) prespecified reports (e.g., exception reporting), and (3) structured analyses (rules-based procedures, multivariate statistics, neural networks, or other models).

Decision tree: Rule-based model consisting of *nodes* (decision points) and branches (connections between nodes) that reaches multiple outcomes based on passing through two or more nodes.

Decompositional method: Perceptual mapping method associated with multidimensional scaling (MDS) techniques in which the respondent provides only an overall evaluation of *similarity* or *preference* between *objects*. This set of overall evaluations is then "decomposed" into a set of "dimensions" that best represent the objects' differences.

Decompositional model: Class of multivariate models that "decompose" the respondent's preference. This class of models present the respondent with a predefined set of independent variables, usually in the form of a hypothetical or actual product or service, and then asks for an overall evaluation or preference of the product or service. Once given, the preference is "decomposed" by relating the known attributes of the product (which become the independent variables) to the evaluation (dependent variable). Principal among such models is conjoint analysis and some forms of multidimensional scaling.

Defect: Any instance or occurrence where the product or service fails to meet customer requirements, however defined (we prefer the use of nonconformity). Also, any output of an opportunity that does not meet a defined specification, or a failure to meet an imposed requirement (intended or reasonable expectation for use, including safety considerations) on a single quality characteristic or a single instance of nonconformance to the specification. If the defect is in relation to safety then, they are often classified as *Class 1*, Critical, leads directly to severe injury or catastrophic economic loss; *Class 2*, Serious, leads directly to significant injury or significant economic loss; *Class 3*, Major, is related to major problems with respect to intended normal or reasonably foreseeable use; and *Class 4*, Minor, is related to minor problems with respect to intended normal or reasonably foreseeable use (see also Blemish, Imperfection, and Nonconformity).

Defect opportunity: A type of potential defect (nonconformity) on a unit of throughput (output) that is important to the customer; example: specific fields on a form that create an opportunity for error that would be important to the customer.

Defective unit: A sample (part) that contains one or more defects, making it unacceptable for its intended, normal usage. See also *Nonconforming unit*.

Defective: Any unit with one or more nonconformities (defects). See also *Defects*.

Defects per million opportunities: Calculation used in Six Sigma Process Improvement initiatives indicating the amount of defects in a process per one million opportunities; number of defects divided by (the number of units times the number of opportunities) = DPO, times 1 million = DPMO. See also DPO, Six Sigma, Defect Opportunity.

Defects per opportunity: Calculation used in Process Improvements to determine the amount of defects per opportunity; number of defects divided by (the number of units times the number of opportunities) = DPO. *See* also Defect, Defect Opportunity.

Define: First DMAIC phase defines the problem/opportunity, process, and customer requirements; because the DMAIC cycle is iterative, the process problem, flow, and requirements should be verified and updated for clarity, throughout the other phases. See also Charter, Customer requirements, Process map, VOC.

Definitive estimate (−5%, +10%): A definitive ESTIMATE is prepared from well-defined data, specifications, drawings, and so on. This category covers all estimate ranges from a minimum to maximum definitive type. These estimates are used for BID proposals, bid evaluations, contract changes, extra work, legal claims, permit, and government approvals. Other terms associated with a Definitive Estimate include check, lump sum, tender, postcontract changes, and so on.

Deflection: The act of transferring all or part of a *risk* to another party, usually by some form of contract.

Deformation: The bending or distorting of an object due to forces applied to it. Deformation can contribute to errors in measurement if the measuring instrument applies enough force.

Degenerate solution: MDS solution that is invalid because of (1) inconsistencies in the data or (2) too few objects compared with the dimensionality of the solution. Even though the computer program may indicate a valid solution, the researcher should disregard the degenerate solution and examine the data for the cause. This type of solution is typically portrayed as a circular pattern with illogical results.

Degradation: Degradation is the undesirable change, over time or usage, in a function of the system or component.

Degrees of freedom (dof *or df or* ν): Value calculated from the total number of observations minus the number of estimated *parameters.* These parameter estimates are restrictions on the data because, once made, they define the population from which the data are assumed to have been drawn. For example, in estimating a regression model with a single independent variable, we estimate two parameters, the intercept (b_0) and a *regression coefficient* for the independent variable (b_1). In estimating the random error, defined as the sum of the prediction errors (actual minus predicted dependent values) for all cases, we would find $(n-2)$ degrees of freedom. Degrees of freedom provide a measure of how restricted the data are to reach a certain level of prediction. If the number of degrees of freedom is small, the resulting prediction may be less generalizable because all but a few observations were incorporated in the prediction. Conversely, a large degrees-of-freedom value indicates that the prediction is fairly "robust" with regard to being representative of the overall sample of respondents.

Yet another way to look at the *df* is the total number of independent pieces of information contributing to the component of variation, minus the number of pieces required to measure it. Analysis of variance is always reported with two values of degrees of freedom. The first informs on the number of test samples, and the second informs on the number of independent and random replicates available for calibrating the test effect against the background "error" variation. For example, a result $F_{2,12} = 3.98$, $p < 0.05$ indicates a significant effect at a threshold Type-I error of $\alpha = 0.05$. This outcome applies to a design with $a = 3$ samples or treatment levels, giving 2 test *df* $(=a - 1$, since one grand mean is required to test variation of *a* sample means). The *a* samples were allocated amongst a total of $N = 15$ sampling units, giving 12 error *df* (=$N - a$, since *a* sample means are required to test within-sample variation of N observations).

F-ratio: The ratio of explained to unexplained Mean Squares (MS), where the numerator is the MS explained by the model and the denominator is the error MS left unexplained by the model. A significant effect is indicated by a value of F that exceeds the critical F for a predefined Type I error. The *Mean Square* is the variance measured as variation per degree of freedom.

Delegating: The process by which authority is distributed from the Project Manager to an individual working on the project.

Deliverable: A report or product of one or more tasks that satisfy one or more objectives and must be delivered to satisfy contractual requirements.

Deleted residual: Process of calculating *residuals* in which the influence of each observation is removed when calculating its residual. This is accomplished by omitting the *i*th observation from the regression equation used to calculate its predicted value.

Demerit chart: A control chart for evaluating a process in terms of a demerit (or quality score), such as a weighted sum of counts of various classified nonconformities.

Deming cycle: Also known as Deming Shewhart cycle. See Plan-do-check-act cycle and plan-do-study-act cycle.

Deming W. Edwards: W. Edwards Deming was an eminent American quality consultant (he prefers consultant in statistical studies) whose work guided Japanese industry toward new principles of management and a new focus upon quality and productivity. He was at the forefront of applying statistical methods to quality management and, among other things, helped to formalize the PDSA (a.k.a. PDCA) cycle.

Demographics: Variables among buyers in the consumer market, which include geographic location, age, sex, marital status, family size, social class, education, nationality, occupation, and income.

Dendrogram: Graphical representation (tree graph) of the results of a *hierarchical procedure* in which each *object* is arrayed on one axis, and the other axis portrays the steps in the hierarchical procedure. Starting with each object represented as a separate cluster, the dendrogram shows graphically how the clusters are combined at each step of the procedure until all are contained in a single cluster.

Dependability failure: Failures that cause customers to be stranded or lose product function.

Dependability: The degree to which a product is operable and capable of performing its required function at any randomly chosen time during its specified operating time, provided that the product is available at the start of that period. (Non-operation-related influences are not included.) Dependability can be expressed by the ratio: [time available]/[time available + time required].

Dependence technique: Classification of statistical techniques distinguished by having a variable or set of variables identified as the *dependent variable(s)* and the remaining variables as *independent*. The objective is prediction of the dependent variable(s) by the independent variable(s). A common method to do this is a regression analysis.

Dependent variable (Y): Variable being predicted or explained by the set of in-dependent variables. In other words, a presumed effect of, or response to, a change in the *independent variable(s)*.

Deployment (to spread around): Used in strategic planning to describe the process of cascading plans throughout an organization.

Derived measures: Procedure to obtain respondents' perceptions of *similarities data*. Derived similarities are typically based on a series of "scores" given to stimuli by respondents, which are then combined in some manner. The semantic differential scale is frequently used to elicit such "scores."

Design (general): The creation of final approach for executing the project's work.

Design: Specific set of conjoint stimuli created to exhibit the specific statistical properties of *orthogonality* and *balance*.

Design contract: A Contract for Design.

Design control: A system for monitoring project scope, schedule, and cost during the project's design stage.

Design efficiency: Degree to which a *design* matches an *orthogonal* design. This measure is primarily used to evaluate and compare *nearly orthogonal* designs. Design efficiency values range from 0 to 100, which denotes an *optimal design*.

Design of experiment: A branch of applied statistics dealing with planning' conducting, analyzing, and interpreting controlled tests to evaluate the factors and noises that control the value of a parameter or group of parameters. It is a systematic approach to estimate the effects of various factors and levels on the performance of products or services or the outputs of processes. There are two approaches to DOE. Classical and the Taguchi approach. In both cases, however, the planning of an experiment to minimize the cost of data obtained and maximize the validity range of the results is the primary concern. Requirements for a good experiment include clear treatment comparisons, controlled fixed and experimental variables, and maximum freedom from systematic error. The experiments should adhere to the scientific principles of statistical design and analysis. Each experiment should include three parts: the experimental statement, the design, and the analysis. Examples of experimental designs include: single/multifactor block, factorial, Latin square, and nested arrangements.

Design review: A meeting of a cross-functional group of technical people assembled to probe and demonstrate the thoroughness and competency of a proposed design's ability to maximize customer's perceived value. It is, therefore, a formal documented, comprehensive, and systematic examination of a design to evaluate its capability to fulfill the requirements for quality.

Design verification (DV): A testing/evaluation discipline that is used to verify that prototype components, subsystems, and/or systems made to print and assembled with simulated or actual production processes are capable of meeting functional, quality, reliability, and durability requirements.

Design verification method (DVM): A method that is used to verify that a Requirement has been met.

Design verification planning (DVP) team: Members of a Process Action Team (PAT) responsible for planning and implementing design verification.

Designing in quality vs. inspecting in quality: See Prevention versus Detection.

Desired quality: Refers to the additional features and benefits a customer discovers when using a product or service that lead to increased customer satisfaction. If missing, a customer may become dissatisfied.

Detail schedule: A schedule used to communicate the day-to-day activities to working levels on the project.

Development (phase): The second of four sequential phases in the generic Project Life Cycle. Also known as Planning Phase.

Deviation: A nonconformance or departure of a characteristic from specified product, process, or system requirements.

DFSS: Acronym for "Design for Six Sigma." Describes the application of Six Sigma tools to product development and Process Design efforts with the goal of "designing in" Six Sigma performance capability.

DFBETA: Measure of the change in a regression coefficient when an observation is omitted from the regression analysis. The value of DFBETA is in terms of the coefficient itself; a standardized form (SDFBETA) is also available. No threshold limit can be established for DFBETA, although the researcher can look for values substantially different from the remaining observations to assess potential influence. The SDFBETA values are scaled by their standard errors, thus supporting the rationale for cutoffs of 1 or 2, corresponding to confidence levels of 0.10 or 0.05, respectively. (Source: http://www.mvstats.com/Downloads/Supplements/Advanced_Regression_Diagnostics.pdf

DFFIT: Measure of an observation's impact on the overall model fit, which also has a standardized version (SDFFIT). The best rule of thumb is to classify as influential any standardized values (SDFFIT) that exceed $2/\sqrt{p/n}$,, where p is the number of independent variables +1, and n

is the sample size. There is no threshold value for the DFFIT measure. (Source: http://www.mvstats.com/Downloads/Supplements/Advanced_Regression_Diagnostics.pdf

Diagnostic journey and remedial journey: A two-phase investigation used by teams to solve chronic quality problems. In the first phase, the diagnostic journey, the team moves from the symptom of a problem to its cause. In the second phase, the remedial journey, the team moves from the cause to a remedy.

Dimension: Attribute of a *data element,* such as age or gender of a customer or date of a historical record.

Dimensions: Features of an *object.* A particular object can be thought of as possessing both *perceived,* or *subjective,* dimensions (e.g., expensive, fragile) and *objective* dimensions (e.g., color, price, features).

Dimensions of quality: Refers to different ways in which quality may be viewed, for example, meaning of quality, characteristics of quality, drivers of quality, and so on.

Direct project costs: The costs directly attributable to a project, including all personnel, goods, and/or services together with all their associated costs, but not including indirect project costs, such as any overhead and office costs incurred in support of the project.

Direct estimation: Process whereby a model is estimated directly with a selected estimation procedure and the confidence interval (and standard error) of each parameter estimate is based on sampling error.

Disaggregate analysis: Approach to MDS in which the researcher generates *perceptual maps* on a respondent-by-respondent basis. The results may be difficult to generalize across respondents. Therefore, the researcher may attempt to create fewer maps by some process of *aggregate analysis,* in which the results of respondents are combined.

Discrete: A characteristic of attribute data, such as R/1000 or CPU. Discrete data can only take on certain values. For instance, there can only be 1, 2, 3, :.. and so on repairs—it makes no sense to talk about 2.2 repairs (except as an average). Warranty and defect data are nearly continuous, since although the repair counts are discrete, the high production volumes mean that the repair rates and CPU data can take on a very large number of values. See also *Continuous.*

Discrete data: Any data *not* quantified on an infinitely divisible scale. Includes a count, proportion, or percentage of a characteristic or category (e.g., gender, loan type, department, location, etc); also referred to as "attribute data."

Discrete probability distribution: Means that the measured process variable takes on a finite or limited number of values; no other possible values exist.

Discrete variable: A variable that assumes only integer values; for example, the number of people in a room is a discrete variable.

Discriminant coefficient: See *Discriminant weight.*

Discriminant function: A variate of the independent variables selected for their discriminatory power used in the prediction of group membership. The predicted value of the discriminant function is the *discriminant Z-score,* which is calculated for each object (person, firm, or product) in the analysis. It takes the form of the linear equation: $Z_{jk} = a + W_iX_ik + W_2X_{2k} + \ldots + W_{ni}X_{nk}$; where: Z_{jk} = discriminant Z-score of discriminant function j for object k; a = intercept; W_1 = discriminant weight for independent variable i; X_{ik} = independent variable i for object k.

Discriminant loadings: Measurement of the simple linear correlation between each independent variable and the discriminant Z-score for each discriminant function; also called *structure correlations.* Discriminant loadings are calculated whether or not an independent variable is included in the discriminant function.

Discriminant Z-score: Score defined by the *discriminant function* for each object in the analysis and usually stated in standardized terms. Also referred to as the Z-score. It is calculated for each object on each discriminant function and used in conjunction with the cutting score to determine predicted group membership this is different from the z-score terminology used in standardized variables.

Discrimination: The requirements imposed on the organization and the procedures implemented by the organization to assure fairness in hiring and promotion practices.

Discussion: Dialogue explaining implications and impacts on objectives. The elaboration and description of facts, findings, and alternatives.

Disordinal interaction: Form of *interaction effect* among independent variables that invalidates interpretation of the *main effects* of the treatments. A disordinal interaction is exhibited graphically by plotting the means for each group and having the lines intersect or cross. In this type of interaction the mean differences not only vary, given the unique combinations of independent variable levels, but the relative ordering of groups changes as well.

Disparities: Differences in the computer-generated distances representing *similarity* and the distances provided by the respondent.

Dispersion (of a statistical sample)**:** The tendency of the values of the elements in a sample to differ from each other. Dispersion is commonly expressed in terms of the range of the sample (difference between the lowest and highest values) or by the standard deviation.

Dispersion analysis diagram: A cause-and-effect diagram for analysis of the various contributions to variability of a process or product. The main factors contributing to the process are first listed, then the specific causes of variability from each factor are enumerated. A systematic study of each cause can then be performed.

Display: A pictorial, verbal, written, tabulated, or graphical means of transmitting findings, results and conclusions.

Disposition of nonconformity: Action taken to deal with an existing nonconformity; action may include: correction (repair), rework, regrade, scrap, obtain a concession, or amendment of a requirement.

Dispute: Disagreements not settled by mutual consent, which could be decided by litigation or arbitration.

Disruptive: A project manager who tends to break apart the unity of a group; one who tends to be an agitator and causes disorder on a project.

Dissatisfiers: Those features or functions that the customer or employee has come to expect and, if they were no longer present, would result in dissatisfaction.

Distribution (of communications): The dissemination of information for the purpose of communication, approval or decision-making.

Distribution: A statistical term that helps describe the output of an unchanging common-cause system of variation, in which individual values are not exactly predictable, but when assembled it describes the amount of potential variation in outputs of a process; it is usually described in terms of its shape, average, and standard deviation.

Divergent thinking: Divergent thinking means generating many answers.

Divisive method: Clustering procedure that begins with all *objects* in a single cluster, which is then divided at each step into two clusters that contain the most dissimilar objects. This method is the opposite of the *agglomerative method.*

DMAIC: The acronym pertains to the methodology (breakthrough strategy) used in the classical Six Sigma approach. It stands for: define, measure, analyze, improve, control. It lends structure to Process Improvement, and Design or Redesign applications, through a systematic process improvement/management system. Whereas the DMAIC model is a process-oriented approach, it must be separated from the DCOV model, which pertains to methodology used in the design for Six Sigma approach and it stands for: Define, characterize, optimize, and verify.

Document control: A system for controlling and executing project documentation in a uniform and orderly fashion.

Documentation: The collection of reports, user information, and references for distribution and retrieval, displays, back-up information, and records pertaining to the project.

Dodge–Romig sampling plans: Plans for acceptance sampling developed by Harold E. Dodge and Harry G. Romig. Four sets of tables were published in 1940: single-sampling lot tolerance tables, double-sampling lot tolerance tables, single-sampling average outgoing quality limit tables, and double-sampling average outgoing quality limit tables.

DOE: Design of experiments.

Downsizing: The planned reduction in workforce due to economics, competition, merger, sale, restructuring, or reengineering.

Downstream: Processes (activities) occurring after the task or activity in question.

DPMO, or defects per million opportunities: Calculation used in Six Sigma Process Improvement initiatives indicating the amount of defects in a process per one million opportunities; number of defects divided by (the number of units times the number of opportunities) = DPO, times 1 million = DPMO. See also *DPO, Six Sigma, Defect opportunity.*

DPO, or defects per opportunity: Calculation used in Process Improvements to determine the amount of defects per opportunity; number of defects divided by (the number of units times the number of opportunities) = DPO. *See* also Defect, Defect opportunity,

DPU: Defects per unit—the number of defects counted, divided by the number of "products" or "characteristics" (units) produced.

Drill down: Accessing more detailed data used to create *aggregate data.* For example, after reviewing weekly sales figures for a set of stores, the researcher would drill down to view daily sales reports for sales or reports for separate stores. One can drill down to the lowest level of detail (*primitive data*) for available data elements.

Drivers of quality: Include customers, products/services, employee satisfaction, total organizational focus.

Dummy activity: An activity, always of zero duration, used to show logical dependency when an activity cannot start before another is complete, but which does not lie on the same *path* through the *network.* Normally, these dummy activities are graphically represented as a dashed line headed by an arrow.

Dummy variable: Nonmetrically measured variable transformed into a metric variable by assigning a 1 or a 0 to a subject (a binary metric), depending on whether it possesses a particular characteristic. Furthermore, one may surmise that a dummy variable is a special metric variable used to represent a single category of a nonmetric variable. To account for L levels of an nonmetric variable, $L - 1$ dummy variables are needed. For example, gender is measured as male or female and could be represented by two dummy variables (X_1 and X_2). When the respondent is male, $X_1 = 1$ and $X_2 = 0$. Likewise, when the respondent is female, $X_1 = 0$ and $X_2 = 1$. However, when $X_1 = 1$, we know that X_2 must equal 0. Thus we need only one variable, either X_1 or X_2, to represent the variable gender. If a nonmetric variable has three levels, only two dummy variables are needed. We always have one dummy variable less than the number of levels for the nonmetric variable.

Durability: It is the ability of a component, subsystem, or system to perform its intended function satisfactorily without requiring overhaul or rebuild due to wearout during its projected design life.

DV: Design verification.

DVP&R: Design verification plan and Report.

DVP: Design verification plan.

Effect Plot: A line graph that visually depicts the difference between the response average at each factor-level setting.

Effectiveness: Measures related to how well the process output(s) meets the needs of the customer (e.g., on-time delivery, adherence to specifications, service experience, accuracy, value-added features, customer satisfaction level); links primarily to customer satisfaction.

Effects: General term that encompasses both main effects and interaction effects. It is a measure of the impact that a factor or interaction has on the response when levels are changed.

Effect size: Standardized measure of group differences used in the calculation of statistical power. Calculated as the difference in group means divided by the standard deviation, it is then comparable across research studies as a generalized measure of effect (i.e., differences in group means). Yet another way to think of the effect size is to think of it as an estimate of the degree to which the phenomenon being studied (e.g., correlation or difference in means) exists in the population.

Effects coding: Method for specifying the reference category for a set of *dummy variables* where the reference category receives a value of minus one (–1) across the set of dummy variables. With this type of coding, the dummy variable coefficients represent group deviations from the mean of all groups. This is in contrast to *indicator coding,* in which the reference category is given the value of zero across all dummy variables and the coefficients represent group deviations from the reference group.

Efficiency: Measures related to the quantity of resources used in producing the output of a process (e.g., costs of the process, total cycle time, resources consumed, cost of defects, scrap, and/or waste); links primarily to company profitability.

Effort: The application of human energy to accomplish an objective.

Eighty–twenty (80–20) rule: A term referring to the Pareto principle, which suggests that most effects come from relatively few causes; that is, 80% of the effects come from 20% of the possible causes.

Eigenvalue: Measure of the amount of variance contained in the correlation matrix so that the sum of the eigenvalues is equal to the number of variables. In essence, it is the Column sum of squared loadings for a factor; also referred to as the *latent root.* It represents the amount of variance accounted for by a factor. Also known as the latent root or characteristic root. See *Canonical roots.*

Employee involvement: A practice within an organization whereby employees regularly participate in making decisions on how their work areas operate, including making suggestions for improvement, planning, objectives setting, and monitoring performance.

Employee relations: Those formal activities and procedures used by an organization to administer and develop its workforce.

Empowerment: A condition whereby employees have the authority to make decisions and take action in their work areas, within stated bounds, without prior approval.

Encoding: Conversion of the categories of a nonmetric variable to a series of binary variables, one for each category. Similar to the dummy variable coding, however, without the deletion of the reference category.

Endogenous construct: *Construct* or variable that is the dependent or outcome variable in at least one *causal relationship.* In terms of a path diagram, there are one or more arrows leading *into* the endogenous construct or variable.

Endorsement: Written approval. Endorsement signifies personal understanding and acceptance of the thing endorsed and recommends further endorsement by higher levels of authority if necessary. Endorsement of *commitment* by a person invested with appropriate authority signifies authorization. See Approve, Authorize.

End users: External customers who purchase products/services for their own use.

Engineered system: See *P Diagram.*

Engineering confidence: Confidence that our DV tests actually expose products to the critical customer usage conditions that they will experience in the field at the right frequency, and that by meeting functional performance targets in these tests, our products demonstrate they will operate reliably in the real world.

English system: The system of measurement units based on the foot, the pound, and the second.

Entity (or item): That which can be individually described and considered: process, product, and organization, a system or a person, or any combination thereof. Totality of characteristics of an entity that bear on its ability to satisfy stated and implied needs.

Entropy group: Group of *objects* independent of any cluster (i.e., they do not fit into any cluster) that may be considered outliers and possibly eliminated from the cluster analysis.

Environment (framework): The combined internal and external forces, both individual and collective, which assist or restrict the attainment of the project objectives. These could be business or project related or may be due to political, economic, technological, or regulatory conditions (Communications Management). The circumstances, objects, or conditions by which one is surrounded.

Environmental correlation: See *Interattribute correlation.*

Equivalent models: Comparable models that have the same number of degrees of freedom (*nested models*) but differ in one or more paths. The number of equivalent models expands very quickly as model complexity increases.

Error variance: Variance of a variable due to errors in data collection or measurement.

Ethical: A project manager who is honest, sincere, able to motivate and to press for the best and fairest solution; one generally goes "by the books."

Ethics: A code of conduct that is based on moral principles and which tries to balance what is fair for individuals with what is right for society.

Euclidean distance: Most commonly used measure of the *similarity* between two *objects*. Essentially, it is a measure of the length of a straight line drawn between two objects.

Event: An event is an identifiable single point in time on a project, task, or group of tasks. It may be the starting or the ending point of a task.

Evolutionary operations (*EVOP*): *A* procedure to optimize the performance of a process by making small, known variations in the parameters entering the process and observing the effects of the variation on the product. This seems identical to Response Surface Methodology, except that it is done in a production situation rather than during process development, and the variations must, therefore, be kept small enough to meet product tolerances. In other words, the process of adjusting variables in a process in small increments in search for a more optimum point on the response surface.

Exception reporting: The process of documenting those situations where there are significant deviations from the quality specifications of a project. The assumption is made that the project will be developed within established boundaries of quality. When the process falls outside of those boundaries, a report is made on why this deviation occurred. Sometimes it is also the documentation that focuses its attention on variations of key control parameters that are critical rather than on those that are progressing as planned.

Excited quality: The additional benefit a customer receives when a product or service goes beyond basic expectations. Excited quality "wows" the customer and distinguishes the provider from the competition. If missing, the customer will still be satisfied.

Exogenous construct: *Construct* or variable that acts only as a predictor or "cause" for other constructs or variables in the model. In path diagrams, the exogenous *constructs* have only causal arrows leading out of them and are not predicted by any other *constructs* in the model.

Expected quality: Also known as basic quality, the minimum benefit a customer expects to receive from a product or service.

Experimental design: Research plan in which the researcher directly manipulates or controls one or more predictor variables (see *Treatment)* and assesses their effect on the dependent variables. Common in the physical sciences, it is gaining in popularity in business and the social sciences. For example, respondents are shown separate advertisements that vary systematically on a characteristic, such as different appeals (emotional versus rational) or types of presentation (color versus black-and-white) and are then asked their attitudes, evaluations, or feelings toward the different advertisements. Also experimental design (ED) is very popular in the quality field especially with Six Sigma and lean methodologies.

Experimental plan: This is determined by the number of factors, levels, and experimental runs identified by the team. It involves choosing an orthogonal array on which to base the experiment's design. This, in turn, gives the combination of factor-level settings to be used for each experimental run.

Experimental run: A trial of the system with the factors fixed at one combination of levels.

Explicit knowledge: Represented by the captured and recorded tools of the day, for example, procedures, processes, standards, and other like documents.

Exploration thinking: Exploration thinking seeks to generate new information.

Exploratory analysis: Analysis that defines possible relationships in only the most general form and then allows the multivariate technique to estimate relationship(s). The opposite of *confirmatory analysis,* the researcher is not looking to "confirm" any relationships specified prior to the analysis, but instead lets the method and the data define the nature of the relationships. An example is stepwise multiple regression, in which the method adds predictor variables until some criterion is met.

Exponential Distribution: A probability distribution mathematically described by an exponential function. Used to describe the probability that a product survives a length of time t in service, under the assumption that the probability of a product failing in any small time interval is independent of time. A continuous distribution where data are more likely to occur below the average than above it. Typically used to describe the break-in portion of the "bathtub" curve.

External customer: A person or organization who receives a product, a service, or information but is not part of the organization supplying it (see also Internal customer).

External failure costs: Costs associated with defects found during or after delivery of the product or service.

External failure: When defective units pass all the way through a process and are received by the customer.

Face validity: See *Content validity*.

F-distribution: The distribution of *F*, the ratios of variances for pairs of samples. Used to determine whether or not the populations from which two samples were taken have the same standard deviation. The *F*-distribution is usually expressed as a table of the upper limit below which *F* can be expected to lie with some confidence level, for samples of a specified number of degrees of freedom.

F-test: Test of whether two samples are drawn from populations with the same standard deviation, with some specified confidence level. The test is performed by determining whether *F*, as defined above, falls below the upper limit given by the *F*-distribution table.

Facilitating: The project manager is available to answer questions and give guidance when needed; he does not interfere with day to day tasks, but rather maintains that status quo.

Facilitator: An individual who is responsible for creating favorable conditions that will enable a team to reach its purpose or achieve its goals by bringing together the *necessary tools*, information, and resources to get the job done.

Facilities/product life cycle: A life cycle that encompasses phases of operation and disposal, as well as, and following, the Project Life Cycle.

Factor analysis: A statistical technique that examines the relationships between a single dependent variable and multiple independent variables. For example, it is used to determine which questions on a questionnaire are related to a specific question such as "Would you buy this product again?"

Factors: Parameters or variables that may influence the performance of the product or process under investigation. Those factors believed to have the most impact are identified by the team and become the focus of attention for the experiment. Simply put, these factors become the "things to change" during the experiment.

Factor indeterminacy: Characteristic of *common factor analysis* such that several different factor scores can be calculated for a respondent, each fitting the estimated factor model. This means the factor scores are not unique for each individual.

Factor loadings: Correlation between the original variables and the factors, and the key to understanding the nature of a particular factor. Squared factor loadings indicate what percentage of the variance in an original variable is explained by a factor.

Factor matrix: Table displaying the *factor loadings* of all variables on each factor. Factor pattern matrix: One- or two-*factor matrices* found in an *oblique rotation* that is most comparable to the factor matrix in an *orthogonal rotation*.

Factor rotation: Process of manipulation or adjusting the factor axes to achieve a simpler and pragmatically more meaningful factor solution.

Factor score: Composite measure created for each observation on each factor extracted in the factor analysis. The factor weights are used in conjunction with the original variable values to calculate each observation's score. The factor score then can be used to represent the factor(s) in subsequent analyses. Factor scores are standardized to have a mean of 0 and a standard deviation of 1.

Factor structure matrix: A factor *matrix* found in an *oblique rotation* that represents the simple correlations between variables and factors, incorporating the unique variance and the correlations between factors. Most researchers prefer to use the *factor pattern matrix* when interpreting an oblique solution.

Factor: Linear combination (variate) of the original variables. Factors also represent the underlying dimensions (constructs) that summarize or account for the original set of observed variables. A factor is also a nonmetric independent variable, which is referred to as a *treatment* or experimental variable. In simple terms a factor is a variable the researcher manipulates that represents a specific attribute. In conjoint analysis, the factors (independent variables) are nonmetric. Factors must be represented by two or more values (also known as *levels*), which are also specified by the researcher. One may also think of a factor as a source of variance in the response. A categorical factor is measured in categorical levels, whereas a covariance factor is measured on a scale of continuous (or sometimes ordinal) variation. A statistical model might be constructed to test the influence of a factor as the sole explanation ($Y = A + \varepsilon$) or as one of many factors variously crossed with each other or nested within each other.

A *fixed* factor has levels that are fixed by the design and could be repeated without error in another investigation. The factor has a significant effect if sample means differ by considerably more than the background variation, or for a covariate, if the variation of the regression line from horizontal greatly exceeds the variation of data points from the line.

A *random* factor has levels that sample at random from a defined population. A random factor will be assumed to have a normal distribution of sample means, and homogenous variance of means, if its MS is the error variance for estimating other effects (e.g., in nested designs). The random factor has a significant effect if the variance among its levels is considerably greater than zero.

Factorial design: Design with more than one *factor* (treatment). Factorial designs examine the effects of several factors simultaneously by forming groups based on all possible combinations of the levels (values) of the various treatment variables. It is a method of designing *stimuli* for evaluation by generating *all* possible combinations of *levels*. For example, a three-factor conjoint analysis with three levels per factor (3 x 3 x 3) would result in 27 combinations that could act as stimuli.

Failure-free life: The period of time after manufacture and before the customer gets the product when you expect no failure to.

Failure mode analysis (FMA): A procedure to determine which malfunction symptoms appear immediately before or after a failure of a critical parameter in a system. After all the possible causes are listed for each symptom, the product is designed to eliminate the problems.

Failure Mode and Effect Analysis (FMEA): An analysis of a product or process that considers the effects of all potential ways that the design could be affected by external and internal factors and develops countermeasures to control or to reduce the effects of those factors. The FMEA is an inductive (bottom-up) analysis of a product or process that (1) considers the effects of all potential ways that the design could be impacted by external and internal factors and (2) develops countermeasures to control or reduce the effects of those factors. FMEAs are performed early in the design concept phase for a new product or process cycle in order to receive maximum benefits. FMEA should be a cross-functional team process. By contrast the FTA is a top-down approach.

Failure mode effects and criticality analysis (FMECA): A procedure that is performed after a failure mode effects analysis to classify each potential failure effect according to its severity and probability of occurrence.

Failure rate: The ratio of the number of units failed to the total number of units that can potentially fail during a specified interval of life units. When the life unit interval is very small, this is referred to as Instantaneous Failure Rate and is a very useful indicator for identifying various patterns of failure. Used for assessing reliability of a product in service.

Failure: A failure occurs when a product ceases to function in the way in which the customer expects.

False Xbar Causes: For *Xbar* control charts: changes in the *Xbar* control chart that are not due to changes in the process mean, but to changes in the corresponding R-chart.

Fast Track: The starting or implementation of a project by overlapping activities, commonly entailing the overlapping of design and construction (manufacturing) activities.

Fault tree analysis (FTA): A deductive (top-down) analysis depicting the functional relationship of a design or a process to study potential failures. It is a technique for evaluating the possible causes that might lead to the failure of a product. For each possible failure, the possible causes of the failure are determined; then the situations leading to those causes are determined, and so forth, until all paths leading to possible failures have been traced. The result is a flow chart for the failure process. Plans to deal with each path can then be made.

Feasibility studies: The methods and techniques used to examine technical and cost data to determine the economic potential and the practicality of project applications. It involves the use of techniques such as the time value of money so that projects may be evaluated and compared on an equivalent basis. Interest rates, present worth factors, capitalization costs, operating costs, depreciation, and so on, are all considered.

Feasibility: The assessment of capability of being completed; the possibility, probability, and suitability of accomplishment.

Feasible project alternatives: Reviews of available alternate Procurement actions that could attain the objectives.

Feather diagram: see cause-and-effect diagram.

Feedback—general: Information (data) extracted from a process or situation and used in controlling (directly) or in planning or modifying immediate or future inputs (actions or decisions) into the process or situation.

Feedback—process: Using the results of a process to control it. The feedback principle has wide application. An example would be using control charts to keep production personnel informed on the results of a process. This allows them to make suitable adjustments to the process. Some form of feedback on the results of a process is essential in order to keep the process under control.

Feedback—teams: The return of information in interpersonal communication; it may be based on fact or feeling and helps the party who is receiving the information judge how well he or she is being understood by the other party. More generally, information about the interaction process that is used to make decisions about its performance and to adjust the process when necessary.

Feedback loops: Pertains to open-loop and closed-loop feedback.

Feedback: The return of information in interpersonal communication; it may be based on fact or feeling and helps the party who is receiving the information judge how well he or she is being understood by the other party. More generally, information about a process that is used to make decisions about its performance and to adjust the process when necessary.

Field cost: Costs associated with the project site rather than with the home office.

Figure of merit: Generic term for any of several measures of product reliability, such as MTBF, mean life, and so on.

Filters: Relative to human to human communication, those perceptions (based on culture, language, demographics, experience, etc.) that affect how a message is transmitted by the sender and how a message is interpreted by the receiver.

Final completion: When the entire work has been performed to the requirements of the contract, except for those items arising from the provisions of warranty, and is so certified.

First-time-through capability: The percentage of units that complete a process and meet quality guidelines without being scrapped, rerun, retested, returned, or diverted into an off-line repair area. FTT = [Units entering the process − (scrap + rerun + retests + repa ired off-line + returns)]/Units entering the process.

Fishbone diagram: See Cause-and-effect diagram.

Fitness for use: A term used to indicate that a product or service fits the customer's defined purpose for that product or service.

Five (5)-Ss: Five practices for maintaining a clean and efficient workplace (Japanese).

Five whys: A persistent questioning technique to probe deeper to surface the root cause of a problem.

Flow chart (for programs, decision making, process development): A pictorial representation of a process indicating the main steps, branches, and eventual outcomes of the process. Also called "process map," "activity flow." A graphical *representation of* the steps in a process, flowcharts are drawn to better understand processes. The flowchart is one of the seven basic tools of quality.

FMA: An analysis of the ways a component or subsystem had failed to meet customer expectations in the past. FMA provides an overall data-supported assessment of a commodity from current or past production. FMA quantifies top concerns reported by the customer and associated planned actions.

FMEA: Failure Mode and Effects Analysis.

Focus group: A qualitative discussion group consisting of 8 to 10 participants, invited from a segment of the customer base to discuss an existing or planned product or service, led by a facilitator working from predetermined questions (focus groups may also be used to gather information in a context other than customers).

Force-field analysis: A technique for analyzing the forces that aid or hinder an organization in reaching an objective. Identifies forces/factors supporting or working against an idea; "restraining" factors listed on one side of the page, "driving forces" listed on the other; used to reinforce the strengths (positive ideas) and overcome the weaknesses or obstacles.

Forcing: The project manager uses his power to direct the solution. This is a type of win-lose agreement where one side gets its way and the other does not.

Forecast final cost: The anticipated cost of a project or component when it is complete. The sum of the committed cost to date and the estimated cost to complete.

Forecast: An *estimate* and prediction of future conditions and events based on information and knowledge available at the time of the forecast.

Forecasting: The work performed to estimate and predict future conditions and events. Forecasting is an activity of the management function of planning. Forecasting is often confused with budgeting, which is a definitive allocation of resources rather than a prediction or estimate.

Formal communication: The officially sanctioned data within an organization, which includes publications, memoranda, training materials/events, public relations information, and company meetings.

Formative quality evaluation: The process of reviewing the project data at key junctures during the Project Life Cycle for a comparative analysis against the preestablished quality specifications. This evaluation process is ongoing during the life of the project to ensure that timely changes can be made as needed to protect the success of the project.

Forward pass: Network calculations that determine the earliest start/earliest finish time (date) of each activity. These calculations are from Data Date through the logical flow of each activity.

Fourteen (14) points: W. Edwards Deming's 14 management practices to help organizations increase their quality and productivity: (1) Create constancy of purpose for improving products and services; (2) adopt a new philosophy; (3) cease dependence on inspection to achieve quality; (4) end the practice of awarding business on price alone; instead, minimize total cost by working with a single supplier; (5) improve constantly and forever every process for planning, production, and service; (6) institute training on the job; (7) adopt and institute leadership; (8) drive out fear; (9) break down barriers between staff areas; (10) eliminate slogans, exhortations, and targets for the workforce; (11) eliminate numerical quotas for the workforce and numerical goals for management; (12) remove barriers that rob people of pride of workmanship and eliminate the annual rating or merit system; (13) institute a vigorous program of education and self-improvement for everyone; and (14) put everybody in the company to work to accomplish the transformation.

Forward addition: Method of selecting variables for inclusion in the regression model by starting with no variables in the model and then adding variables based on their contribution to prediction.

FRACAS: Failure reporting analysis and corrective action system is a data repository of product and systems failure history that acquires data and analyses and records corrective actions for all failures occurring during reliability test efforts.

Fraction defective chart (p-chart): An attribute control chart used to track the proportion of defective units.

Fractional factorial design: Approach, as an alternative to a *factorial design*, which uses only a subset of the possible stimuli needed to estimate the results based on the assumed composition rule. Its primary task is to reduce the number of evaluations collected while still maintaining *orthogonality* among the *levels* and subsequent *part-worth* estimates. The simplest design is an *additive model,* in which only *main effects* are estimated. If selected *interaction terms* are included, then additional stimuli are created. The design can be created either by referring to published sources or by using computer programs that accompany most conjoint analysis packages.

Frequency distribution: For a sample drawn from a statistical population, the number of times each outcome was observed.

FTA (fault tree analysis): It focuses on single combination of events to describe how the top (undesirable event) may occur. Useful for identifying appropriate corrective actions.

FTT: First-time-through capability.

Full Factorial Experimentation: An experimental design that explores all the possible combinations of all the levels of all the factors.

Full-profile method: Method of presenting *stimuli* to the respondent for evaluation that consists of a complete description of the stimuli across all attributes. For example, let us assume that a candy was described by three factors with two levels each: price (15 or 25 cents), flavor (citrus or butterscotch), and color (white or red). A full-profile stimulus would be defined by one level of each factor. One such full-profile stimulus would be a red butterscotch candy costing 15 cents.

Function (PM function): The series of processes by which the project objectives in that particular area of Project Management, for example, scope, cost, time, and so on, are achieved.

Function–quality integration: The process of actively ensuring that quality plans and programs are integrated, mutually consistent, necessary, and sufficient to permit the Project Team to achieve the defined product quality.

Functional block diagrams: Functional block diagrams are used to break a system into its smaller elements and to show the functional and physical relationships between the elements.

Functional organization: An organization organized by discrete functions, for example, marketing/sales, engineering, production, finance, and human resources.

Functional partitioning: Defines the physical systems and subsystems, by function, which will be affected by the new design. Programs can partition any product architecture in multiple ways.

Funnel experiment: An experiment that demonstrates the effects of tampering. Marbles are dropped through a funnel in an attempt to hit a flat-surfaced target below. The experiment shows that adjusting a stable process to compensate for an undesirable result or an extraordinarily good result will produce output that is worse than if the process had been left alone.

Future reality tree: A technique used in the application of Goldratt's Theory of Constraints.

Gainsharing: A type of program that rewards individuals financially on the basis of organizational performance.

Gantt chart: A type of bar chart used in process/project planning and control to display planned work and finished work in relation to time. Also called a "milestone chart."

Gap analysis: A technique that compares a company's existing state to its desired state (as expressed by its long-term plans) to help determine what needs to be done to remove or minimize the gap.

Gatekeeping: The role of an individual (often a facilitator) in a group meeting in helping ensure effective interpersonal interactions (e.g., someone's ideas are not ignored due to the team moving on to the next topic too quickly).

Gauge repeatability and reproducibility (GR&R): The evaluation of a gauging instrument's accuracy by determining whether the measurements taken with it are repeatable (i.e., there is close agreement among a number of consecutive measurements of the output for the same value of the input under the same operating conditions) and reproducible (i.e., there is close agreement among repeated measurements of the output for the same value of input made under the same operating conditions over a period of time).

General conditions: General definition of the legal relationships and responsibilities of the parties to the contract, and how the contract is to be administered. They are usually standard for a corporation and/or project.

General linear model (GLM): A linear model is a statistical model with linear (additive) combinations of parameter constants describing effect sizes and variance components. Linear models can describe nonlinear trends in covariates, for example, by transformation of the data or fitting a polynomial model. GLM is a generic term for parametric analyses of variance that can accommodate combinations of factors and covariates, and unbalanced and nonorthogonal designs. GLMs generally use an unrestricted model for analyzing combinations of fixed and random factors. Significant effects are tested with the F-statistic, which is constructed from sums of squared deviations of observations from means, adjusted for any nonorthogonality. This statistic assumes random sampling of independent replicates, homogeneous within-sample variances, a normal distribution of the residual error variation around sample means, and a linear response to any covariate. Transformations may be necessary to the response and/or covariate to meet these assumptions. A further generalization of GLM, the Generalized Linear Model (GLIM), accommodates nonnormally distributed response variables, and partitions the components of variation using maximum likelihood rather than sums of squares. (Source: http://www.soton.ac.uk/~cpd/anovas/datasets/GLM.htm)

Genetic algorithm: Learning-based models developed on the principles of evolution. Partial solutions to a problem "compete" with one another, and then the best solutions are selected and combined to provide the basis for further problem solving.

Geometric dimensioning and tolerancing (GDT): A method to minimize production costs by considering the functions or relationships of part features in order to define dimensions and tolerances.

Global 8D: An orderly team-oriented approach to problem solving, developed by FORD Motor company and used practically in all industries. It has become the "default" process of problem-solving methodology in many industries.

Global problem solving (GPS): A problem-solving methodology very similar to the Global 8D.

Goodness-of-fit: Degree to which the actual or observed input matrix (covariances or correlations) is predicted by the estimated model. *Goodness-of-fit* measures are computed only for the total input matrix, making no distinction between exogenous and endogenous constructs or indicators.

Go/no-go: State of a unit or product. Two parameters are possible: go conforms to specifications, and no-go does not conform to specifications.

Goal statement: Description of the intended target or desired results of Process Improvement or Design/Redesign activities; usually included in a team charter and supported with actual numbers and details once data have been obtained.

Goal: A statement of general intent, aim, or desire; it is the point toward which management directs its efforts and resources; goals are often nonquantitative.

Goodness-of-Fit: Any measure of how well a set of data matches a proposed distribution. Chi-square is the most common measure for frequency distributions. Simple visual inspection of a histogram is a less quantitative, but equally valid, way to determine goodness of fit.

Goodwill: Special repairs or refunds not covered by warranty, owner notification, or service recall program. The intent of goodwill payments is to maintain customer satisfaction/owner loyalty related to parts failures that are perceived by the customer/dealer as "unexpected" and, therefore, the company's responsibility.

Grade: A planned or recognized difference in requirements for quality.

Grand average: Overall average of data represented on an Xbar chart at the time the control limits were calculated. Is denoted as ($\bar{\bar{X}}$).

Grapevine: The informal communication channels over which information flows within an organization, usually without a known origin of the information and without any confirmation of its accuracy or completeness (sometimes referred to as the "rumor mill").

Graph (quality mgt.): A visual comparison of variables that yield data in numerical facts. Examples include: trend graphs, histograms, control charts, frequency distributions, and scatter diagrams (Time Management). The display or drawing that shows the relationship between activities. Pictorial representation of relative variables.

Green belt: Green belts are Six Sigma team leaders capable of forming and facilitating Six Sigma teams and managing Six Sigma projects from concept to completion. Typically, green-belt training consists of 5 days of classroom training and is conducted in conjunction with Six Sigma team projects.

Greatest characteristic root (gcr): Statistic for testing the null hypothesis in MANOVA. It tests the first *discriminant function* of the dependent variables for its ability to discern group differences.

Group communication: The means by which the project manager conducts meetings, presentations, negotiations, and other activities necessary to convey the *project's* needs and concerns to the project team and other groups.

Guideline: A document that recommends methods to be used to accomplish an objective.

Half-normal probability plots (Categorized): This type of graph is used to evaluate the normality of the distribution of a variable, that is, whether and to what extent the distribution of the variable follows the normal distribution. The selected variable will be plotted in a scatterplot against the values "expected from the normal distribution." The half-normal probability plot is constructed in the same way as the standard normal probability plot except that only the positive half of the normal curve is considered. Consequently, only positive normal values will be plotted on the Y-axis. This plot is used when one wants to ignore the sign of the residual, that is, when one is mostly interested in the distribution of absolute residuals, regardless of the sign. The most used half-normal probability is the Daniel half plot and they are primarily used with the Taguchi methodology in parameter design.

Hammock: An aggregate or summary activity. All related activities are tied as one summary activity and reported at the summary level.

Handoff: Any time in a process when one person (or job title) passes on the item moving through the process to another person; potential to add defects, time, and cost to a process.

Hat($\wedge$): Symbol used to represent the estimated standard deviation given by the formula $Rbar/d_2$. The estimated standard deviation may only be used if the data are normally distributed and the process is in control.

Hat matrix: Matrix that contains values for each observation on the diagonal, known as *hat values,* which represent the impact of the observed dependent variable on its predicted value. If all cases have equal influence, each would have a value of p/n, where p equals the number of independent variables +1, and n is the number of cases. If a case has no influence, its value would be $-1 \pm n$, whereas total domination by a single case would result in a value of $(n - 1)/n$. Values exceeding $2\,p/n$ for larger samples, or $3\,p/n$ for smaller samples $(n \leq 30)$, are candidates for classification as *influential observations*.

Hat value: See *H matrix*.

Hazard Rate: The instantaneous failure rate. The probability of failure during the current interval.

Heteroscedasticity: Is a condition where the variance of Y changes as the independent variable changes location. Another way to say it is that the variance of Y is correlated to the mean of Y. This is an observed condition generally when the independent variable of interest interacts with some other independent variable. As the interaction increases, so does the heteroscedasticity. See *Homoscedasticity*.

Heywood cases: A common type of *offending estimate*, which occurs when the estimated error term for an indicator becomes negative, which is a nonsensical value. The problem is remedied either by deleting the indicator or by constraining the measurement error value to be a small positive value.

Hidden layer: Series (layer) of *nodes* in a *multilayer perceptron neural network* that are between the input and output nodes. The researcher may or may not control the number of nodes per hidden layer or the number of hidden layers. The hidden layers provide the capability to represent nonlinear functions in the neural network system.

Hierarchical procedures: Stepwise clustering procedures involving a combination (or division) of the *objects* into clusters. The two alternative procedures are *the agglomerative and divisive methods. The result is the construction of a hierarchy,* or treelike structure *(dendrogram)*, depicting the formation of the clusters. Such a procedure produces $N - 1$ cluster solutions, where N is the number of objects. For example, if the agglomerative procedure starts with five objects in separate clusters, it will show how four clusters, then three, then two, and finally one cluster are formed.

Hierarchy block diagrams: Hierarchy block diagrams break the product into natural and logical elements and become more detailed at each level down.

Hierarchy structure: Describes an organization that is organized around functional departments/product lines or around customers/customer segments and is characterized by top-down management (also referred to as a bureaucratic model or pyramid structure).

Histogram: A graphic representation of a frequency distribution. The range of the variable is divided into a number of intervals of equal size (called cells) and an accumulation is made of the number of observations falling into each cell. The histogram is essentially a bar graph of the results of this accumulation. A graphic summary of variation in a set of data. The pictorial nature of the histogram lets people see patterns that are difficult to see in a simple table of numbers. When raw numbers are plotted in a histogram, we refer to it as a frequency plot or distribution and we are concerned to see the "centeredness" of the sample or population as the case may be. The pictorial nature of the histogram lets people see patterns that are difficult to see in a simple table of numbers. The histogram is one of the seven basic tools of quality. Used to make a visual comparison to the *normal distribution*.

Histogram or frequency plot: Chart used to graphically represents the frequency, distribution, and "centeredness" of a population.

Historic records: Project documentation that can be used to predict trends, analyze feasibility, and highlight problem areas/pitfalls on future similar projects.

Historical data banks: The data stored for future reference and referred to on a periodic basis to indicate trends, total costs, unit costs and technical relationships, and so on. Different applications require different database information. These data can be used to assist in the development of future estimates.

Hold point: A point defined in an appropriate document beyond which an activity must not proceed without the approval of a designated organization or authority.

Holdout stimuli: See *Validation stimuli*.

Homoscedasticity: When the variance of the error terms (e, or more commonly known as epsilon = ε) appears constant over a range of predictor variables, the data are said to be homoscedastic. The assumption of equal variance of the population error E (where E is estimated from ε) is critical to the proper application of linear regression. When the error terms have increasing or modulating variance, the data are said to be *heteroscedastic*. Analysis of *residuals* best illustrates this point.

Horizontal structure: Describes an organization that is organized along a process or value-added chain, eliminating hierarchy and functional boundaries (also referred to as a systems structure).

Hoshin kanri, hoshin planning: Japanese-based strategic planning/policy deployment process that involves consensus at all levels as plans are cascaded throughout the organization, resulting in actionable plans and continual monitoring and measurement.

House of quality: A diagram (named for its house-shaped appearance) that clarifies the relationship between customer needs and product features. It helps correlate market or customer requirements and analysis of competitive products with higher-level technical and product characteristics and makes it possible to bring several factors into a single figure. Also known as Quality Function Deployment (QFD).

Hotelling's P: Test to assess the statistical significance of the difference on the means of two or more variables between two groups. It is a special case of MANOVA used with two groups or levels of a treatment variable.

Hybrid model: See *Adaptive model.*

Hypergeometric distribution: A probability distribution for the probability of drawing exactly n objects type from a sample of N objects of which r are of the desired type. A discrete (probability) distribution defining the probability of r occurrences in n trials of an event, when there are a total of d occurrences in a population of N.

Hypothesis: A testable explanation for observations. Science in general proceeds by an incremental process of refuting null hypotheses. Evidence is then presented persuasively in the form of a pattern that has been calibrated against unmeasured variation.

- The *null hypothesis* is the refutable hypothesis of negligible systematic difference or pattern, in other words that nothing interesting is going on beyond random variation.

- The *test hypothesis* is the alternative hypothesis of pattern, in the form of one or more real treatment effects, as defined by the statistical model.

- A statistical test will reject the null hypothesis with probability α of doing so mistakenly (and thereby making a "Type I error"), or it will accept the null hypothesis with probability β of doing so mistakenly (and thereby making a "Type II error"). Data collection should be designed with a view to maximizing the power to detect true effects at a given α, with power defined by the probability $1 - \beta$.

Hygiene factors: A term used by Frederick Herzberg to label "dissatisfiers" (see Dissatisfiers).

Hypothesis statement: A complete description of the suspected cause(s) of a process problem.

I-Chart: Individuals chart or X-chart. A control charting methodology developed for continuous, variables; data that use individuals' data (not averages). Even though R/1000 and CPU data are technically discrete attribute data, I-charts work well for most warranty data applications. This is because of the large divisors involved, which mitigate the discreteness.

Ideal function: The ideal relationship between signal and response (e.g., Y-Beta*M). It is based on the energy transformation that achieves what the product or process is primarily or fundamentally supposed to do.

Ideal point: Point on a perceptual map that represents the most preferred combination of perceived attributes (according to the respondents).

A major assumption is that the position of the ideal point (relative to the other objects on the perceptual map) would define relative *preference* such that objects farther from the ideal point should be preferred less.

Identification: Degree to which there is a sufficient number of equations to "solve for" each of the coefficients (unknowns) to be estimated. Models can be *underidentified* (cannot be solved), *just-identified* (number of equations equals number of estimated coefficients with no degrees of freedom), or *overidentified* (more equations than estimated coefficients and degrees of freedom greater than zero). The researcher desires to have an overidentified model for the most rigorous test of the proposed model. Also see *Degrees of freedom.*

Ignorable missing data: *Missing data process* that is explicitly identifiable and/or is under the control of the researcher. Ignorable missing data do not require a remedy because the missing data are explicitly handled in the technique used.

Impact analysis: The mathematical examination of the nature of individual RISKS on the PROJECT as well as potential structures of interdependent risks. It includes the quantification of their respective impact severity, probability, and sensitivity to changes in related project variables, including the Project Life Cycle. To be complete, the analysis should also include an examination of the external "status quo" prior to project implementation as well as the project's internal intrinsic worth as a reference baseline. A determination should also be made as to whether all risks identified are within the scope of the project's risk response planning process.

Impact interpretation: Clarification of the significance of a variance with respect to overall objectives.

Imperfection: A quality characteristic's departure from its intended level or state without any association to conformance to specification requirements or to the usability of a product or service (see also Blemish, Defect, and Nonconformity).

Implementation (phase): The third of four sequential phases in the Project Life Cycle. Also known as Execution or Operation Phase.

Implementation, completion of: Also known as Closeout Phase. Completion of implementation means that the Project Team has successfully completed the project and has transfer the knowledge of operating and maintaining the integrity of the project to the customer.

Improve: The fourth element of the DMAIC phase where solutions and ideas are creatively generated and decided upon. Once a problem has been fully identified, measured, and analyzed, potential solutions can be determined to solve the problem in the problem statement and support the goal statement. See also *Charter.*

Importance–performance grid: Two-dimensional approach for assisting the researcher in labeling dimensions. The first axis is the respondents' perceptions of the importance (e.g., as measured on a scale of "extremely important" to "not at all important"). The opposing axis is performance (e.g., as measured on a scale of "highly likely to perform" to "highly unlikely to perform") for each brand, or product or service, on various attributes. Each object is represented by its values on importance and performance.

Imputation methods: Process of estimating the *missing data* of an observation based on valid values of the other variables. The objective is to employ known relationships that can be identified in the valid values of the sample to assist in representing or even estimating the replacements for missing values.

In-control process: A process in which the statistical measure being evaluated is in a state of statistical control; that is, the variations among the observed sampling results can be attributed to a constant system of chance/common causes (see also Out-of-control process).

In-control: Used to characterize the state of a process. A process is said to be "in-control" if its data are falling randomly about a fixed process average, and within the control limits of a run chart. In other words, there are no run rule violations. This means that only an unchanging mix of common causes are in effect. In warranty, this translates to "no improvement," a highly undesirable state if the competition is doing better. See also *Out-of-control, Run Chart, Random, Run rules.*

In-progress activity: An activity that has been started but is not completed on a given date.

Incremental fit measure: Measure of *goodness-of-fit* that compares the current model to a specified *null model* to determine the degree of improvement over the null model. Complements the other two types of goodness-of-fit measures, the *absolute fit* and *parsimonious fit measures.*

Independence: Critical assumption of ANOVA or MANOVA that requires that the dependent measures for each respondent be totally uncorrelated with the responses from other respondents in the sample. A lack of independence severely affects the statistical validity of the analysis unless corrective action is taken.

Independent variable: Variable(s) selected as predictors and potential explanatory variables of the dependent variable. It is presumed to be a cause of any change in the *dependent variable.*

Index of fit: Squared correlation index (R^2) that may be interpreted as indicating the proportion of variance of the *disparities* (optimally scaled data) that can be accounted for by the MDS procedure. It measures how well the raw data fit the MDS model. This index is an alternative to the *stress measure* for determining the number of dimensions.

Similar to measures of covariance in other multivariate techniques, measures of 0.60 or greater are considered acceptable.

Indicator coding: Method for specifying the reference category for a set of *dummy variables* where the reference category receives a value of zero across the set of dummy variables. The dummy variable coefficients represent the category differences from the reference category. Also see *Effects coding*.

Indicator: An observation that has a disproportionate influence on one or more aspects of the regression estimates. This influence may be based on extreme values of the independent or dependent variables, or both. Influential observations can either be "good," by reinforcing the pattern of the remaining data, or "bad," when a single or small set of cases unduly affect the regression estimates. It is not necessary for the observation to be an *outlier*, although many times outliers can be classified as influential observations as well.

Indicator: Single variable used in conjunction with one or more other variables to form a *composite measure. Also, it is an* observed value *(manifest variable)* used as a measure of a concept or *latent construct* that cannot be measured directly. The researcher must specify which indicators are associated with each latent construct.

Individuals chart with moving range: A control chart used when working with one sample per subgroup. The individual samples are plotted on the x-chart rather than subgroup averages. The individuals chart is always accompanied by a moving range chart, usually using two subgroups (two individual readings) to calculate the moving range points.

Individuals outside the project: Refers to all those individuals who impact the project work but who are not considered members of the Project Team.

Infant mortality: Early failures attributable to defects in design, manufacturing, or construction. They typically exist until debugging eliminates faulty components. These failures often happen during burn-in or during early usage.

Influential observation: Observation with a disproportionate influence on one or more aspects of the regression estimates. This influence may have as its basis: (a) substantial differences from other cases on the set of independent variables, (b) extreme (either high or low) observed values for the criterion variables, or (c) a combination of these effects. Influential observations can either be "good," by reinforcing the pattern of the remaining data, or "bad," when a single or small set of cases unduly affect (biases) the regression estimates.

Information: Data transferred into an ordered format that makes them usable and allows one to draw conclusions.

Informal communication: The unofficial communication that takes place in an organization as people talk freely and easily; examples include impromptu meetings and personal conversations (verbal or e-mail).

Information flow (distribution list): A list of individuals that would receive information on a given subject or project.

Information gathering: Researching, organizing, recording, and comprehending pertinent.

Information system: Technology-based systems used to support operations, aid day to-day decision making, and support strategic analysis (other names often used include: management information system, decision system, information technology—IT, data processing).

Information systems: A structured, interacting, complex of persons, machines, and procedures designed to produce information that is collected from both internal and external sources for use as a basis for decision making in specific contract/procurement activities.

Initial dimensionality: A starting point in selecting the best spatial configuration for data. Before beginning an MDS procedure, the researcher must specify how many *dimensions* or features are represented in the data.

Input: Any product, service, or piece of information that comes into the process from a supplier. A supplier may be internal or external to the organization.

Input limits: Imposition of limitations to the resources through which the plan will be executed.

Input measures: Measures related to and describing the input into a process; predictors of output measures.

Input milestones: Imposed target dates or target events that are to be accomplished and which control the plan with respect to time.

Input priorities: Imposed priorities or sequence desired with respect to the scheduling of activities within previously imposed constraints.

Input restraints: Imposed external restraints, such as dates reflecting input from others and target dates reflecting output required by others, and such items as float allocation and constraints.

Inspection: Examination or measurement of work to verify whether an item or activity conforms to a specific requirement. That means, measuring, examining, testing, and gauging one or more characteristics of a product or service and comparing the results with specified requirements to determine whether conformity is achieved for each characteristic.

Inspection accuracy: The percentage of defective units that are correctly identified by an inspector. The percentage is determined by having a second inspector review both the accepted and rejected units.

Inspection studies: A study to better define and improve understanding of troubles and areas of product dissatisfaction reported by owners through actual inspection of these owners' vehicles.

Instability (of a process): A process is said to show instability if it exhibits variations larger than its control limits or shows a systematic pattern of variation.

Institutionalization: Fundamental changes in daily behaviors, attitudes, and practices that make changes "permanent", cultural adaptation of changes implemented by Process Improvement, Design, or Redesign, including complex business systems such as HR, MIS, Training, and so on.

Intended function: Typically a noun–verb description of what the product or process is supposed to do (i.e., create dimension, turn vehicle). It generally only includes the response.

Interaction: Mutual action or reciprocal action or influence.

Interaction effect: In factorial designs, the joint effects of two *treatment* variables in addition to the individual *main effects.* This means that the difference between groups on one treatment variable varies depending on the level of the second treatment variable. Another way of explaining is to think of a condition in which the impact of a factor on the system changes depending on the level of another factor. If this is the case, it is said that the two factors interact with each other. These are also known as two-way interactions or second order effects. There are higher-order interaction effects, such as third-order effects that involve three factors, but frequently, higher-order interactions are insignificant. A simple example: Assume that respondents were classified by income (three levels) and gender (males vs. females). A significant interaction would be found when the differences between males and females on the independent variable(s) varied substantially across the three income levels.

Interaction plot: A line graph that depicts the interaction between factors.

Interattribute correlation: Correlation among attributes, also known as *environmental correlation,* which makes combinations of attributes unbelievable or redundant. A negative correlation depicts the situation in which two attributes are naturally assumed to operate in different directions, such as horsepower and gas mileage. As one increases, the other is naturally assumed to decrease. Thus, because of this correlation, all combinations of these two attributes (e.g., high gas mileage and high horsepower) are not believable. The same effects can be seen for positive correlations, where perhaps price and quality are assumed to be positively correlated. It may not be believable to find a high-price, low-quality product in such a situation. The presence of strong interattribute correlations requires that the researcher closely examine the stimuli presented to respondents and avoid unbelievable combinations that are not useful in estimating the part-worths.

Intercept (b_0): Value on the *Y-axis* (dependent-variable axis) where the line defined by the regression equation $y = b_0 + b_1X_1$ crosses the axis. It is described by the constant term b_0 in the regression equation. In addition to its role in prediction, the intercept may have a managerial interpretation. If the complete absence of the independent variable has meaning, then the intercept represents that amount. For example, when estimating sales from past advertising expenditures, the intercept represents the level of sales expected if advertising is eliminated. But in many instances the constant has only predictive value because there is no situation in which all independent variables are absent. An example is predicting product preference based on consumer attitudes. All individuals have some level of attitude, so the intercept has no managerial use, but it still aids in prediction.

Interdependence: Shared dependence between two or more items.

Interdependence technique: Classification of statistical techniques in which the variables are not divided into *dependent* and *independent sets* (e.g., factor analysis); rather, all variables are analyzed as a single set.

Interface activity: An activity connecting anode in one subnet with a node in another subnet, representing logical interdependence. The activity identifies points of interaction or commonality between the PROJECT activities and outside influences.

Interface management: The management of communication, coordination, and responsibility across a common boundary between two organizations, phases, or physical entities that are interdependent.

Intermediate customers: Distributors, dealers, or brokers who make products and services available to the end user by repairing, repackaging, reselling, or creating finished goods from components or subassemblies.

Internal audit: An audit conducted within an organization by members of the organization to measure its strengths or weaknesses against its own procedures and/or external standards a "first-party audit."

Internal capability analysis: A detailed view of the internal workings of organization (e.g., determine how well the capabilities of the match to strategic needs).

Internal customer: The recipient, person, or department of another person's or department's output (product, service, or information) within an organization (see also External customer).

Internal failure costs: Costs associated with defects found before the product or service is delivered.

Internal project sources: Intrafirm sources and records including historical data on similar procurements, cost, and performance data on various suppliers and other data that could assist in proposed procurements.

Interobject similarity: The correspondence or association of two *objects* based on the variables of the *cluster variate*. Similarity can be measured in two ways. First is a measure of association, with higher positive correlation coefficients representing greater similarity. Second, "proximity," or "closeness," between each pair of objects can assess similarity, where measures of distance or difference are used, with smaller distances or differences representing greater similarity.

Interpret: Present in understandable terms.

Interpretation: Reduction of information to appropriate and understandable terms and explanations.

Interfacing requirement: A Requirement that a specific DVP Team must classify, but they are not responsible for the Design Verification Plan.

Interrelationship digraph: A management and planning tool that displays the relationship between factors in a complex situation. It identifies meaningful categories from a mass of ideas and is useful when relationships are difficult to determine.

Intervention: An action taken by a leader or a facilitator to support the effective functioning of a team or work group.

Intervention intensity: Refers to the strength of the intervention by the intervening person; intensity is affected by words, voice inflection, and nonverbal behaviors.

Intimidating: A project manager that frequently reprimands employees for the sake of an image as a "tough guy," at the risk of lowering department morale.

Intraction effect: A condition in which the impact of a factor on the system changes depending on the level of another factor. If this is the case, it is said that the two factors interact with each other. These are also known as two-way interactions or second-order effects. There are higher-order interaction effects, such as third order-effects that involve three factors, but frequently, higher-order interactions are insignificant.

Ishikawa diagram: See *Cause-and-effect Diagram*.

ISO 14000 series: A set of standards and guidelines relevant to developing and sustaining an environmental management system.

ISO 9000 series standards: A set of individual but related international standards and guidelines on quality management and quality assurance developed to help companies effectively document the quality system elements to be implemented to maintain an efficient quality system. The standards, initially published in 1987, revised in 1994, 2000, and 2008, are not specific to any particular industry, product, or service. The standards were developed by the International Organization for Standardization, a specialized international agency for standardization composed of the national standards bodies of more than 100 countries.

ISO: "Equal" (Greek). A prefix for a series of standards published by the International Organization for Standardization.

ISO 9000: Standard and guideline used to certify organizations as competent in defining and adhering to documented processes; mostly associated with quality assurance systems, not quality improvement. A quality systems orientation.

Jackknife: Procedure for drawing repeated samples based on omitting one observation when creating each sample. Allows for the empirical estimation of parameter confidence levels rather than parametric estimation.

Jidoka: Japanese method of autonomous control involving the adding of intelligent features to machines to start or stop operations as control parameters are reached and to signal operators when necessary.

Jitter: Small amounts of random noise added to the values used in *training* a neural network that smooth the error function and assist in obtaining a global optimum solution.

Job aid: Any device, document, or other media that can be provided a worker to aid in correctly performing their tasks (e.g., laminated setup instruction card hanging on machine, photos of product at different stages of assembly, metric conversion table).

Job description: A narrative explanation of the work, responsibilities, and basic requirements of the job.

Job descriptions (scope management): Documentation of a project participant's job title, supervisor, job summary, responsibilities, authority, and any additional job factors (Human Resources Management). Written outlines of the skills, responsibilities, knowledge, authority, environment, and interrelationships involved in an individual's job.

Job enlargement: Increasing the variety of tasks performed by an employee.

Job enrichment: Increasing the worker's authority in work to be done.

Job specification: A list of the important functional and quality attributes (knowledge, skills, aptitudes, and personal characteristics) needed to succeed in the job.

Joint planning meetings: A meeting involving representatives of a key customer and the sales and service team for that account to determine how better to meet the customer's requirements and expectations.

Judgment sampling: Approach that involves making educated guesses about which items or people are representative of a whole, generally to be avoided.

Judicial: A project manager that exercises the use of sound judgment or is characterized by applying sound judgment to most areas of the project.

Juran's trilogy. See *Quality trilogy.*

Just-identified model: Structural model with zero *degrees of freedom* and exactly meeting the *order condition*. This corresponds to a perfectly fitting model but has no generalizability.

Just-in-time manufacturing (JIT): An optimal material requirement planning system for a manufacturing process in which there is little or no manufacturing material inventory on hand at the manufacturing site and little or no incoming inspection. It is time-dependent. That means, time manufacturing coordinates inventory and production to get away from the batch mode of production in order to improve quality.

Just-in-time-training: Providing job training coincidental with or immediately prior to its need for the job.

K: For process capability studies: a measure of difference between the process mean and the specification mean (nominal). It is represented by (Mean – Midpoint)/(Tolerance/2).

Kaikaku (Japanese): A breakthrough improvement in eliminating waste.

Kaizen blitz/event: An intense, short time-frame, team approach to employ the concepts and techniques of continuous improvement (e.g., to reduce cycle time, increase throughput).

Kaizen: A Japanese term that means gradual unending improvement by doing little things better and setting and achieving increasingly higher standards. The term was made famous by Masaaki Imai in his book *Kaizen: The Key to Japan's Competitive Success.*

Kanban: A system inspired by Tauchi Ohno's (Toyota) visit to a US supermarket. The system signals the need to replenish stock or materials or to produce more of an item (also called a "pull" approach).

Kano model: A representation of the three levels of customer satisfaction defined as dissatisfaction, neutrality, and delight.

Key event schedule: A schedule comprised of key events or milestones. These events are generally critical accomplishments planned at time intervals throughout the project and used as a basis to monitor overall project performance. The format may be either network or bar chart and may contain minimal detail at a highly summarized level. This is often referred to as a milestone schedule.

Key life tests: A test that duplicates the stresses under customer operation. It includes the specified extremes of loads, usage, manufacturing/assembly variability, environmental stresses, and their interactions to address a single critical failure mode or a group of failure modes that result from the same stress patterns. It may be accelerated to reduce test time.

KJ method: See *Affinity diagram.*

Knowledge discovery in databases (KDD): Extraction of new information from databases through a variety of knowledge discovery processes.

Knowledge management: Involves transforming data into information, the acquisition or creation of knowledge, as well as the processes and technology employed in identifying, categorizing, storing, retrieving, disseminating, and using information and knowledge for the purposes of improving decisions and plans.

Kohonen model: *Neural network* model that is designed for clustering problems and operates in an unsupervised *learning* mode.

KPIV: Key Process INPUT Variable—an independent material or element, with descriptive characteristic(s), which is either an object (going into) or a parameter of a process (step) and which has a significant (Key) effect on the output of the process.

KPOV: Key Process OUTPUT Variable—a dependent material or element, with descriptive characteristic(s), which is the result of a process (step) that either is, or significantly affects, the customers' CTQ.

Kurtosis: Measure of the peakedness or flatness of a distribution when compared with a normal distribution. A positive value indicates a relatively peaked distribution, and a negative value indicates a relatively flat distribution. Therefore, it is a measure of the shape of a distribution. If the distribution has longer tails than a normal distribution of the same standard deviation, then it is said to have positive kurtosis (platykurtosis); if it has shorter tails, then it has negative kurtosis (leptokurtosis).

L_8 Array: An orthogonal array that is used for an eight-run experiment. It can be used to test a maximum of seven factors.

L_{12} Array: An orthogonal array that is used for a twelve run experiment. It can be used to test a maximum of 11 factors. This array is especially handy for a screening experiment involving many two-level factors because main effects are confounded with fractions of interactions (rather than the entire interaction.) This array cannot be used to measure interactions because the interactions cannot be separated from the main effects.

L_{16} Array: An orthogonal array that is used for a 16-run experiment. It can be used to test a maximum of 15 factors.

L_{18} Array: A mixed orthogonal array. The first column accommodates two levels and the rest of the columns are for three levels. It is one of the preferred arrays for screening.

Latent construct or variable: Operationalization of a *construct* in structural equation modeling. A latent variable cannot be measured directly but can be represented or measured by one or more variables *(indicators)*. For example, a person's attitude toward a product can never be measured so precisely that there is no uncertainty, but by asking various questions we can assess the many aspects of the person's attitude. In combination, the answers to these questions give a reasonably accurate measure of the latent construct (attitude) for an individual.

Latent root: See *Eigenvalue*.

Latin squares: These are balanced variants of the randomized complete block design, with treatment factor(s) replicated in two cross-factored blocks. The two blocking factors each have the same number of blocks as there are levels of the treatment factor(s). The defining feature of a Latin square is that treatment factor levels are randomly allocated to cells within the square grid of column and row blocks in such a way as to have each factor level represented once in every column block and every row block (making it a balanced and complete design). These designs are particularly useful for field experiments that benefit from blocking random variation in two dimensions.

They are also used in crossover trials, where the orthogonal blocking factors are Subject and Order of treatment. Unless replicated, Latin squares do not allow testing of interactions with blocks that must be assumed to have negligible effects.

LCL (Lower Control Limit): For control charts: the limit above which the process subgroup statistics (Xbar, R, sigma) remain when the process is in control.

Leader: An individual, recognized by others, as the person to lead an effort. One cannot be a "leader" without one or more "followers." The term is often used interchangeably with "manager" (see Manager). A "leader" may or may not hold an officially designated management-type position. Key ingredients of any leader are that he or she (a) must be willing to be different (rock the boat); (b) must be willing to get involved in the tranches to find the real problems—the problem is always closer to the source; (c) must be willing to delegate authority and responsibility for task at hand; (d) know when to stop and when to ignore advisers and consultants; (e) must be able to develop and cultivate selective amnesia for the success of the project at hand—this means change opinion and strategy in the middle of the campaign; (f) must be able to prioritize and be a master of the 80/20 rule; (g) must be able to bring closure and quit in his or her terms; and (h) ALWAYS remember that leadership is an issue of responsibility.

Leadership—general: An essential part of a quality improvement effort. Organization leaders must establish a vision, communicate that vision to those in the organization, and provide the tools, knowledge, and motivation necessary to accomplish the vision.

Leadership—PM: The process by which the Project Manager influences the Project Team to behave in a manner that will facilitate project goal achievement.

Lean approach/lean thinking: ("lean" and "agile" may be used interchangeably). The focus is on reducing cycle time and waste using a number of different techniques and tools, for example, value stream mapping, and identifying and eliminating "monuments" and non-value-added steps. You know that a lean culture is present when (a) you have less fire fighting; (b) you meet all your operating targets and operations are stable; (c) your entire organization shares the same vision; (d) everyone is always looking for a way to improve, but the improvements come from a standardized foundation; (e) you can walk out on your manufacturing floor on any day, at any time, and know the actual condition of your operation; and (f) you hear "we" and "us" more often than "you" and "they."

Lean manufacturing: Applying the lean approach to improving manufacturing and nonmanufacturing operations.

Learner-controlled instruction (also called "self-directed learning"): The process when the learner is working without an instructor, at their own pace, building mastery of a task (computer-based training is a form of LCI).

Learning: Sequential processing of large samples of observations (known as the *training* sample) in which the prediction or classification errors are used to recalibrate the weights to improve estimation.

Learning curve: The time it takes to achieve mastery of a task or body of knowledge. Or another way of saying it, it is a concept that recognizes the fact that productivity by workers improves as they become familiar with the sequence of activities involved in the production process.

Learning objectives (also called "terminal objectives"): The objectives to be met upon completion of a course of study or the learning of a skill.

Learning organization: An organization that has as a policy to continue to learn and improve its products, services, processes, and outcomes; "an organization that is continually expanding its capacity to create its future" (Senge).

Least squares: Estimation procedure used in simple and multiple regression whereby the regression coefficients are estimated so as to minimize the total sum of the squared *residuals.*

Leptokurtosis: For frequency distributions: a distribution which that a higher peak and shorter "tails" than a normal distribution with the same standard deviation.

Level: Specific value describing a *factor.* Each factor must be represented by two or more levels, but the number of levels typically never exceeds four or five. If the factor is metric, it must be reduced to a small number of levels. For example, the many possible values of size and price may be represented by a small number of levels: size (10, 12, or 16 ounces), or price ($1.19, $1.39, or $1.99). If the variable is nonmetric, the original values can be used as in these examples: color (red or blue), brand (X, Y, or Z), or fabric softener additive (present or absent).

Level of detail: A policy expression of content of plans, schedules, and reports in accordance with the scale of the breakdown of information.

Leverage point: An observation that has substantial impact on the regression results due to its differences from other observations on one or more of the independent variables. The most common measure of a leverage point is the *hat value,* contained in the *hat matrix.*

Leverage points: Type of *influential observation* defined by one aspect of influence termed *leverage.* These observations are substantially different on one or more independent variables, so that they affect the estimation of one or more *regression coefficients.*

Life cycle costing: The concept of including all costs within the total life of a project from concept, through implementation, startup to dismantling. It is used for making decisions between alternatives and is a term used principally by the government to express the total cost of an article or system. It is also used in the private sector by the real estate industry.

Life cycle: A product life cycle is the total time frame from product concept to the end of its intended use; a project life cycle is typically divided into five stages: concept, planning, design, implementation, evaluation, and close-out.

Limited life component: A component that is designed to have a specific life time at which point it is discarded/replaced.

Linear composites: See *Canonical variates.*

Linear regression: The mathematical application of the concept of a scatter diagram where the correlation is actually a cause-and-effect relationship.

Linear responsibility matrix: A matrix providing a three-dimensional view of project tasks, responsible person, and level of relationship.

Linearity: The extent to which a measuring instrument's response varies with the measured quantity. It is used to express the concept that the model possesses the properties of additivity and homogeneity. In a simple sense, linear models predict values that fall in a straight line by having a constant unit change (slope) of the dependent variable for a constant unit change of the independent variable. In the population model $Y = b_0 + b_1X_1 + E$, the effect of a change of 1 in X_1 is to add b_1 (a constant) units of Y.

Linear structural relations (LISREL): A software model that is used to evaluate structural equation modeling. The software makes it possible to combine structural equation modeling and confirmatory factor analysis.

Line/functional manager: Those responsible for activities in one of the primary functions of the organization, such as production or marketing, with whom the Project Manager must relate in achieving the project's goals.

Listening post data: customer data and information gathered from designated "listening posts."

Lists, project: The tabulations of information organized in meaningful fashion.

Location: Refers to the central tendency, commonly expressed by the average (or mean), or by the median. Spread—refers to dispersion, commonly expressed by the standard deviation or variance (which is the standard deviation squared). Shape—refers to characteristics such as symmetry and peakedness. The Normal Distribution, for instance, is often described as "bell shaped."

Logic: The interdependency of the activities in a network. As applied to warranty data, this term refers to an algorithm (formula) that is used to select claims and vehicles to be included in R/1000 and CPU calculations.

Long-term goals: Refers to goals that an organization hopes to achieve in the future, usually in 3 to 5 years. They are commonly referred to as strategic goals.

Loop: A path in a network closed on itself passing through any node more than once on any given path. The network cannot be analyzed, as it is not a logical network.

Lot formation, The process of collecting units into lots for the purpose of acceptance sampling: The lots are chosen to ensure, as much as possible, that the units have identical properties, that is, that they were produced by the same process operating under the same conditions.

Lot tolerance percent defective (LTPD): For acceptance sampling: expressed in percent defective units; the poorest quality in an individual lot that should be accepted. Commonly associated with a small consumer's risk (see Consumer's risk).

Lot: A defined quantity of product accumulated under conditions that are considered uniform for sampling purposes.

Lower control limit (LCL): Control limit for points below the central line in a control chart.

LSL (lower specification limit): The lowest value of a product dimension or measurement which is acceptable.

Macro environment: Consideration, interrelationship, and action of outside changes such as legal, social, economic, political, or technological, which may directly or indirectly influence specific project actions.

Macro processes: Broad, far-ranging processes that often cross-functional boundaries.

Mahalanobis distance (D²): Standardized form of *Euclidean distance. It is a* measure of the uniqueness of a single observation based on differences between the observation's values and the mean values for all other cases across all independent variables. The source of influence on regression results is for the case to be quite different on one or more predictor variables, thus causing a shift of the entire regression equation. The scaling responses are in terms of standard deviations that standardize the data, with adjustments made for intercorrelations between the variables.

Main effect: A measure of the influence of varying the factor levels on the system response (or quality characteristic). In factorial designs, the individual effect of each *treatment* variable on the dependent variable. May be complemented by interaction effects in specific situations.

Maintainability: The probability that a given maintenance action for an item under given usage conditions can be performed within a stated time interval when the maintenance is performed under stated conditions using stated procedures and resources. Maintainability has two categories: serviceability—the ease of conducting scheduled inspections and servicing, and repairability—the ease of restoring service after a failure. The basic measure is Mean Time To Repair (MTTR).

Management: The process of planning, organizing, executing, coordinating, monitoring, forecasting, and exercising control.

Management plan: Document that describes the overall guidelines within which a project is organized, administered, and managed to assure the timely accomplishment of project objectives.

Management styles: A manager may adopt several different management styles, according to circumstances, in the process of leadership and team motivation. These include: authoritarian, combative, conciliatory, disruptive, ethical, facilitating, intimidating, judicial, promotional, and secretive. Management styles include the following:

> **Authoritarian:** Lets individuals know what is expected of them, gives specific guidance as to what should be done, makes his part of the group understood, schedules work to be done, and asks group members to follow standard rules and regulations.
>
> **Combative:** A project manager who is marked by an eagerness to fight or be disagreeable over any given situation.
>
> **Conciliatory:** A project manager who is friendly and agreeable; one who attempts to assemble and unite all project parties involved to provide a compatible working team.
>
> **Disruptive:** A project manager who tends to break apart the unity of a group; one who tends to be an agitator and causes disorder on a project.
>
> **Ethical:** A project manager who is honest, sincere, able to motivate, and to press for the best and fairest solution; one generally goes "by the books."
>
> **Facilitating:** The project manager is available to answer questions and give guidance when needed; he does not interfere with day-to-day tasks, but rather maintains that status quo.
>
> **Intimidating:** A project manager who frequently reprimands employees for the sake of an image as a "tough guy," at the risk of lowering department morale.
>
> **Judicial:** A project manager who exercises the use of sound judgment or is characterized by applying sound judgment to most areas of the project.
>
> **Promotional:** Encourages subordinates to realize their full potential, cultivates a team spirit, and lets subordinates know that good work will be rewarded.
>
> **Secretive:** A project manager who is not open or outgoing in speech activity, or purpose much to the detriment of the overall project.

Management time: Manhours related to the project management team.

Management training: Usually refers to training and/or education provided to any management- or professional-level person from front-line supervision up to, but not including executives.

Management-by-fact: Decision making using criteria and facts; supporting "intuition" with data; tools used include process measurement, process management techniques, and rational decision-making tools (e.g., criteria matrix).

Manager: An individual who manages and is responsible for resources (people, material, money, time). A person officially designated with a management-type position title. A "manager" is granted authority from above, whereas a "leader's" role is derived by virtue of having followers. However, the terms "manager" and "leader" are often used interchangeably.

Managerial grid: A management theory developed by Robert Blake and Jane Mouton, which maintains that a manager's management style is based on his or her mindset toward people, focuses on attitudes rather than behavior. The theory uses a grid to measure concern with production and concern with people.

Managerial quality administration: The managerial process of defining and monitoring policies, responsibilities, and systems necessary to retain quality standards throughout the project.

Managerial reserves: The reserve accounts for allocating and maintaining funds for contingency purposes on over- or under-spending on project activities. These accounts will normally accrue from the contingency and other allowances in the Project Budget Estimate.

Manpower planning: The process of projecting the organization's manpower needs overtime, in terms of both numbers and skills, and obtaining the human resources required to match the organization's needs. See Human resources management.

Manifest variable: Observed value for a specific item or question, obtained either from respondents in response to questions (as in a questionnaire) or from observations by the researcher. Manifest variables are used as the *indicators* of *latent constructs* or *variables.*

Market-perceived quality: The customer's opinion of an organization's products or services as compared to those of the competitors.

Master black belt: This is the highest level of technical and organizational proficiency in the six sigma methodology. Because master black belts train black belts, they must know everything the black belts know, as well as understand the mathematical theory on which the statistical methods are based. Masters must be able to assist black belts in applying the methods correctly in unusual situations. Whenever possible, statistical training should be conducted only by master black belts. If it's necessary for black belts and green belts to provide training, they should only do so under the guidance of master black belts. Because of the nature of the master's duties, communications and teaching skills should be judged as important as technical competence in selecting candidates.

Master schedule: An executive summary-level schedule that identifies the major components of a project; usually also identifies the major milestones.

Materials review board (MRB): A quality-control committee or team, usually employed in manufacturing or other materials-processing installations, which has the responsibility and authority to deal with items or materials that do not conform to fitness-for-use specifications. An equivalent, error review board, is sometimes used in software development.

Matrix chart/diagram: A management and planning tool that shows the relationships among various groups of data; it yields information about the relationships and the importance of task/method elements of the subjects.

Matrix organization: A two-dimensional organizational structure in which the horizontal and vertical intersections represent different staffing positions with responsibility divided between the horizontal and vertical authorities.

Matrix structure: Describes an organization that is organized into a combination of functional and product departments; it brings together teams of people to work on projects and is driven by product scope.

Matrix: A two-dimensional structure in which the horizontal and vertical intersections form cells or boxes. In each cell may be identified a block of knowledge whose interface with other blocks is determined by its position in the structure.

Maximum likelihood estimation (MLE): Estimation method commonly employed in structural equation models, including LISREL and EQS. An alternative to ordinary least squares used in multiple regression, MLE is a procedure that iteratively improves parameter estimates to minimize a specified fit function.

MDS: See Multidimensional scaling.

Mean time between failures (MTBF): A reliability measure that estimates the average time interval between failures for repairable product for a defined unit of measure (e.g., operating hours, cycles, or miles). MTBF is calculated as the total operating time on all units divided by the total number of failures.

Mean time to failure (MTTF): The average time of operation to failure of a nonrepairable system. MTTF is calculated as a total operating time on all units divided by the total number of failures. It is also known as Expected life or Mean Life.

Mean time to repair (MTTR): The average time to complete a repair. MTTR is the basic measure of maintainability.

Mean: A measure of central tendency and is the arithmetic average of all measurements in a data set. Also known as the average. There are two means: (1) **Mean** (of a population) (μ)—the true arithmetic average of all elements in a population. Xbar approximates the true value of the population mean, and (2) **Mean** (of a statistical sample) ($\overline{X}$-Xbar)—the arithmetic average value of some variable. The mean is given by the formula, where "x" is the value of each measurement in the sample. All *x*s are added together and divided by the number of elements (*n*) in the sample.

Means (in the Hoshin planning usage): The step of identifying the ways by which multiyear objectives will be met, leading to the development of action plans.

Measure: The second element of the DMAIC model, where key measures are identified, and data are collected, compiled, and displayed. A quantified evaluation of specific characteristics and/or level of performance based on observable data.

Measure of sampling adequacy (MSA): Measure calculated both for the entire correlation matrix and each individual variable evaluating the appropriateness of applying factor analysis. Values above 0.50 for either the entire matrix or an individual variable indicate appropriateness.

Measurement: Refers to the reference standard or sample used for the comparison of properties.

Measurement accuracy: The extent to which the average result of a repeated measurement tends toward the true value of the measured quantity. The difference between the true value and the average measured value is called the instrument bias and may be due to such things as improper zero-adjustment, nonlinear instrument response, or even improper use of the instrument.

Measurement error: The difference between the actual and measured value of a measured quantity. Degree to which the data values do not truly measure the characteristic being represented by the variable. In other words, inaccuracies in measuring the "true" variable values due to the fallibility of the measurement instrument (i.e., inappropriate response scales), data entry errors, or respondent errors. For example, when asking about total quality, there are many sources of measurement error (e.g., errors in machinery, instruments, observations, and process) that make the data values imprecise. A more detailed definition may be the degree to which the variables we can measure (the *manifest variables*) do not perfectly describe the *latent construct(s)* of interest. Sources of measurement error can range from simple data entry errors to definition of constructs (e.g., abstract concepts such as patriotism or loyalty that mean many things to different people) that are not perfectly defined by any set of manifest variables. For all practical purposes, all constructs have some measurement error, even with the best *indicator variables*. However, the researcher's objective is to minimize the amount of measurement error. SEM can take measurement error into account in order to provide more accurate estimates of the *causal relationships*.

Measurement model: Submodel in SEM that (1) specifies the *indicators* for each *construct,* and (2) assesses the *reliability* of each construct for estimating the *causal relationships.* The measurement model is similar in form to factor analysis; the major difference lies in the degree of control provided the researcher. In factor analysis, the researcher can specify only the number of factors, but all variables have loadings (i.e., they act as indicators) for each factor. In the measurement model, the researcher specifies which variables are indicators of each construct, with variables having no loadings other than those on its specified construct.

Measurement precision: The extent to which a repeated measurement gives the same result. Variations may arise from the inherent capabilities of the instrument, from variations of the operator's use of the instrument, from changes in operating conditions, and so on.

Measurement system capability: The method and tools used to collect and measure the results of the experiment must be carefully verified to avoid introducing distortion or bias.

Median (of a statistical sample): The middle number or center value of a set of data when all the data are arranged in an increasing sequence. For a sample of a specific variable, the median is the point $\tilde{X}$ such that half the sample elements are below and the other half are above of the median. That is: the midpoint or fiftieth percentile.

Median chart: For variables data: a control chart of the median of subgroups.

Metadata: Complete description of a data element, not only defining its storage location and data type but also any attributes, dimensions, or associations with any other data elements. Simply put, "data about data," which are stored in a centralized *repository* for common access to all database users.

Method: The manner or way in which work is done. When formalized into a prescribed manner of performing specified work, a method becomes a procedure.

Metric: A standard of measurement.

Metric data: Also called *quantitative data, interval data,* or *ratio data,* these measurements identify or describe subjects (or objects) not only on the possession of an attribute but also by the amount or degree to which the subject may be characterized by the attribute. For example, a person's age and weight are metric data.

Metrology: The science and practice of measurement.

Micro environment: Consideration of company, project, or client imposed policies and procedures applicable to project actions.

Micromanaging: Managing every little detail (e.g., approving requisition for post-it notes).

Micro processes: Narrow processes made up of detailed steps and activities that could be accomplished by a single person.

Micro procurement environment: Consideration of firm, project, or client imposed policies and procedures applicable in the procurement actions.

Milestone chart: Another name for a Gantt chart.

Milestone schedule: See *Summary schedule.*

Milestone: A significant event in the project (key item or key event). A point in time when a critical event is to occur; a symbol placed on a milestone chart to locate the point when a critical event is to occur.

Milestones for control: Interim objectives, points of arrival in terms of time for purposes of progress management.

MIL-STD: Military standard.

MIL-STD-105D: A set of specifications for acceptance sampling plans based on acceptable quality level (AQL). For a given AQL, lot size, and level of inspection, the specification lists the number of defective units that is acceptable, and the number that requires rejection of the lot. The specifications allow for tightening or loosening of inspection requirements based on previous inspection.

MIL-STD-2000: A set of standard requirements for soldered electrical and electronic assemblies. Standards address design, production, process control, and inspection of these assemblies.

MIL-STD-414: A set of specifications for acceptance sampling plans based on acceptable quality level (AQL) for variables data using the assumption that the variable is normally distributed.

Mind mapping: A technique for creating a visual representation of a multitude of issues or concerns by forming a map of the interrelated ideas.

Minimum feature configurations: Identify, using Marketing Features Availability List codes, the systems, subsystem, or components needed to verify a specific requirement.

M.I.S: Management information systems.

M.I.S. quality requirements: The process of organizing a project's objectives, strategies, and resources for the M.I.S. data systems.

Missing at random (MAR): Classification of *missing data* applicable when missing values of Y depend on X, but not on Y. When missing data are MAR, observed data for Y are a truly random sample for the X values in the sample, but not a random sample of all Y values due to missing values of X.

Missing completely at random (MCAR): Classification of *missing data* applicable when missing values of Y are not dependent on X. When missing data are MCAR, observed values of Y are a truly random sample of all Y values, with no underling process that lends bias to the observed data.

Missing data process: Any systematic event external to the respondent (such as data entry errors or data collection problems) or any action on the part of the respondent (such as refusal to answer a question) that leads to *missing data.*

Missing data: Information not available for a subject (or case) about which other information is available. Missing data often occur when a respondent fails to answer one or more questions in a survey or the researcher cannot obtain the data or he or she has lost the data.

Mission statement: An explanation of purpose or reasons for existing as an organization; it provides the focus for the organization and defines its scope of business.

Mitigation: The act of revising the project's scope, budget, schedule, or quality, preferably without material impact on the project's objectives, in order to reduce uncertainty on the project.

Mixed model: A model with random and fixed factors.

Mixture: A combination of two distinct populations. On control charts a mixture is indicated by an absence of points near the centerline.

Mode (of a statistical sample): The value of the sample variable that occurs most frequent.

Mode: Value that occurs most often or point of highest probability.

Model 1: In designs without full replication, an ANOVA model that assumes the presence of block-by-treatment interactions, even though the design has not allowed for their estimation. Randomized-block designs may be analyzed by Model 1 or Model 2. Repeated measures designs are generally analyzed by Model 1.

Model 2: In designs without full replication, an ANOVA model that assumes the absence of block by treatment interactions, even though the design has not allowed for any direct test of this assumption. Randomized-block designs may be analyzed by Model 1 or 2. Split-plot designs are generally analyzed by Model 2.

Model development strategy: Structural modeling strategy incorporating *model respecification* as a theoretically driven method of improving a tentatively specified model. This allows exploration of alternative model formulations that may be supported by theory. It does not correspond to an *exploratory approach* in which model respecifications are made atheoretically.

Model respecification: Modification of an existing model with estimated parameters to correct for inappropriate parameters encountered in the estimation process or to create a *competing model* for comparison.

Model—generic: Specified set of dependence relationships that can be tested empirically—an operationalization of a *theory*. The purpose of a model is to concisely provide a comprehensive representation of the relationships to be examined. The model can be formalized in a path diagram or in a set of structural equations.

Model—in an ANOVA: The hypothesized effect(s) on a response, which can be tested with ANOVA for evidence of pattern in the data. An ANOVA model contains one or more terms, each having an effect on the response that is tested against unmeasured error or residual variation. A model with a single factor (whether categorical or covariate) is written: $Y = A + \varepsilon$, and the ANOVA tests the term A against the residual ε. A fully replicated model with two crossed-factors is written: $Y = A + B + B^*A + \varepsilon$, and the two-way ANOVA tests each main effect A and B, and the interaction B^*A, against the residual ε. Models with multiple factors require care with declaring all terms in a statistical package. For example, the cross-factored nesting model: $Y = C|B'(A) + \varepsilon$ is analyzed by declaring the terms: $C|A + C|B(A)$. The two-factor randomized block model: $Y = S'|B|A$ is analyzed by declaring the terms: $S|B|A - S^*B^*A$ for a Model-1 analysis, or the terms: $S + B|A$ for a Model-2 analysis.

Moderator effect: Effect in which a third independent variable (the moderator variable) causes the relationship between a dependent/independent variable pair to change, depending on the value of the moderator variable. It is also known as an interactive effect and similar to the interaction effect seen in analysis of variance methods.

Modification indices: Values calculated for each unestimated relationship possible in a specified model. The modification index value for a specific unestimated relationship indicates the improvement in overall model fit (the reduction in the chi-square statistic) that is possible if a coefficient is calculated for that untested relationship. The researcher should use modification indices only as a guideline for model improvements of those relationships that can theoretically be justified as possible modifications.

Modified control limits: Control limits calculated from information other than the process's statistical variation, such as tolerances. Must be used cautiously, because the process could be working within its normal variation but show up on the control chart as out of control if limits do not account for that variation.

Moment of truth (MOT): Any event or point in a process when the external customer has an opportunity to form an opinion (positive, neutral, or negative) about the process or organization. A MOT is described by Jan Carlzon, former CEO of Scandinavian Air Services, in the 1980s as: "Any episode where a customer comes into contact with any aspect of your company, no matter how distant, and by this contact, has an opportunity to form an opinion about your company."

Monitoring actuals versus budget: One of the main responsibilities of Cost Management is to continually measure and monitor the actual cost versus the budget in order to identify problems, establish the variance, analyze the reasons for variance, and take the necessary corrective action. Changes in the forecast final cost are constantly monitored, managed, and controlled.

Monitoring: The capture, analysis, and reporting of actual performance compared to planned performance.

Monte Carlo simulation: A computer modeling technique to predict the behavior of a system from the known random behaviors and interactions of the system's component parts. A mathematical model of the system is constructed in the computer program, and the response of the model to various operating parameters, conditions, and so on can then be investigated. The technique is useful for handling systems whose complexity prevents analytical calculation

Monument: The point in a process that necessitates a product must wait in a queue before processing further; a barrier to continuous flow. Sometimes it is called "bottleneck" or "constraint."

Motivating: The process of inducing an individual to work toward achieving the organization's objectives while also working to achieve personal objectives.

Moving average moving range charts: A control chart that combines rational subgroups of data and the combined subgroup averages and ranges are plotted. Often used in continuous process industries, such as chemical processing, where single samples are analyzed.

MTBF (mean time between failures): Mean time between successive failures of a repairable product. This is a measure of product reliability.

Muda (Japanese): An activity that consumes resources but creates no value; seven categories are part of the Muda concept. They are correction, processing, inventory, waiting, overproduction, internal transport, and motion.

Multi attribute evaluation: Simpler than QFD, this process rank orders and weights customer requirements relative to the competition. In addition, it estimates the cost of each requirement in order to prioritize improvement actions.

Multi collinearity: Extent to which a variable can be explained by the other variables in the analysis. As multicollinearity increases, it complicates the interpretation of the *variate* as it is more difficult to ascertain the effect of any single variable, owing to their interrelationships. See also *Collinearity.*

Multidisciplinary team: Sometimes known as *Multiplication Table.* A team of people are brought together to work on the experiment. Each member has expertise or knowledge that is directly applicable to the achievement of the experiment's objective.

Multivoting: Narrowing and prioritization tool used in decision making. It enables a group to sort and narrow through a long list of ideas to identify priorities. Faced with a list of ideas, problems, causes, and so on, each member of a group is given a set number of "votes." Those receiving the most votes get further attention/consideration

Multivariate control chart: A control chart for evaluating the stability of a process in terms of the levels of two or more variables or characteristics.

Multidimensional data: *Data element* with multiple *dimensions* (attributes).

Multidimensional scaling (MDS): It is also known as perceptual mapping. It is a procedure that allows a researcher to determine the perceived relative image of a set of objects (firms, ideas, products, or other items associated with commonly held perception).

Multilayer perceptron: Most popular form of *neural network,* contains at least one *hidden layer* of *nodes* between input and output nodes.

Multiple correspondence analysis: Form of *correspondence analysis* that involves three or more categorical variables related in a common perceptual space.

Multiple regression: Regression model with two or more independent variables.

Multivariate analysis: Analysis of multiple variables in a single relationship or set of relationships.

Multivariate graphical display: Method of presenting a multivariate profile of an observation on three or more variables. The methods include approaches such as glyphs, mathematical transformations, and even iconic representations (e.g., faces).

Multivariate measurement: Use of two or more variables as *indicators* of a single *composite measure.* For example, a specific quality test may provide the answer(s) to a series of individual questions (indicators), which are then combined to form a single score *(summated scale)* representing the particular item that is being tested.

Multivariate normal distribution: Generalization of the univariate normal distribution to the case of p variables. A multivariate normal distribution of sample groups is a basic assumption required for the validity of the significance tests in MANOVA.

Myers-Briggs-type indicator/MBTI: A method and instrument for assessing personal "type" based on Carl Jung's theory of personality preferences.

Mystery shopper: A person who pretends to be a regular shopper in order to get an unencumbered view of how a company's service process works.

N: Population sample size.

n: Sample size (the number of units in a sample).

Natural team: A work group having responsibility for a particular process.

NDE: Nondestructive evaluation (see Nondestructive testing and Evaluation).

Nearly orthogonal: Characteristic of a stimuli *design* that is not *orthogonal,* but the deviations from orthogonality are slight and carefully controlled in the generation of the stimuli. This type of design can be compared with other stimuli designs with measures of *design efficiency.*

Near-term activities: Activities that are planned to begin, be in process, or be completed during a relatively short period of time, such as 30, 60, or 90 days.

Negotiating: The process of bargaining with individuals concerning the transfer of resources, the generation of information, and the accomplishment of activities.

Nested models: Models that have the same constructs but differ in terms of the number or types of *causal relationships* represented. The most common form of nested model occurs when a single relationship is added to or deleted from another model. Thus, the model with fewer estimated relationships is "nested" within the more general model.

Nesting: One factor is nested within another when each of its levels are tested in (or belong to) only one level of the other. For example, a response measured per leaf for a treatment factor applied across replicate bushes must declare the bushes as a random factor nested in the treatment levels. The sampling unit of Leaf, L, is then correctly nested in Bush, B, nested in Treatment, T. The model is: $Y = B(T) + \varepsilon$, where the residual error term ε refers to $L(B(T))$. This model is called in a statistics package by requesting the terms: $T + B(T)$ and declaring B as a random factor

Network diagram: A schematic display of the sequential and logical relationship of the activities that comprise the project. Two popular drawing conventions or notations for scheduling are arrow and precedence diagramming.

Networking: The exchange of information or services among individuals, groups, or institutions.

Neural network: Nonlinear predictive model that learns through *training*. It resembles the structure of biological neural systems.

Next operation as customer: Concept that the organization is comprised of service/product providers and service/product receivers or "internal customers."

Node: Basic "building block" of *neural networks* that can act as an input, output, or processing/analysis function. In project management it is defined as one of the defining points of a network; a junction point joined to some or all of the others by dependency lines.

Noise factors: Factors that cannot be controlled or which the engineer decides not to control due to reasons such as difficulty or expense.

Nominal: For a product whose size is of concern: the desired mean value for the particular dimension, the target value.

Nominal chart: A control chart that plots the deviation from the nominal value. Often used when individual samples are taken in short-run, low-volume processes. Allows multiple part numbers manufactured by similar processes to be plotted on the same control charts.

Nominal group technique: A technique similar to brainstorming, used by teams to generate ideas on a particular subject. Team members are asked to silently come up with as many ideas as possible, writing them down. Each member is then asked to share one idea, which is recorded. After all the ideas are recorded, they are discussed and prioritized by the group.

Noncausal part: Other parts replaced as a result of the causal part repair.

Nonconformance: A deficiency in characteristics, documentation, or procedure that renders the quality of material/service unacceptable or indeterminate.

Nonconforming unit: A sample (part) that has one or more nonconformities, making the sample unacceptable for its intended use. See also *Defective Unit*.

Nonconformity: A departure of a quality characteristic from its intended level or state. The nonfulfillment of a specified requirement. See also *defect, blemish, and imperfection*.

Nondestructive testing and evaluation (MDT): Testing and evaluation methods that do not damage or destroy the product being tested.

Nondestructive testing and evaluation (NDT&E): Testing and evaluation methods that do not damage or destroy the product being tested.

Nonhierarchical procedures: Procedures that produce only a single-cluster solution for a set of cluster seeds. Instead of using the tree-like construction process found in the *hierarchical procedures, cluster seeds* are used to group *objects* within a prespecified distance of the seeds. For example, if four cluster seeds are specified, only four clusters are formed. Nonhierarchical procedures do not produce results for all possible numbers of clusters as is done with a *hierarchical procedure*.

Nonlinearity (of a measuring instrument): The deviation of the instrument's response from linearity.

Nonmetric data: Also called *qualitative* data, these are attributes, characteristics, or categorical properties that identify or describe a subject or object. They differ from *metric data* by indicating the presence of an attribute, but not the amount. Examples are occupation (engineer, laboratory technician, operator, trainer) or buyer status (buyer, non-buyer). Also called *nominal data* or *ordinal data*.

Nonrecursive: Relationship in a path diagram indicating a mutual or reciprocal relationship between two constructs. Each construct in the pair has a causal relationship with the other construct. Depicted by a two-headed straight arrow between both constructs.

Non-value-added: Refers to tasks or activities that can be eliminated with no deterioration in product or service functionality, performance, or quality in the eyes of the customer.

Non-value-adding activities: Steps/tasks in a process that do not add value to the external customer and do not meet all three criteria for value adding; includes rework, handoffs, inspection/control, wait/delays, and so on. See also *Value-adding Activities*.

Nonverbal communication: Involving minimal use of the spoken language: gestures, facial expressions, and verbal fragments that communicate emotions without the use of words; sometimes known as body language.

Nonwork unit: A calendar unit during which work may not be performed on an activity, such as weekends and holidays.

Normal distribution: Purely theoretical continuous probability distribution in which the horizontal axis represents all possible values of a variable and the vertical axis represents the probability of those values occurring. The scores on the variable are clustered around the mean in a symmetrical, unimodal pattern known as the bell-shaped, or normal, curve. Furthermore, it is the most commonly used probability distribution in the shape of a bell. It is because of this shape that sometimes it is referred as the "bell-shape" distribution. The normal distribution is a good approximation for a large class of situations. It is a distribution for continuous data and where most of the data are concentrated around the average, and it is equally likely that an observation will occur above or below the average. One example is the distribution resulting from the random additions of a large number of small variations. The Central Limit Theorem expresses this for the distribution of means of samples; the distribution of means results from the random additions of a large number of individual measurements, each of which contributes a small variation of its own. The normal distribution (or Gaussian distribution) is exhibited by many naturally occurring variables and its predictive properties are used extensively in statistical analysis. It is significant to know that in this kind of distribution the average (mean), the middle, and the mode are the same.

Normal plots: These are a useful graphical tool for (among other uses) helping to identify which factor effects are likely to be real (or active). Normal plots are used to determine the normality of a set of data. Factor effects that do not follow the expected normal distribution are assumed to have resulted from changing factor levels (as opposed to random fluctuations).

Normal probability plot: Graphical comparison of the form of the distribution to the *normal distribution*. In the normal probability plot, the normal distribution is represented by a straight line angled at 45 degrees. The actual distribution is plotted against this line, so that any differences are shown as deviations from the straight line, making identification of differences quite apparent and interpretable.

Normality: Degree to which the distribution of the sample data corresponds to a *normal distribution*.

Normal score: These are expected ordered values for the standard normal distribution (i.e., having a mean of 0 and a standard deviation of 1). They form the x-axis locations of a normal plot. The scores can be obtained from existing tables.

Normalized distance function: Process that converts each raw data score to a standardized variate with a mean of 0 and a standard deviation of 1, to remove the bias introduced by differences in scales of several variables.

Norms: Behavioral expectations, mutually agreed upon rules of conduct, protocols to be followed, social practice.

Notation: In the statistical world there are some basic notations for simple communication for intermediate and advanced meanings. Here we present some of those notations.

Symbol	Meaning
Y	Continuous response variable.
A, B, C	Fixed factor (e.g., Treatment A of watering regime).
A′, B′, C′	Random factor (e.g., Treatment B′ of crop genotype).
a, b, c	Number of sample levels of factor A, B, C (e.g., factor A may have $a = 2$ levels, corresponding to 'low' and 'high').
S′, P′, Q′, R′	Random factor representing randomly selected subjects/blocks (S′), plots (P′), subplots (Q′), or sub-sub-plots (R′), to which treatments are applied.
$S_i, P_i, Q_i, R_i,$	Independent and randomly chosen subject/block, plot, subplot or sub-sub-plot that provides a replicate observation of the response.
n	The size of each sample, given by the number of measures of the response in each combination of factor levels (including any repeated measures), or by the number of measures across all values of a covariate.
N	Total number of measures of the response across all factor levels.
B′(A)	Hierarchical nesting of one factor in another (here, B′ is nested in A).
B*A	Interaction between factors in their effects on the response (here, interaction of B with A).
B∣A	Cross-factoring, with analysis of interaction(s) and main effects (here, B cross-factored with A, and analysis of A + B + B*A).
ε	Residual variation left unexplained by the model, taking the form S′(...), P′(...), Q′(...) or R′(...).
Y = C∣B∣A + ε	Full model (here, variation in Y around the grand mean partitions amongst the three main effects A, B, C plus the three two-way interactions B*A, C*A, C*B plus the one three-way interaction C*B*A, plus the unexplained residual (error) variation ε = S′(C*B*A) around each sample mean. This would not be a full model if only main effects were tested, or only main effects and two-way interactions).

np-Chart: For attributes data: a control chart of the number of defective units in a subgroup. Assumes a constant subgroup size.

NPV (net present value): A discounted cash flow technique for finding the present value of each future year's cash flow.

Null hypothesis: Hypothesis that samples come from populations with equal means for either a dependent variable (univariate test) or a set of dependent variables (multivariate test). The null hypothesis can be accepted or rejected depending on the results of a test of statistical significance.

Null model: Baseline or comparison standard used in *incremental fit indices.* The null model is hypothesized to be the simplest model that can be theoretically justified. The most common example is a single *construct* model related to all *indicators* with no *measurement error.*

Null plot: Plot of residuals versus the predicted values that exhibits a random pattern. A null plot is indicative of no identifiable violations of the assumptions underlying regression analysis.

Number of affected units chart (np-chart): A control chart for evaluating the stability of a process in terms of the total number of units in a sample in which frequency of an event of a given classification occurs.

NVH: Noise, vibration, harshness

Object: Any stimulus, including tangible entities (product or physical object), actions (service), sensory perceptions (smell, taste, sights), or even thoughts (ideas, slogans), which can be compared and evaluated by the respondent. It may also be a person, product, or service, firm, or any other entity that can be evaluated on a number of attributes.

Objective (time management): A predetermined result; the end toward which *effort* is directed (Contract/Procurement Management). To define the method to follow and the service to be contracted or resource to be procured for the performance of work.

Objective: The primary reason for attempting the experiment, that is, a description of the outcome the experiment is designed to achieve. This may be the elimination of an existing problem or the achievement of a desirable goal. On the other hand, it may also be a quantitative statement of future expectations and an indication of when the expectations should be achieved. It flows from goals and clarifies what people must accomplish.

Objective dimension: Physical or tangible characteristics of an *object* that have an objective basis of comparison. For example, a product has size, shape, color, weight, and so on.

Objective evidence: Verifiable qualitative or quantitative observations, information, records, or statements of fact pertaining to the quality of an item or service or to the existence and implementation of a quality system element.

Observation: An item of objective evidence found during an audit.

Oblique factor rotation: *Factor rotation* computed so that the extracted factors are correlated. Rather than arbitrarily constraining the factor rotation to an *orthogonal* solution, the oblique rotation identifies the extent to which each of the factors are correlated.

OC (operating characteristic) curve: For a sampling plan, the OC curve indicates the probability of accepting a lot based on the sample size to be taken and the fraction defective in the batch.

Offending estimates: Any value that exceeds its theoretical limits. The most common occurrences are *Heywood cases* with negative error variances (the minimum value should be zero, indicating no measurement error) or very large standard errors. The researcher must correct the offending estimate with one of a number of remedies before the results can be interpreted for overall model fit and the individual coefficients can be examined for statistical significance.

Off-the-job training: Training that takes place away from the actual work site.

One factor at a time testing: The traditional approach to experimentation where different levels of a factor are compared under fixed conditions. That is, with all other factors kept at a constant level.

One-to-one marketing: The concept of knowing customers' unique requirements and expectations and marketing to these (see also Customer relationship management).

Online analytical processing (OLAP): Database application that allows users to view, manipulate, and analyze *multidimensional data.*

On-the-job-training (OJT): Training conducted usually at the workstation, typically done one-on-one.

Open book management: An approach to managing that exposes employees to the organization's financial information, provides instruction in business literacy, and enables employees to better understand their role and contribution and its impact on the organization.

Operating characteristic curve: For acceptance sampling: a curve showing the probability of accepting a lot versus the percentage of defective units in the lot.

Operation: The operation of a new facility is described by a variety of terms, each depicting an EVENT in its early operating life. These are defined below, in chronological order: (1) *Initial Operation:* the Project Milestone date on which material is first introduced into the system for the purpose of producing products, and (2) *Normal Operation:* the project milestone date on which the facility has demonstrated the capability of sustained operations at design conditions and the facility is accepted by the client.

Operational data: Data used by operational systems (e.g., accounting, inventory management) that reflect the current status of the organization that is necessary for day-to-day business functions.

Operational definition: A clear, precise description of the factor being measured or the term being used; ensures a clear understanding of terminology and the ability to operate a process or collect data consistently.

Operational systems: Information systems that service the basic functions of the organization (accounting, inventory management, order processing) for their day-to-day operations.

Opportunity: Any event that generates an output (product, service, or information).

Optimal design: Stimuli *design* that is *orthogonal* and *balanced.*

Optimization: Refers to achieving planned process results that meet the needs of the customer and supplier alike and minimize their combined costs. Also, it is the third element of the DCOV model.

Optimizing procedure: *Nonhierarchical clustering* procedure that allows for the reassignment of *objects* from the originally assigned cluster to another cluster on the basis of an overall optimizing criterion.

Order condition: Requirement for *identification* that the model's *degrees of freedom* must be greater than or equal to zero.

Order of magnitude (–25, + 75 Percent): This is an approximate *estimate* made without detailed data, which is usually produced from cost capacity curves, scale up or down factors that are appropriately escalated and approximate cost–capacity ratios. This type of estimate is used during the formative stages of an expenditure program for initial evaluation of the project. Other terms commonly used to identify an Order of Magnitude estimate are preliminary, conceptual, factored, quickie, feasibility, and SWAG.

Ordinal interaction: Acceptable type of *interaction effect* in which the magnitudes of differences between groups vary, but the groups' relative positions remain constant. It is graphically represented by plotting mean values and observing nonparallel lines that do not intersect.

Organization development (OD): An organization-wide (usually) planned effort, managed from the top, to increase organization effectiveness and health through interventions in the organization's processes; using behavioral science knowledge, technology, research, and theory to change an organization's culture to meet predetermined objectives involving participation, joint decision making, and team building.

Organization Structure: Identification of participants and their hierarchical relationships.

Organizational Politics: The informal process by which personal friendships, loyalties, and enmities are used in an attempt to gain an advantage in influencing project decisions.

Original duration: The first estimate of work time needed to execute an activity. The most common units of time are hours, days, and weeks.

Orthogonal array: A matrix of levels arranged in rows and columns. Each row represents the combination of factor-level settings in a given experimental run. Each column represents a specific factor that can be changed from experiment to experiment. When a factor has not been allocated to a column, the column can be used to estimate an interaction effect. The array is called orthogonal because the effect of each factor on the experimental results can be separated.

Orthogonal contrasts for analysis of variance are independent comparisons between the groups of a factor with at least three fixed levels. The sum of squares for a factor A with a levels is partitioned into a set of $a - 1$ orthogonal contrasts each with two levels (so each has $p = 1$ test degree of freedom), to be tested against the same error MS as for the factor. Each contrast is assigned a coefficient at each level of A such that its a coefficients sum to zero, with coefficients of equal value indicating pooled levels of the factor, coefficients of opposite sign indicating factor levels to be contrasted, and a zero indicating an excluded factor level. With this numbering system, two contrasts are orthogonal to each other if the products of their coefficients sum to zero.

Orthogonal factor rotation: Factor rotation in which the factors are extracted so that their axes are maintained at 90 degrees. Each factor is independent of, or *orthogonal* to, all other factors. The correlation between the factors is determined to be 0.

Orthogonal: Statistical independence or absence of association. This means that there is a mathematical independence (no correlation) of factor axes to each other (that is at right angles, or 90 degrees). Orthogonal *variates* explain unique variance, with no variance explanation shared between them. Orthogonal *contrasts* are *planned comparisons* that are statistically independent and represent unique comparisons of group means. In conjoint analysis, orthogonality refers to the ability to measure the effect of changing each attribute level and to separate it from the effects of changing other attribute levels and from experimental error.

Orthogonality: A cross-factored design is orthogonal if each of its factors are independent of each other. Two categorical cross factors are orthogonal by design if each level of one is measured at every level of the other. Orthogonal designs partition total variation in the response straightforwardly into testable components using sequential sums of squares for each effect in turn. Although a balanced design generally (but not inevitably) ensures orthogonality, this can be difficult to achieve in practice, especially with covariates. Two covariates are only orthogonal if they have a correlation coefficient of zero. Loss of orthogonality can reduce or enhance the power of a design to detect effects, and usually requires analysis with the aid of adjusted sums of squares calculated in a General Linear Model (GLM).

Outlier: In strict terms, an observation that has a substantial difference between the actual value for the dependent variable and the predicted value. Cases that are substantially "different," with regard to either the dependent or independent variables, are often termed outliers as well. In all instances, the objective is to identify observations that are inappropriate representations of the population from which the sample is drawn, so that they may be discounted or even eliminated from the analysis as unrepresentative. At issue is its representativeness of the population.

Out-of-control process: A process in which the statistical measure being evaluated is not in a state of statistical control (i.e., the variations among the observed sampling results cannot all be attributed to a constant system of chance causes; special or assignable causes exist. See *also In-control process.*

Out-of-control: Used to characterize the state of a process. In other words, if a process exhibits variations larger than the control limits is said to be out of control. Also, a process is said to be "out-of-control" if its data violate one or more of the run rules on a run chart. In warranty there may be special causes or common cause shifts that are in effect. These need to be understood and addressed (if they are in the unfavorable direction). Usually, "out-of-control" signals are first noted by in-plant or other upstream indicators, in advance of receiving the warranty data. A special note here for the reader. The process may be out of control but be in specification and vice versa. The reason for this is that specifications are defined by the customer whereas the control limits are the voice of the process.

Output: Any product, service, or piece of information coming out of, or resulting from, the activities in a process.

Output measures: Measures related to and describing the output of the process; total figures/overall measures.

Outsourcing: A strategy to relieve an organization of processes and tasks in order to reduce costs, improve quality, reduce cycle time (e.g., by parallel processing), reduce the need for specialized skills, and increase efficiency.

Overall quality philosophy: The universal belief and performance throughout the company, based on established quality policies and procedures. Those policies and procedures become the basis for collecting facts about a project in an orderly way for study (statistics).

Overidentified model: Structural model with a positive number of *degrees of freedom*, indicating that some level of generalizability may be possible. The objective is to achieve the maximum model fit with the largest number of degrees of freedom.

Owned requirement: A requirement assigned to a specific DVP Team that is responsible for creating the Design Verification Plan.

p-Chart: Fraction defective chart (also called a proportion chart; percent defective). For attributes data: a control chart of the percentage of defective units (or fraction defective) in a subgroup.

P Diagram: A schematic representation of the relationship among signal factors, control factors, noise factors, and the measured response.

Pairwise comparison method: Method of presenting a pair of stimuli to a respondent for evaluation, with the respondent selecting one stimuli as preferred.

PALS: Product attribute leadership strategy.

Panels: Groups of customers recruited by an organization to provide *ad hoc* feedback on performance or product development ideas.

Parallel structure: Describes an organizational module in which groups, such as quality circles or a quality council, exist in the organization in addition to and simultaneously with the line organization (also referred to as collateral structure).

Parallel threshold method: *Nonhierarchical clustering* procedure that selects the cluster seeds simultaneously in the beginning. *Objects* within the threshold distances are assigned to the nearest seed. Threshold distances can be adjusted to include fewer or more objects in the clusters. This method is the opposite of the *sequential threshold method.*

Parameter: Quantity (measure) characteristic of the population. For example, μ, and σ^2 are the symbols used for the population parameters mean (μ) and variance (σ^2). These are typically estimated from sample data in which the arithmetic average of the sample is used as a measure of the population average and the variance of the sample is used to estimate the variance of the population.

Parameter design (Taguchi): The use of orthogonal arrays in the design of experiments for identifying the major contributors to variation. Emphasis is on factor contribution rather than interactions.

Parameter design: Parameter design achieves robustness with no increase in the product/manufacturing cost by running a statistical experiment with lowest-cost components, selecting optimum values for control factors that will reduce the variability of the response and shift the mean of the response toward target, and by selecting the lowest cost setting for factors that have minimal effect on the response.

Parametric cost estimating: An estimating methodology using statistical relationships between historical costs and other project variables such as system physical or performance characteristics, contractor output measures, or manpower loading, and so on. Also referred to as "top down" estimating.

Pareto analysis: An analysis of the frequency of occurrence of various possible concerns. This is a useful way to decide quality control priorities when more than one concern is present. The underlying "Pareto Principle" states that a very small number of concerns is usually responsible for most quality problems.

Pareto chart: A graphical display of the relative contribution of various factors to the total; used to prioritize the most important factors. Typically a few causes account for most of the cost (or variation), so problem-solving efforts are best prioritized to concentrate on the vital few causes, temporarily ignoring the trivial many. It is one of the basic quality tools used to graphically rank causes from most significant to least significant. It utilizes a vertical bar graph in which the bar height reflects the frequency or impact of causes. Sometimes it is also referred to as Pareto Diagram.

Pareto diagram: See *Pareto chart*

Pareto diagrams (quality): A graph, particularly popular in nontechnical projects, to prioritize the few change areas (often 20% of the total) that cause most quality deviations (often 80% of the total) in a descending order.

Pareto principle: The 80/20 rule; based on Alfredo Pareto's research stating that the vital few (20%) of causes have a greater impact than the trivial many (80%) causes with a lesser impact.

Parsimonious fit measure: Measure of overall *goodness-of-fit* representing the degree of model fit per estimated coefficient. This measure attempts to correct for any "overfitting" of the model and evaluates the *parsimony* of the model compared to the *goodness-of-fit*.

Parsimony: Degree to which a model achieves *goodness-of-fit* for each estimated coefficient. The objective is not to minimize the number of coefficients or to maximize the fit, but to maximize the amount of fit per estimated coefficient and avoid "overfitting" the model with additional coefficients that achieve only small gains in model fit.

Part correlation: Value that measures the strength of the relationship between a dependent and a single independent variable when the predictive effects of the other independent variables in the regression model are removed. The objective is to portray the *unique* predictive effect due to a single independent variable among a set of independent variables. Differs from the *partial correlation coefficient,* which is concerned with incremental predictive effect.

Partial correlation coefficient: Value that measures the strength of the relationship between the criterion or dependent variable and a single independent variable when the effects of the other independent variables in the model are held constant. For example, r_{y,x_2,x_1} measures the variation in Y associated with X_2 when the effect of X_1 on both X_2 and Y is held constant. This value is used in sequential variable selection methods of regression model estimation to identify the independent variable with the greatest incremental predictive power beyond the independent variables already in the regression model.

Partial F (or t) values: The partial F-test is simply a statistical test for the additional contribution to prediction accuracy of a variable above that of the variables already in the equation. When a variable (X_a) is added to a regression equation after other variables are already in the equation, its contribution may be very small even though it has a high correlation with the dependent variable. The reason is that X_a is highly correlated with the variables already in the equation. The partial F-value is calculated for all variables by simply pretending that each, in turn, is the last to enter the equation. It gives the additional contribution of each variable above all others in the equation. A low or insignificant partial F-value for a variable not in the equation indicates its low or insignificant contribution to the model as already specified. A t-value may be calculated instead of F-values in all instances, with the t-value being approximately the square root of the F-value.

Partial regression plot: Graphical representation of the relationship between the dependent variable and a single independent variable. The scatterplot of points depicts the partial correlation between the two variables, with the effects of other independent variables held constant (see Partial correlation coefficient). This portrayal is particularly helpful in assessing the form of the relationship (linear versus nonlinear) and the identification of *influential observations.*

Participative management: A style of managing whereby the manager tends to work from theory Y assumptions about people, involving the workers in decisions made.

Partnership/alliance: A strategy leading to a relationship with suppliers or customers aimed at reducing costs of ownership, maintenance of minimum stocks, just-in-time deliveries, joint participation in design, exchange of information on materials and technologies, new production methods, quality improvement strategies, and the exploitation of market synergy.

Part-worth: Estimate from conjoint analysis of the overall preference or *utility* associated with each *level* of each *factor* used to define the product or service.

Path analysis: Method that employs simple bivariate correlations to estimate the relationships in a system of structural equations. The method is based on specifying the relationships in a series of regression-like equations (portrayed graphically in a *path diagram*) that can then be estimated by determining the amount of correlation attributable to each effect in each equation simultaneously. When employed with multiple relationships among *latent constructs* and a *measurement model,* it is then termed *structural equation modeling.*

Path diagram: Graphical portrayal of the complete set of relationships among the model's constructs. *Causal relationships* are depicted by straight arrows, with the arrow emanating from the predictor variable and the arrowhead "pointing" to the dependent *construct* or variable. Curved arrows represent correlations between constructs or indicators, but no causation is implied.

Payback period: The number of years it will take the results of a project or capital investment to recover the investment from net cash flows.

PDF: Probability density function.

PDM finish to finish relationship: This relationship restricts the finish of the work activity until some specified duration following the finish of another work activity.

PDM finish to start relationship: The relationship where the work activity may start just as soon as another work activity is finished.

PDM start to finish felationship: The relationship restricts the finish of the work activity until some duration following the start of another work activity.

PDM start to start relationship: This relationship restricts the start of the work activity until some specified duration following the start of the preceding work activity.

PDM: See *Precedence diagram method.*

PDSA Cycle: PDSA is short for Plan, Do, Study, and Act (sometimes called the PDCA cycle for Plan, Do, Check, and Act). It is a cyclical methodology that can be applied to the process of gaining more knowledge about a system. The cycle was formally defined by the American quality consultant, W. Edwards Deming and the statistician, Walter A. Shewhart.

Perceived dimension: A respondent's subjective attachment of features to an *object* that represents its intangible characteristics. Examples include "quality," "expensive," and "good-looking." These perceived dimensions are unique to the individual respondent and may bear little correspondence to actual *objective dimensions.*

Percent complete: A ratio comparison of the completion status to the current projection of total work.

Percent defective: For acceptance sampling: the percentage of units in a lot that are defective, that is, of unacceptable quality.

Percentile customer: A level of customer usage that exceeds that generated by *n* percent of customers. In other words, *n* percent of customers will use or stress products less than the nth percentile level. Typical values of *n* are 90 and 95. Sometimes this term is used generically to describe severe customer usage conditions under which designs should be tested for survival. (It must be emphasized that designing for the nth percentile customer is generally appropriate only for wear-out-type failures. Care should be taken to identify the nth percentile customer within the market where a particular failure mode is likely to occur, for example, hot or cold climates, high or low humidity regions, etc.)

Perceptual map: Visual representation of a respondent's perceptions of *objects* on two or more dimensions. Usually this map has opposite levels of *dimensions* on the ends of the X- and Y-axis, such as "sweet" to "sour" on the ends of the X-axis and "high-priced" to "low-priced" on the ends of the Y-axis. Each *object* then has a spatial position in the perceptual map that reflects the relative *similarity* or *preference* to other objects with regard to the dimensions of the perceptual map. This may be a good tool to evaluate suppliers.

Performance appraisal: A formal method of measuring employees' progress against performance standards and providing feedback to them.

Performance control: Control of work during contract execution.

Performance evaluation: The formal system by which managers evaluate and rate the quality of subordinates' performance over a given period of time.

Performance management system: A system that supports and contributes to the creation of high-performance work and work systems by translating behavioral principles into procedures.

Performance plan: A performance management tool that describes desired performance and provides a way to assess the performance objectively.

Performance study: Analysis of a process to determine the distribution of a run. The process may or may not be in statistical control.

Performance: The calculation of achievement used to measure and manage project quality.

Personal recognition: The public acknowledgment of an individual's performance on the project.

Personal rewards: Providing an individual with psychological or monetary benefits in return for his or her performance.

Personnel training: The development of specific job skills and techniques required by the individual to become more productive.

Persuade: To advise, to move by argument, entreaty, or expostulation to a belief, position, or course of action.

PERT: Program evaluation and review technique. An event- and probability-based network analysis system generally used in the research and development field where, at the planning stage, activities and their durations between events are difficult to define. Typically used on large programs where the projects involve numerous organizations at widely different locations.

Phase: See *Project phase*.

Pilot: Trial implementation of a solution, on a limited scale, to ensure its effectiveness and test its impact; an experiment verifying a root cause hypothesis.

Plan: An intended future course of action.

Plan development: Stage of planning during which the plan is initially created.

Plan matrix: In a designed experiment, the combination of factors and levels are assigned to a matrix that specifies the setting of factors for particular experimental runs. This matrix is known as the Plan Matrix. Alternative names given to this matrix are the Experimental Design and Experimental Layout.

Plan-do-check-act cycle (PDCA): A four-step process for quality improvement. In the first step (plan), a plan to effect improvement is developed. In the second step (do), the plan is carried out, preferably on a small scale. In the third step (check), the effects of the plan are observed. In the last step (act), the results are studied to determine what was learned and what can be predicted. The plan-do-check-act cycle is sometimes referred to as the Shewhart cycle because Walter A. Shewhart discussed the concept in his book *Statistical Method from the Viewpoint of Quality Control* and as the Deming cycle because W. Edwards Deming introduced the concept in Japan. The Japanese subsequently called it the Deming cycle.

Plan-do-study (check)-act, or (PDS(C)A): a Variation of the basic model or set of steps in continuous improvement; also referred to as "Shewhart Cycle," PDCA or "Deming Cycle."

Plan-do-study-act (PDSA): The same basic model as the PDCA except instead of "Check," Deming in the early 1980s replaced it with "study" to denote action rather than complacency.

Planned activity: An activity that has not started or finished prior to the Data Date.

Planned comparison: *A priori test* that tests a specific comparison of group mean differences. These tests are performed in conjunction with the tests for *main* and *interaction effects* by using a *contrast.*

Platykurtosis: For frequency distributions: a distribution that has longer "tails" than a normal distribution with the same standard deviation.

Point estimate (Statistics): A single-value estimate of a population parameter. Point estimates are commonly referred to as the points at which the interval estimates are centered; these estimates give information about how much uncertainty is associated with the estimate.

Poisson distribution: A probability distribution for the number of occurrences of an event; n = number of trials; p = probability that the event occurs for a single trial; r = the number of trials for which the event occurred. The Poisson distribution is a good approximation of the binomial distribution for a case where p is small. A simpler way to say this is: A distribution used for discrete data is applicable when there are many opportunities for occurrence of an event but with a low probability (less than 0.10) on each trial.

Poka-yoke: A term that means to mistake-proof a process by building safeguards into the system that avoid or immediately find errors. It comes from *poka,* which means "error," and *yokeru,* which means "to avoid." Also known as mistake proofing and error proofing.

PM function: See *Function.*

PM: See *Project Management.*

Policies/procedures: See *Project policies.*

Policy: Directives issued by management for guidance and direction where uniformity of action is essential. Directives pertain to the approach, techniques, authorities, and responsibilities for carrying out the management function.

Polychoric correlation: Measure of association employed as a replacement for the product–moment correlation when both variables are ordinal measures with three or more categories.

Polynomial: Transformation of an independent variable to represent a curvilinear relationship with the dependent variable. By including a squared term (X^2), a single inflection point is estimated. A cubic term estimates a second inflection point. Additional terms of a higher power can also be estimated.

Polyserial correlation: Measure of association used as a substitute for the product–moment correlation when one variable is measured on an ordinal scale and the other variable is metrically measured.

PONC: Price of nonconformance. The cost of not doing things right the first time.

Pooling: The construction of an error term from more than one source of variance in the response. *A priori* pooling occurs in designs without full replication, where untestable interactions with random factors are pooled into the residual variation. The analysis then proceeds on the assumption that the interactions are either present (Model 1) or absent (Model 2). Planned *post hoc* pooling is applied to mixed models by pooling a nonsignificant error term with its own error term. The design is thereby influenced by the outcome of the analysis (in terms of whether or not an error term is itself significant). More generally, pooling can describe the process of joining together samples, for example, in calculating a main effect Mean Square by pooling across levels of a cross factor.

Population (general): A group of people, objects, observations, or measurements about which one wishes to draw conclusions.

Population (statistical): The set of all possible outcomes of a statistical determination. The population is usually considered as an essentially infinite set from which a subset called a sample is selected to determine the characteristics of the population, that is, if a process were to run for an infinite length of time, it would produce an infinite number of units. The outcome of measuring the length of each unit would represent a statistical universe, or population. Any subset of the units produced (say, a hundred of them collected in sequence) would represent a sample of the population. Also known as universe.

Population: A group of people, objects, observations, or measurements about which one wishes to draw conclusions.

***Post hoc* test:** Statistical test of mean differences performed after the statistical tests for *main effects* have been performed. Most often, *post hoc* tests do not use a single *contrast*, but instead test for differences among all possible combinations of groups. Even though they provide abundant diagnostic information, they do inflate the overall *Type I* error rate by performing multiple statistical tests and thus must use very strict confidence levels.

Post processing: Processing of data done after it is collected, usually by computer.

Postproject analysis and report: A formal analysis and documentation of the project's results including cost, schedule, and technical performance versus the original plan.

Postproject evaluation: An appraisal of the costs and technical performance of a completed project and the development of new applications in Project Management methods to overcome problems that occurred during the project life to benefit future projects.

Power: Probability of correctly rejecting the null hypothesis when it is false, that is, correctly finding a hypothesized relationship when it exists. Power is defined as $(1 - \beta)$ and is determined as a function of (1) the statistical significance level (alpha, α) set by the researcher for a *Type I error*, (2) the sample size used in the analysis, and (3) the *effect size* being examined. See *Beta*.

P_{pk}: Potential process capability used in the validation stage of a new product launch (uses the same formula as C_{pk}, but a higher value is expected due to the smaller time span of the samples).

PPM: Parts per million.

Practical significance: Means of assessing multivariate analysis results based on their substantive findings rather than their statistical significance. Whereas statistical significance determines whether the result is attributable to chance, practical significance assesses whether the result is useful (i.e., substantial enough to warrant action).

Precedence diagram method (PDM): A method of constructing a logic network using nodes to represent the activities and connecting them by lines that show dependencies.

Precedence diagram method arrow: A graphical symbol in PDM networks used to represent the LAG describing the relationship between work activities.

Precision (of measurement): The extent to which repeated measurement of a standard with a given instrument yields the same result. A characteristic of measurement that addresses the consistency or repeatability of a measurement system when the identical item is measured a number of times.

Precision: A characteristic of measurement that addresses the consistency or repeatability of a measurement system when the identical item is measured a number of times. Another way of looking at it is the accuracy of the measure you plan to do. This links to the type of scale or detail of your operational definition, but it can have an impact on your sample size, too.

Precontrol: A method of controlling a process based on the specification limits. It is used to prevent the manufacture of defective units, but does not work toward minimizing variation of the process. The area between the specifications are split into zones (green, yellow, and red) and adjustments made when a specified number of points fall in the yellow or red zones. A control process, with simple rules, based on tolerances. It is effective for any process where a worker can measure a quality characteristic (dimension, color, strength, etc.) and can adjust the process to change that characteristic, and where there is either continuous output or discrete output totaling three or more pieces.

Predecessor activity: Any activity that exists on a common path with the activity in question and occurs before the activity in question.

Prediction equation: An equation that can predict an estimate of a response with factors set at predetermined levels.

Predictive validity: See *Criterion validity*.

Predictor variable (X_n): See *Independent variable*.

Preference structure: Representation of both the relative importance or worth of each *factor* and the impact of individual *levels* in affecting utility.

Preference: Implies that *objects* are judged by the respondent in terms of dominance relationships; that is, the stimuli are ordered in preference with respect to some property. Direct ranking, paired comparisons, and preference scales are frequently used to determine respondent preferences.

Preliminary plan: Used when developing milestones for team activities related to process improvement; includes key tasks, target completion dates, responsibilities, potential problems, obstacles and contingencies, and communication strategies.

Prerequisite tree: A technique used in the application of Goldratt's Theory of Constraints.

Prescribe: To direct specified action. To prescribe implies that action must be carried out in a specified fashion.

PRESS statistic: Validation measure obtained by eliminating each observation one at a time and predicting this dependent value with the regression model estimated from the remaining observations.

Prevention costs: Costs incurred to keep internal and external failure costs and appraisal costs to a minimum.

Prevention versus detection: A term used to contrast two types of quality activities. Prevention refers to those activities designed to prevent nonconformances in products and services. Detection refers to those activities designed to detect nonconformances already in products and services. Another term used to describe this distinction is "designing in quality vs. inspecting in quality."

Prevention: A strategy for preventing concerns from occurring by using disciplined methods to assure that specific past concerns are not repeated and the processes (procedures, acceptance criteria, test/analytical methods) that allowed those concerns to occur are corrected to permanently prevent other similar concerns from occurring.

Preventive action: Action taken to eliminate the causes of a potential non-conformity; defect, or other undesirable situation in order to prevent occurrence.

Primary reference standard: For measurements: a standard maintained by the National Bureau of Standards for a particular measuring unit. The primary reference standard duplicates as nearly as possible the international standard and is used to calibrate other (transfer) standards, which in turn are used to calibrate measuring instruments for industrial use. It is the source for traceability.

Primary: Process that refers to the basic steps or activities that will produce the output without the "nice-to-haves."

Primitive data: Data elements maintained at their lowest level of detail, such as individual transactions.

Priorities: The imposed sequences desired with respect to the scheduling of activities within previously imposed constraints.

Priorities matrix: A tool used to choose between several options that have many useful benefits, but where not all of them are of equal value.

Probability: Refers to the likelihood of occurrence. The chance of an event happening (the number of ways a particular event can happen relative to the total number of possible outcomes).

Probability (Mathematical): The likelihood that a particular occurrence (event) has a particular outcome. In mathematical terms, the probability that outcome X occurs is expressed by the formula: $P(x) = $ (number of trials giving outcome x/total number of trials). Note that, because of this definition, summing up the probabilities for all values of X always gives a total of 1: This is another way of saying that each trial must have exactly one outcome.

Probability distribution: A relationship giving the probability of observing each possible outcome of a random event. The relationship may be given by a mathematical expression, or it may be given empirically by drawing a frequency distribution for a large enough sample. In essence, it is a mathematical formula that relates the values of characteristics to their probability of occurrence in a population.

Probit analysis: Probit analysis is a type of regression used to analyze binomial response variables. It transforms the sigmoid dose-response curve to a straight line that can then be analyzed by regression either through least squares or maximum likelihood. It can be conducted by one of three techniques: (a) Using tables to estimate the probits and fitting the relationship by eye; (b) hand-calculating the probits, regression coefficient, and confidence intervals; or (c) using a software statistical package such as Minitab, SPSS, and others to do it all for you.

Problem resolution: The interaction between the Project Manager and an individual team member with the goal of finding a solution to a technical or personal problem that affects project accomplishment.

Problem solving: A rational process for identifying, describing, analyzing, and resolving situations in which something has gone wrong without explanation.

Problem/need statement/goal: Documentation to define the problem, to document the need to find a solution, and to document the overall aim of the sponsor.

Problem/opportunity statement: Description of the symptoms or the "pain" in the process; usually written in noun–verb structure: usually included in a team charter and supported with numbers and more detail once data have been obtained. See also *Charter*.

Procedure: A prescribed method of performing specified work. A document that answers the questions: What has to be done? Where is it to be done? When is it to be done? Who is to do it? Why do it? (contrasted with a work instruction that answers: How is it to be done? With what materials and tools is it to be done?); in the absence of a work instruction, the instructions may be embedded in the procedure.

Process: There are many ways to define a process. However, the most common are (a) the combination of people, equipment, materials, methods, measurement, and environment; (b) an activity or group of activities that takes an input, adds value to it, and provides an output to an internal or external customer; (c) a planned and repetitive sequence of steps by which a defined product or service is delivered; and (d) a process where the energy transformation (or value added) takes place. A process can involve any aspect of the business from Manufacturing and Assembly to Management Decision Making. A key tool for managing processes is statistical process control. In manufacturing we talk about a process being a task that includes singularly or in combination one of the following: Manpower, Machine, Method, Material, Measurement, and Environment. In service we talk about a process being a task that includes singularly or in combination one of the following: Manpower, Environment, Procedure, Policy, Method, and Material.

Process (framework): The set of activities by means of which an output is achieved. A series of actions or operations that produce a result (especially, a continuous operation).

Process analysis diagram: A cause-and-effect diagram for a process. Each step of the process and the factors contributing to it are shown, indicating all cause-and-effect relationships. This allows systematic tracing of any problems that may arise, to identify the source of the problem.

Process capability: Determination of whether a process, with normal variation, is capable of meeting customer requirements; measure of the degree a process is/is not meeting customer requirements, compared to the distribution of the process. See also *Control and Control Charts*. That is, the level of uniformity of product that a process is capable of yielding. Process capability may be expressed by the percent of defective products, the range or standard deviation of some product dimension, and so on. Process capability is usually determined by performing measurements on some (or all) of the product units produced by the process. A statistical measure of the inherent process variability for a given characteristic (see C_p, C_{pk}, and P_{pk}).

Process capability index: The value of the tolerance specified for the characteristic divided by the process capability. There are several types of process capability indexes, including the widely used C_p and C_{pk}.

Process characterization: Is concerned with the identification and benchmarking of key product characteristics. This is done by way of gap analysis.

Process control: Maintaining the performance of a process at its capability level. Process control involves a range of activities such as sampling the process product, charting its performance, determining causes of any excessive variation, and taking corrective actions.

Process decision program chart (PDPC): A management and planning tool that identifies all events that can go wrong and the appropriate countermeasures for these events. It graphically represents all sequences that lead to a desirable effect.

Process design: Creation of an innovative process needed for newly introduced activities, systems, products, or services

Process improvement (PI): A strategy and methodology for improving the process of performing the work (in other words, on reducing common causes). This includes any process at any level; for example, manufacturing, assembly, management decision-making, and so on. In general, it refers to the act of changing a process to reduce variability and cycle time and make the process more effective, efficient, and productive. Specifically, it is focused on incremental changes/solutions to eliminate or reduce defects, costs, or cycle time; leaves basic design and assumptions of a process intact. See also *Process redesign*.

Process improvement team (PIP): A natural work group or cross-functional team whose responsibility is to achieve needed improvements in existing processes. The life span of the team is based on the completion of the team purpose and specific goals.

Process management: The collection of practices used to implement and improve process effectiveness; it focuses on holding the gains achieved through process improvement and assuring process integrity. Also, defined and documented processes, monitored on an ongoing basis, which ensure that measures are providing feedback on the flow/function of a process; key measures include financial, process, people, innovation. See also *Control*.

Process map, or flowchart: Graphic display of the process flow that shows all activities, decision points, rework loops, delays, inspection, and handoffs.

Process mapping: The flowcharting of a work process in detail, including key measurements.

Process measures: Measures related to individual steps as well as to the total process; predictors of output measures.

Process optimization: Is aimed at the identification and containment of those process variables that exert undue influence over the key product characteristics.

Process organization: A form of departmentalization where each department specializes in one phase of the process.

Process owner: The manager or leader who is responsible for ensuring that the total process is effective and efficient.

Process quality audit: An analysis of elements of a process and appraisal of completeness, correctness of conditions, and probable effectiveness.

Process redesign: Method of restructuring process flow elements eliminating handoffs, rework loops, inspection points, and other non-value-adding activities; typically means "clean slate" design of a business segment and accommodates major changes or yields exponential improvements (similar to reengineering). See also *Process Improvement; Reengineering.*

Process village: Refers to machines grouped by type of operation (contrast with a cell layout).

Producer's risk: The maximum probability of saying a process or lot is unacceptable when, in fact, it is acceptable. For a sampling plan, refers to the probability of not accepting a lot, the quality of which has a designated numerical value representing a *level* that is generally desirable. Usually the designated value will be the acceptable quality *level* (also called alpha risk and Type I error).

Product analysis, reliability, and service evaluation report (PARSER): Product Analysis, Reliability, and Service Evaluation Reports = a PC-based early warranty reporting tool and is used primarily in the automotive industry. All claims approved for payment appear in PARSER at the end of the week they are submitted. PARSER provides detail and customer-defined aggregate reporting. Numerous standard format report requests are available, as well as flexible *ad hoc* capabilities. Provides earlier access to claims information but does not have sample size information.

Product attribute leadership strategy (PALS): A process to identify broad areas of product attribute leadership in the long term to position the company at competitive advantage (product strategy, vision, market, and target customer).

Product buybacks: Costs incurred to repurchase or replace products that have experienced repeated repairs that are either "not fixable" in the customer's mind or are substantial enough that the customer has pursued (or threatened to pursue) legal action to dispose of the vehicle. The primary concern code indicates the primary reason the product was brought back.

Product design: A specific term that refers to an Engineering Specification. It can also mean the process of designing.

Product organization: A departmentalization where each department focuses on a specific product type or family.

Product orientation: Refers to a tendency to see customers' needs in terms of a product they want to buy, not in terms of the services, value, or benefits the product will produce.

Product quality audit: A quantitative assessment of conformance to required product characteristics.

Product/service liability: The obligation of a company to make restitution for loss related to personal injury, property damage, or other harm caused by its product or service.

Productivity: The measurement of labor efficiency when compared to an established base. It is also used to measure equipment effectiveness, drawing productivity, and so on.

Professional development plan: An individual development tool for an *employee.* Working together, the employee and his or her supervisor create a plan that matches the individual's career needs and aspirations with organizational demands.

Profile diagram: Graphical representation of data that aids in screening for outliers or the interpretation of the final cluster solution. Typically, the variables of the *cluster variate* or those used for validation are listed along the horizontal axis, and the value scale is used for the vertical axis. Separate lines depict the scores (original or standardized) for individual *objects* or cluster *centroids* in a graphic plane.

Profitability: A measure of the total income of a project compared to the total monies expended at any period of time. The techniques that are utilized are Payout Time, Return on Original Investment (ROI), Net Present Value (NPV), Discounted Cash Flow (DCF), Sensitivity and Risk Analysis.

Profound knowledge, system of: As defined by W. Edwards Deming, states that learning cannot be based on experience only; it requires comparisons of results to a prediction, plan, or an expression of theory. Predicting why something happens is essential to understand results and to continually improve. The four components of the system of profound knowledge are (1) appreciation for a system, (2) knowledge of variation, (3) theory of knowledge, and (4) understanding of psychology.

Program evaluation and review technique (PERT): An event-oriented project management planning and measurement technique that utilizes an arrow diagram or road map to identify all major project events and demonstrates the amount of time (critical path) needed to complete a project. It provides three time estimates: optimistic, most likely, and pessimistic.

Program management: The management of a related series of projects executed over abroad period of time, and which are designed to accomplish broad goals, to which the individual projects contribute.

Program: An endeavor of considerable scope encompassing a number of projects.

Progress analysis (time management): The evaluation of calculated progress against the approved schedule and the determination of its impact (Cost Management). The development of performance indices such as (a) Cost Performance Index = CPI = BCWP/ACWP, (b) Schedule Performance Index = SPI = BCWP/BCWS, and (c) Productivity.

Progress: Development to a more advanced state. Progress relates to a progression of development and, therefore, shows relationships between current conditions and past conditions.

Progress trend: An indication of whether the progress rate of an activity or of a project is increasing, decreasing, or remaining the same (steady) over a period of time.

Project: Any undertaking with a defined starting point and defined objectives by which completion is identified. In practice most projects depend on finite or limited resources by which the objectives are to be accomplished.

Project budget: The amount and distribution of money allocated to a project.

Project change: An approved change to project work content caused by a scope of work change or a special circumstance on the project (weather, strikes, etc.). See also *Project cost changes.*

Project close-out and start-up costs: The estimated extra costs (both capital and operating) that are incurred during the period from the completion of project implementation to the beginning of normal revenue earnings on operations.

Project close-out: A process that provides for acceptance of the PROJECT by the project sponsor, completion of various project records, final revision, and issue of documentation to reflect the "as-built" condition and the retention of essential project documentation. See Project life cycle.

Project cost: The actual costs of the entire project.

Project cost changes: The changes to a project and the initiating of the preparation of detail estimates to determine the impact on project costs and schedule. These changes must then be communicated clearly (both written and verbally) to all participants that approval/rejection of the project changes have been obtained (especially those that change the original project intent).

Project goods: Equipment and/or materials needed to implement a project.

Project information sources: Identification and listing of various available sources, internal as well as external, to provide relevant information on specific procurements.

Project integration: The bringing together of diverse organizations, groups, or parts to form a cohesive whole to successfully achieve project objectives.

Project investment cost: The activity of establishing and assembling all the cost elements (capital and operating) of a project as defined by an agreed scope of work. the estimate attempts to predict the final financial outcome of a future investment program even though all the parameters of the project are not yet fully defined.

Project life cycle: The four sequential phases in time through which any project passes, namely: concept, development, execution (implementation or operation), and finishing (termination or close out). Note that these phases may be further broken down into stages depending on the area of project application. Sometimes these phases are known as: concept, planning, design, implementation, and evaluation.

Project management (PM): The art of directing and coordinating human and material resources throughout the life of a project by using modern management techniques to achieve predetermined objectives of scope, cost, time, quality, and participant satisfaction.

Project manager: The individual appointed with responsibility for Project Management of the project. (In the Six Sigma methodology that person has been designated to be the Black Belt).

Project manual: See *Project Policies/Procedures*.

Project objectives: Project Scope expressed in terms of outputs, required resources and timing.

Project organization: The orderly structuring of project participants.

Project personnel: Those members of a Project Team employed directly by the organization responsible for a project.

Project phase: The division of a project time frame (or Project Life Cycle) into the largest logical collection of related activities.

Project plan—general: All the documents that comprise the details of why the project is to be initiated, what the project is to accomplish, when and where it is to be implemented, who will have responsibility, how the implementation will be carried out, how much it will cost, what resources are required, and how the project's progress and results will be measured.

Project plan—in PM: A management summary document that gives the essentials of a project in terms of its objectives, justification, and how the objectives are to be achieved. It should describe how all the major activities under each Project Management function are to be accomplished, including that of overall project control. The

Project Plan will evolve through successive stages of the project life cycle. Prior to project implementation, for example, it may be referred to as a Project Brief. See also *Baseline and Baseline concept*.

Project planning: The identification of the project objectives and the ordered activity necessary to complete the project. The identification of resource types and quantities required to carry out each activity or task.

Project policies: General guidelines/formalized methodologies on how a project will be managed.

Project preselection meetings: Meetings held to supplement and/or verify qualifications, data, and specifications.

Project Procedures: The methods, practices, and policies (both written and verbal communications) that will be used during the project life.

Project rationale (aka business case): Broad statement defining area of concern or opportunity, including impact/benefit of potential improvements, or risk of not improving a process; links to business strategies, the customer, and/or company values. Provided by business leaders to an improvement team and used to develop problem statement and Project Charter.

Project reporting: A planning activity involved with the development and issuance of (internal) Time Management analysis reports and (external) progress reports.

Project risk analysis: Analysis of the consequences and probabilities that certain undesirable *events* will occur and their impact on attaining the Contract/Procurement Objectives.

Project risk characterization: Identifying the potential external or internal risks associated with procurement actions using estimates of probability of occurrence.

Project risk: The cumulative effect of the chances of uncertain occurrences which will adversely affect Project Objectives. It is the degree of exposure to negative events and their probable consequences. Project Risk is characterized by three factors: Risk Event, Risk Probability and the Amount At Stake.

Project segments: project subdivisions expressed as manageable components.

Project services: Expertise and/or labor needed to implement a project not available directly from a Project Manager's organization.

Project stage: A sub-set of Project Phase.

Project start date/schedule: The earliest calendar start date among all *activities* in the *network*.

Project team (framework): The central management group . . . of the project. The group of people, considered as a group, that shares responsibility for the accomplishment of project goals and who report either part-time or full-time to the project manager.

Projections: Points defined by perpendicular lines from an object to a vector. Projections are used in determining the *preference* order with *vector* representations.

Promotional: Encourages subordinates to realize their full potential, cultivates a team spirit and lets subordinates know that good work will be rewarded.

Proportion defective: Fraction of units with nonconformances (defects); number of defective units divided by the total number of units; translate the decimal figure to a percentage. See also *Defects; Defective.*

Proposal project plan: Usually the first plan issued on a project and accompanies the proposal. it contains key analysis, procurement, and Implementation Milestones; historical data; and any client-supplied information. Usually presented in bar chart form or summary level network, and used for inquiry and contract negotiations.

Psychographic customer characteristics: Variables among buyers in the consumer market that address lifestyle issues and include consumer interests, activities, and opinions' pull system: (see Kanban)

Public relations: An activity designed to improve the environment in which an organization operates in order to improve the performance of that organization.

Public, the (project external): All those that are not directly involved in the project but who have an interest in its outcome. This could include for example, environmental protection groups, Equal Employment Opportunity groups, and others with a real or imagined interest in the project or the way it is managed.

Public, the (project internal): All personnel working directly or indirectly on the project.

Punch list: A list made near to the completion of a project showing the items of work remaining in order to complete the project scope.

Q factor analysis: Forms groups of respondents or cases based on their similarity on a set of characteristics.

QFD: Quality Function Deployment.

Q-Press: Quality Plant Resident Engineering Systems Support. A system which utilizes a Resident Engineer to interface between the Plant Quality Teams and QRTs to reduce customer concerns through product and process assistance at the control and "sister" plant.

Qualifications contractor: A review of the experience, past performance, capabilities, resources and current work loads of potential service resources.

Qualitative measurement: A measure which is classified by type and relies on the observer making a subjective judgment.

Quality adviser: The person (facilitator) who helps team members work together in quality processes and is a consultant to the team. The adviser is concerned about the process and how decisions are made rather than about which decisions are made. In the Six Sigma initiative this person is also called Champion.

Quality assessment: The process of identifying business practices, attitudes, and activities that are enhancing or inhibiting the achievement of quality improvement in an organization.

Quality assurance (contract/procurement mgt.): Planned and systematic actions necessary to provide adequate confidence that the performed service or supplied goods will serve satisfactorily for its intended and specified purpose. (Managerial). The development of a comprehensive program which includes the processes of identifying objectives and strategy, of client interfacing and of organizing and coordinating planned and systematic controls for maintaining established standards. This in turn involves measuring and evaluating performance to these standards, reporting results and taking appropriate action to deal with deviations.

Quality assurance (general): The function of assuring that a product or service will satisfy given needs. The function includes necessary verification, audits, and evaluations of quality factors affecting the intended usage and customer satisfaction. This function is normally the responsibility of one or more upper management individuals overseeing the quality assurance program; primary tools are statistical analysis, inspection and SPC. It is also a discipline (or department) that is charged with maintaining product or service conformance to customer specifications.

Quality assurance (QA): Discipline (or department) of maintaining product or service conformance to customer specifications; primary tools are statistical analysis, inspection and SPC.

Quality assurance/quality control (QA/QC): Two terms that have many interpretations because of the multiple definitions for the words *assurance* and *control*. For example, *assurance* can mean the act of giving confidence, the state of being certain, or the act of making certain; *control* can mean an evaluation to indicate needed corrective responses, the act of guiding, or the state of a process in which the variability is attributable to a constant system of chance causes. (For a detailed discussion on the multiple definitions, see ANSI/ISO/ASQC A35342, Statistics-vocabulary and Symbols-statistical quality control.) One definition of quality assurance is: all the planned and systematic activities implemented within the quality system that can be demonstrated to provide confidence that a product or service will fulfill requirements for quality. One definition for quality control is: the operational techniques and activities used to fulfill requirements for quality. Often, however, *quality assurance* and *quality control* are used interchangeably, referring to the actions performed to ensure the quality of a product, service, or process. The focus of *assurance* is planning and that of *control* is appraising.

Quality audit: A systematic, independent examination and review to determine whether quality activities and related results comply with planned arrangements and whether these arrangements are implemented effectively and are suitable to achieve the objectives.

Quality characteristic: A particular aspect of a product which relates to its ability to perform its intended function. Or, the aspect of a system to be measured in an experiment. It is a measure of the desired response of the system as opposed to error states or failure modes. This is used to determine the success of the experiment's objective. It is also the unique characteristics of products and of services by which customers evaluate their perception of quality.

Quality circles: Quality improvement or self-improvement study groups composed of a small number of employees-10 or fewer-and their supervisor.

Quality control: The process of maintaining an acceptable level of product quality.

Quality control (technical): The planned process of identifying established system requirements and exercising influence through the collection of specific (usually highly technical and itself standardized) data. The basis for decision on any necessary Corrective Action is provided by analyzing the data and reporting it comparatively to System Standards.

Quality cost reports: A system of collecting quality costs that uses a spreadsheet to list the elements of quality costs against a spread of the departments, areas, or projects in which the costs occur and summarizes the data to enable trend analysis and decision making. The reports help organizations review Prevention costs, appraisal costs, and internal and external failure costs.

Quality costs: see *Cost of quality*.

Quality council: (sometimes called "quality steering committee). Is the group driving the quality improvement effort and usually having oversight responsibility for the implementation and maintenance of the quality management system; operates in parallel with the normal operation of the business. (This leadership guides the implementation of quality or Six Sigma within an organization; establishes, reviews, and supports the progress of quality improvement teams.)

Quality culture: Consists of employee opinions, beliefs, traditions, and practices concerning quality.

Quality engineering: The analysis of a manufacturing system at all stages to maximize the quality of the process itself and the products it produces.

Quality evaluation methods: The technical process of gathering measured variables or counted data for decision making in Quality Process Review. Normally these evaluation methods should operate in a holistic context involving proven statistical analysis, referred to previously as statistical process control. A few example methods are: graphs and charts; Pareto Diagrams; and Exception Reporting.

Quality function: The function of maintaining product quality levels; i.e., the execution of quality control. The entire collection of activities through which an organization achieves fitness for use, no matter where these activities are performed.

Quality function deployment (QFD): A structured method in which customer requirements are translated into appropriate technical requirements for each stage of product development and production. The QFD process is often referred to as listening to the voice of the customer. Also see *House of quality.*

Quality function: The entire collection of activities through which an organization achieves fitness for use, no matter where these activities are performed.

Quality improvement: Actions taken throughout the organization to increase the effectiveness and efficiency of activities and processes in order to provide added benefits to both the organization and its customers.

Quality level agreement (QLA): Internal service/product providers assist their internal customers in clearly delineating the level of service/product required in quantitative measurable terms. A QLA may contain specifications for accuracy, timeliness, quality/usability, product life, service availability, responsiveness to needs, etc.

Quality loop: Conceptual model of interacting activities that influence quality at the various stages ranging from the identification of needs to the assessment of whether those needs are satisfied.

Quality loss function: A parabolic approximation (Taylor's Series) of the quality loss that occurs when a quality characteristic deviates from its target value. The quality loss function is expressed in monetary units: The cost of deviating from the target increases as a quadratic function the farther the quality characteristic moves from the target. The formula used to compute the quality loss function depends on the type of quality characteristic being used. The quality loss function was first introduced in this form by Genichi Taguchi.

Quality management: Quality itself is the composite of material attributes (including performance features and characteristics) of the product, processor service that are required to satisfy the need for which the project is launched. Quality policies, plans, procedures, specifications and requirements are attained through the sub-functions of Quality Assurance (Managerial) and Quality Control (Technical). Therefore, QM is viewed as the umbrella for all activities of the overall management function that determine the quality policy, objectives, and responsibilities, and implement them by means such as quality planning, quality control, quality assurance, and quality improvement within the quality system.

Quality manual: document stating the quality policy and describing the quality system of an organization.

Quality metrics: Numerical measurements that give an organization the ability to set goals and evaluate actual performance versus plan.

Quality plan: The document setting out the specific quality practices, resources and sequence of activities relevant to a particular product, project or contract. Also known as Control plan.

Quality planning: The activity of establishing quality objectives and quality requirements.

Quality planning: The activity of establishing quality objectives and quality requirements.

Quality policy: Top management's formally stated intentions and direction for the organization pertaining to quality.

Quality principles: Rules or concepts that an organization believes in collectively. The principles are formulated by senior management with input from others and are communicated and understood at every level of the organization.

Quality probe studies: A study to obtain, in the words of owners, their perceptions of product quality. Findings are the volunteered verbatim responses to open-end questions.

Quality process review: The technical process of using data to decide how the actual project results compare with the quality specifications/ requirements. if deviations occur, this analysis may cause changes in the project design, development, use, etc., depending on the decisions of the client, involved shareholders and the Project Team.

Q-Q plot: A Q-Q plot is used to compare the shapes of distributions, providing a graphical view of how properties such as location, scale and skewness are similar or different in the two distributions. Q-Q plots can be used to compare collections of data or theoretical distributions. The use of Q-Q plots to compare two samples of data can be viewed as a nonparametric approach to comparing their underlying distributions. A Q-Q plot is generally a more powerful approach to doing this than the common technique of comparing histograms of the two samples, but requires more skill to interpret. Q-Q plots are commonly used to compare a data set to a theoretical model.

Quality responsible team (QRT): A cross-functional/organizational team responsible for the quality of a particular group for a particular commodity.

Quality score chart (Q-chart): A control chart for evaluating the stability of a process in terms of a quality score. The quality score is the weighted sum of the count of events of various classifications in which each classification is assigned a weight.

Quality specifications: Particular specifications of the limits within which each quality characteristic of a product is to be maintained.

Quality system audit: A documented activity performed to verify, by examination and valuation of objective evidence, that applicable elements of the quality system are suitable and have been developed, documented, and effectively implemented in accordance with specified requirements.

Quality system: The organizational structure, procedures, processes, and resources needed to implement quality management.

Quality trilogy: A three-pronged approach to managing for quality. The three legs are quality planning (developing the products and processes required to meet customer needs), quality control (meeting product and process goals), and quality improvement (achieving unprecedented levels of performance). Attributed to Joseph M. Juran.

Quality: A broad concept and/or discipline involving degree of excellence; a distinguished attribute or nature; conformance to specifications; measurable standards of comparison so that applications can be consistently directed toward business goals. No matter how quality is defined, in the final analysis it is a subjective term for which each person has his or her own definition. In technical usage, quality can have two meanings: (1) the characteristics of a product or service that bear on its ability to satisfy stated or implied needs and (2) a product or service free of deficiencies.

Quantitative measurement: A numerical measure which is independent of the observer, and is objective.

Quasi *F*-ratio: For some mixed and random models there is no exact *F*-ratio denominator for certain tests. In such cases it may be possible to add and subtract mean squares to construct an error mean square with the appropriate estimated components of variation. This error mean square is then used to obtain a quasi *F*-ratio.

Query: Directed access to a database requesting specific information. Can be used to *drill down* from *aggregated data*.

Questionnaires: See *Surveys*.

Queue processing: Processing in batches (contrast with continuous flow processing).

Queue time: Wait time of product awaiting next step in process.

Quincunx: A tool that creates frequency distributions. Beads tumble over numerous horizontal rows of pins, which force the beads to the right or left. After a random journey, the beads are dropped into vertical slots. After many beads are dropped, a frequency distribution results. In the classroom, quincunxes are often used to simulate a manufacturing process. The quincunx was invented by English scientist Francis Galton in the 1890s.

R factor analysis: Analyzes relationships among variables to identify groups of variables forming latent dimensions *(factors)*.

Radar chart: A visual method to show in graphic form the size of gaps among a number of both current organization performance areas and ideal performance areas; resulting chart resembles a radar screen.

Radial basis function: Alternative form of *neural network* that functions much like the *multilayer perceptron* in being appropriate for prediction and classification problems.

Random: Varying with no discernable pattern.

Random number generator: Used to select a stated quantity of random numbers from a table of random numbers, the resulting selection is then used to pull specific items or records corresponding to the selected numbers to comprise a "random sample."

Random sample: The process of selecting a sample of size n where each part in the lot or batch has an equal probability of being selected.

Random sampling: Method that allows each item or person chosen to be measured is selected completely by chance. In other words every element in the population or sample has an equal chance of being included.

Randomization: The executing of experimental runs in a random order. If the Quality Characteristic, or response, is likely to vary due to testing conditions changes, one precaution is to randomize the run order.

Randomized complete block: Is a design with each block containing each level of a treatment factor, or combination of levels for multiple factors. The randomization of treatment levels is therefore restricted to plots within blocks. The Latin Square is a variant on this design for two blocking factors.

Range: Measure of dispersion. That is, the difference between the highest and lowest of a group of values.

Range chart (R-chart): A control chart in which the subgroup range, R, is used to evaluate the stability of the variability within a process. A control chart of the range of variation among the individual elements of a sample (i.e., the difference between the largest and smallest elements) as a function of time, or lot number, or similar chronological variable.

Rank condition: Requirement for *identification* that each estimated parameter be defined algebraically. Much more laborious than the *order condition* or empirical estimation of identification.

Ratio analysis: The process of relating isolated business numbers, such as sales, margins, expenses, debt, and profits, to make them meaningful.

Rational subgroup: A subgroup which is expected to be as free as possible from assignable causes (usually consecutive items). For control charting: a subgroup of units selected to minimize the differences due to assignable causes. Usually samples taken consecutively from a process operating under the same conditions will meet this requirement

R bar: Average range value displayed on a range control chart. Value is set at the time control limit(s) are calculated.

Real time: The application of external time constraints which might affect the calendar time position of execution of each activity in the schedule.

Reciprocal: See *Nonrecursive*.

Recommend: To offer or suggest for use. Recommendation describes the presentation of plans, ideas, or things to others for adoption. To recommend is to offer something with the option of refusal.

Record: Document or electronic medium which furnishes objective evidence of activities performed or results achieved.

Records management: The procedures established by an organization to manage all documentation required for the effective development and application of its work force.

Record retention: The necessity to retain records for reference for a specified period after Contract Close-Out, in case they are needed.

Recovery schedule: A special schedule showing special efforts to recover time lost compared to the Master Schedule.

Recruitment, selection and job placement: Attracting of a pool of potential employees, determining which of those employees is best suited for work on the project, and matching that employee to the most appropriate task based on his or her skills and abilities.

Red bead experiment: An experiment developed by W. Edwards Deming to illustrate that it is impossible to put employees in rank order of performance for the coming year based on their performance during the past year because performance differences must be attributed to the system, not to employees. Four thousand red and white beads, 20 percent red, in a jar and six people are needed for the experiment. The participants' goal is to produce white beads because the customer will not accept red beads. One person begins by stirring the beads and then, blindfolded, selects a sample of 50 beads. That person hands the jar to the next person, who repeats the process, and so on. When everyone has his or her sample, the number of red beads for each is counted. The limits of variation between employees that can be attributed to the system are calculated. Everyone will fall within the calculated limits of variation that could arise from the system. The calculations will show that there is no evidence one person will be a better performer than another in the future. The experiment shows that it would be a waste of management's time to try to find out why, say, John produced four red beads and Jane produced 15; instead, management should improve the system, making it possible for everyone to produce more white beads.

Redundancy index: Amount of variance in a *canonical variate* (dependent or independent) explained by the other canonical variate in the *canonical function*. It can be computed for both the dependent and the independent canonical variates in each canonical function. For example, a redundancy index of the dependent variate represents the amount of variance in the dependent variables explained by the independent canonical variate.

Reengineering: Completely redesigning or restructuring a whole organization, an organizational component, or a complete process. It's a "start all over again from the beginning" approach, sometimes called a "breakthrough." In terms of improvement approaches, reengineering is contrasted with incremental *improvement (kaizen)*. It is similar to Process Redesign, though in practice usually at a much larger scale or scope.

Refinement: The rework, redefinition or modification of the logic or data that may have previously been developed in the Planning Process as required to properly input milestones, restraints and priorities.

Registration to standards: A process in which an accredited, independent third-party organization conducts an on-site audit of a company's operations against the requirements of the standard to which the company wants to be registered. Upon successful completion of the audit, the company receives a certificate indicating that it has met the standard requirements.

Reference category: The omitted level of a nonmetric variable when a *dummy variable* is formed from the nonmetric variable.

Regression analysis: A study used to understand the relationship between two or more variables, or another way of saying it, a technique for determining the mathematical relation between a measured quantity and the variables it depends on. The relationship can be determined and expressed as a mathematical equation. For example, the method might be used to determine the mathematical form of the probability distribution from which a sample was drawn, by determining which form best "fits" the frequency distribution of the sample. The frequency distribution is the "measured quantity" and the probability distribution is a "mathematical relation". Regression analysis makes it possible not only to explain relationships but also to predict one variable from knowledge about another.

Regression coefficient (b_n): Numerical value of the parameter estimate directly associated with an independent variable; for example, in the model $Y = b_0 + b_1 X_1$, the value b_1 is the regression coefficient for the variable X_1. The regression coefficient represents the amount of change in the dependent variable for a one-unit change in the independent variable. In the multiple predictor model (e.g., $Y = b_0 + b_1 X_1 + b_2 X_2$), the regression coefficients are partial coefficients because each takes into account not only the relationships between Y and X_1 and between Y and X_2, but also between X_1 and X_2. The coefficient is not limited in range, as it is based on both the degree of association and the scale units of the independent variable. For instance, two variables with the same association to Y would have different coefficients if one independent variable was measured on a 7-point scale and another was based on a 100-point scale.

Regression coefficient variance—decomposition matrix: Method of determining the relative contribution of each *eigenvalue* to each estimated coefficient. If two or more coefficients are highly associated with a single eigenvalue *(condition index),* an unacceptable level of *multicollinearity* is indicated.

Regression variate: Linear combination of weighted independent variables used collectively to predict the dependent variable.

Reinforcement: The process of providing positive consequences when an individual is applying the correct knowledge and skills to the job. It has been described as "catching people doing things right and recognizing their behavior," Caution: less than desired behavior can also be reinforced unintentionally.

Rejectable quality level (RQL): For acceptance sampling: expressed as percentage or proportion of defective units; the poorest quality in an individual lot that should be accepted. Commonly associated with a small consumer's risk. See also *LTPD.*

Reliability: The probability that a product will function properly for some specified period of time, under specified conditions. In measurement system analysis, refers to the ability of an instrument to produce the same results over repeated administration-to measure consistently. In reliability engineering it is the probability of a product performing its intended function under stated conditions for a given period of time (see also Mean time between failures). Yet another way of understanding reliability is to think of it as the degree to which a set of *latent construct indicators* are consistent in their measurements. In more formal terms, reliability is the extent to which a set of two or more indicators "share" in their measurement of a construct. This means that: the extent to which a variable or set of variables is consistent in what it is intended to measure. If multiple measurements are taken, the reliable measures will all be very consistent in their values. The indicators of highly reliable constructs are highly intercorrelated, indicating that they all are measuring the same latent construct. As reliability decreases, the indicators become less consistent and thus are poorer indicators of the latent construct. Reliability can be computed as 1.0 minus the *measurement error.* Reliability differs from *validity* in that it relates not to what should be measured, but instead to how it is measured. Also see *Validity.*

Reliability block diagrams: Reliability block diagrams are used to break down a system into smaller elements and to show their relationship from a reliability standpoint.

Reliability mission: Satisfy customer expectations for reliability throughout the product useful, as expected by the customer.

Remaining available resource: The difference between the Resource Availability Pool and the Level Schedule resource requirements. Computed from the Resource Allocation Process.

Remaining duration: The estimated work units needed to complete an activity as of the Data Date.

Remaining float (RF): The difference between the Early Finish and the Late Finish Date.

Remedy: Something that eliminates or counteracts a problem cause; a solution.

Repair: Action taken on a nonconforming product so that it will fulfill the intended usage requirements although it may not conform to the originally specified requirements.

Reparability: Probability that a failed system will be restored to operable condition within a specific repair time.

Repeatability (of a measurement): The extent to which repeated measurements of a particular object with a particular instrument produces the same value. In other words, it is a measurement stability concept in which a single person gets the same results each time he/she measures and collects data; necessary to ensure data consistency and stability. See also *Reproducibility*.

Repeatability and reproducibility (R & R): a measurement validation process to determine how much variation exists in the measurement system (including the variation in product,- the gage used to measure, and the individuals using the gage).

Repeatability: Measurement stability concept in which a single person gets the same results each time he/she measures and collects data; necessary to ensure data consistency and stability. See also *Reproducibility*.

Repeated measures: Use of two or more responses from a single individual in an ANOVA or MANOVA analysis. The purpose of a repeated measures design is to control for individual-level differences that may affect the within-group variance. Repeated measures are a form of respondent's lack of *independence*.

Repetition: Repetition means taking multiple measurements for each row of an experiment before proceeding to the next row. This precaution will combat the lack of repeatability.

Replication: Taking one measurement for each row, then repeating the entire experiment multiple times. In a sense then, replication is the readministration of an experiment with the intent of validating the results in another sample of respondents.

Replication, *n:* is the number of random independent replicates from which an ANOVA model calculates the unmeasured variation that is used to calibrate the significance of effects. A balanced ANOVA model has the same number of random independent observations in each sample. This sample size defines the replication of at least the most nested term in a hierarchical model, and the highest order interaction term in a cross-factored model. Main effects have more replication than interactions (except in randomized-block designs with Model-1 analysis), which generally gives them higher power.

The replication available for any effect in a balanced ANOVA is given by the product of all the variables contributing to the denominator degrees of freedom, q, that do not also contribute to the numerator degrees of freedom, p. The table below shows some examples of designs with treatment factors A, B, C, each with a, b, c levels, and s random independent replicate sampling units S' (e.g., subjects or plots or blocks).

Design	Terms	Effect	Error	p	q	Replication
One-factor fully randomized[1]	S'(A)	A	S'(A)	a	$(s-1)a$	s
Three-factor nested [2]	S'(C'(B'(A)))	A	B'(A)	a	$(b-1)a$	b
Three-factor fully randomized[3]	S'(C|B|A)	C*B*A	S'(C*B*A)	cba	$(s-1)cba$	s
Three-factor fully randomized	S'(C|B|A)	B*A	S'(C*B*A)	ba	$(s-1)cba$	sc
Three-factor fully randomized	S'(C|B|A)	A	S'(C*B*A)	a	$(s-1)cba$	scb
Two-factor randomized block[4]	S'|B|A	B*A	S'*B*A	ba	$(s-1)(b-1)(a-1)$	s
Two-factor randomized block	S'|B|A	A	S'*A	a	$(s-1)(a-1)$	s
Two-factor split plot[5,6]	B|S'(A)	B*A	B*S'(A)	ba	$(b-1)(s-1)a$	s
Two-factor split plot	B|S'(A)	A	S'(A)	a	$(s-1)a$	s
Two-factor split plot	B|S'(A)	B	B*S'(A)	b	$(b-1)(s-1)a$	sa

[1] For example, $s = 6$ subjects nested in each of $a = 3$ levels of factor A, requiring a total of $sa = 18$ observations and giving $n = 6$ replicates for testing the A main effect.

[2] For example, $s = 24$ subjects nested in each of $c = 12$ towns in each of $b = 6$ counties in each of $a = 3$ countries, requiring a total of $scba = 5184$ observations and giving $n = 6$ replicates for testing the effect A of country.

[3] For example, $s = 2$ plots nested in each of $cba = 27$ combinations of three-level factors C, B, A, requiring a total of $scba = 54$ observations and giving $n = 2, 6, 18$ replicates for testing respectively the three-way interaction, two-way interactions, and main effects.

[4] For example, $s = 6$ blocks, each with $ba = 9$ combinations of three-level factors B, A, requiring a total of $sba = 54$ observations and giving $n = 6$ replicates for testing both the interaction and main effects.

[5] For example, $s = 6$ blocks nested in each of $a = 2$ levels of factor A, and each block split into $b = 3$ levels of factor B, requiring a total of $sba = 36$ observations and giving $n = 6, 6, 12$ replicates for testing respectively the interaction, A main effect, and B main effect.

[6] For example, $s = 6$ subjects nested in each of $a = 2$ levels of intervention factor A: treatment and control, and repeated measures on each subject at $b = 2$ levels of time factor B: before and after intervention, requiring a total of $sba = 24$ observations and giving $n = 6$ replicates for testing the treatment effect in the B*A interaction. (Source: http://www.soton.ac.uk/~cpd/anovas/datasets/Replication.htm).

Reporting: Planning activity involved with the development and issuance of (internal) time management analysis reports and (external) progress reports.

Repository: Centralized collection of *metadata* available for access by all functions of the *data warehouse.*

Reproducibility: The variation between individual people taking the same measurement and using the same gauging. In other words, it is a measurement stability concept in which different people get the same results when they measure and collect data using the same methods; necessary to ensure data consistency and stability. See also *Repeatability.*

Resampling: Nonstatistical estimation of a parameter's confidence interval through the use of an empirically derived sampling distribution. The empirical distribution is calculated from multiple samples drawn from the original sample.

Residual (e or E): Error in predicting our sample data. Seldom will our predictions be perfect. We assume that random error will occur, but we assume that this error is an estimate of the true random error in the population (ε), not just the error in prediction for our sample (e). We assume that the error in the population we are estimating is distributed with a mean of 0 and a constant *(homoscedastic)* variance. Furthermore, a residual is a portion of a dependent variable not explained by a multivariate technique. Associated with dependence methods that attempt to predict the dependent variable, the residual represents the unexplained portion of the dependent variable. Residuals can be used in diagnostic procedures to identify problems in the estimation technique or to identify unspecified relationships.

Residual analysis: A means of comparing actual and predicted results of the experiment so as to verify that no bias has been introduced into the experimental procedure.

Residual error: All ANOVA models have residual variation defined by the variation amongst sampling units within each sample. This is always given by the last mean square in ANOVA tables, and denoted 'ε' (epsilon) in the descriptions of fully replicated models where it represents the error variance for at least some of the treatment effects. Models without full replication may have no degrees of freedom for measuring residual variation (e.g., randomised block, split plot, and repeated measures models). By contrast, *error variance* is the random variation in the response against which an effect is tested, containing all of the same components of variation estimated in the population except for the test effect. The validity of ANOVA depends on three assumptions about the error variance: (i) that the random variation around fitted values is the same for all sample means of a factor, or across the range of a covariate; (ii) that the residuals contributing to this variation are free to vary independently of each other; (iii) that the residual variation approximates to a normal distribution.

Resistance to change: Unwillingness to change beliefs, habits, and ways of doing things.

Resolution (of a measurement): The smallest unit of measure which an instrument is capable of indicating.

Resolution (in experimental design): In general, the resolution of a design is one more than the smallest order interaction which is aliased with some main effect. If some main effects are confounded with some 2-level interactions, the resolution is 3. Note: Full factorial designs have no confounding and are said to have resolution "infinity". For most practical purposes, a resolution 5 design is excellent and a resolution 4 design may be adequate. Resolution 3 designs are useful as economical screening designs.

Resolution III: An experimental plan where some of the main effects are confounded with two-way interactions.

Resolution IV: An experimental plan where all the main effects are unconfounded with two-way interactions, but the two-way interactions are confounded with each other.

Resolution V: An experimental plan where all the main effects and two-way interactions are unconfounded with each other.

Resource: Any factors, except time, required or consumed to accomplish an activity. Any substantive requirement of an activity that can be quantified and defined, e.g. manpower, equipment, material, etc.

Response: The continuous variable to be tested for sources of variance by taking observations from independent and randomly assigned sampling units (e.g., from subjects or plots). The response is the variable Y on the left of the equals sign in the model equation: $Y = A + \varepsilon$, etc.

Response data: The results of each run of the designed experiment as a measure of the identified quality characteristic.

Response planning: The process of formulating suitable Risk Management strategies for the project, including the allocation of responsibility to the project's various functional areas. It may involve mitigation, deflection and Contingency Planning. It should also make some allowance, however tentative, for the completely unforeseen occurrence.

Responseplot: The results of each run of the designed experiment as a measure of the identified quality characteristic.

Response surface methodology (RSM): A method of determining the optimum operating conditions and parameters of a process, by varying the process parameters and observing the results on the product. This is the same methodology used in Evolutionary Operations (EVOP), but is used in process development rather than actual production, so that strict adherence to product tolerances need not be maintained. An important aspect of RSM is to consider the relationships among the parameters, and the possibility of simultaneously varying two or more parameters to optimize the process.

Response-style effect: Series of systematic responses by a respondent that reflect a "bias" or consistent pattern. Examples include responding that an object always performs excellently or poorly across all attributes with little or no variation.

Response system: The on-going process put in place during the life of the project to monitor, review, and update Project Risk and make the necessary adjustments. Examination of the various risks will show that some risks are greater in some stages of the project life cycle than in others.

Responsibility charting: The activity of clearly identifying personnel and staff responsibilities for each task within the project.

Responsibility: Charged personally with the duties, assignments, and accountability for results associated with a designated position in the organization. Responsibility can be delegated but cannot be shared.

Restraint: An externally imposed factor affecting when an activity can be scheduled. the external factor may be labor, cost, equipment, or other such resource.

Restricted model: A mixed model (i.e., with random and fixed factors) is termed restricted if a random factor is not allowed to have fixed cross factors amongst its components of variation estimated in the population. This restriction influences the choice of error Mean Square for random effects. It is generally used for balanced and orthogonal designs; nonorthogonal designs requiring analysis by General Linear Model use the unrestricted model.

RETAD: Rapid exchange of tooling and dies

Return on equity (ROE): The net profit after taxes, divided by last year's tangible stockholders' equity, and then multiplied by 100 to provide a percentage (also referred to as return on net worth).

Return on investment (ROI): An umbrella term for a variety of ratios measuring an organization's business performance and calculated by dividing some measure of return by a measure of investment and then multiplying by 100 to provide a percentage. In its most basic form, ROI indicates what remains from all money taken in after all expenses are paid.

Return on net assets (RONA): A measurement of the earning power of the firm's investment in assets, calculated by dividing net profit after taxes by last year's tangible total assets and then multiplying by 100 to provide a percentage.

Reverse scoring: Process of reversing the scores of a variable, while retaining the distributional characteristics, to change the relationships (correlations) between two variables. Used in *summated scale* construction to avoid a "canceling out" between variables with positive and negative *factor loadings* on the same factor.

Review: To examine critically to determine suitability or accuracy.

Revision plans: A mechanism (process) for updating processes, procedures, and documentation.

Rework loop: Any instance in a process when the thing moving through the process has to be corrected by returning it to a previous step or person/organization in the process; adds time, costs, and potential for confusion and more defects. See also *Non-value-adding activities*.

Rework: Action taken on a nonconforming product so that it will fulfill the specified requirements (may also pertain to a service).

Right the first time: A term used to convey the concept that it is beneficial and more cost effective to take the necessary steps up front to ensure a product or service meets its requirements than to provide a product or service that will need rework or not meet customers' needs. In other words, an organization should engage in defect prevention rather than defect detection.

Risk assessment/management: The process of determining what risks are present in a situation (for example, project plan) and what actions might be taken to eliminate or mediate them.

Risk data applications: The development of a data base of Risk Factors both for the current project and as a matter of historic record.

Risk deflection: The act of transferring all or part of a risk to another party, usually by some form of contract.

Risk event: The precise description of what might happen to the detriment of the project.

Risk factor: Any one of Risk Event, Risk Probability or Amount At Stake, as defined above.

Risk identification: The process of systematically identifying all possible Risk Events which may impact on a project. They may be conveniently classified according to their cause or source and ranked roughly according to ability to manage effective responses. Not all risk events will impact all projects, but the cumulative effect of several risk events occurring in conjunction with each other may well be more severe than the examination of the individual risk events would suggest.

Risk management: The art and science of identifying, analyzing and responding to Risk Factors throughout the life of a project and in the best interests of its objectives. In other words, the process of review, examination and judgment whether or not the identified risks are acceptable in the proposed actions.

Risk mitigation: The act of revising the Project's Scope, Budget, Schedule Or Quality, preferably without material impact on the project's objectives, in order to reduce uncertainty on the project.

Risk probability: The degree to which the Risk Event is likely to occur.

Risk response planning: The process of formulating suitable Risk Management strategies for the project, including the allocation of responsibility to the project's various functional areas. It may involve Risk Mitigation, Risk Deflection and Contingency Planning. It should also make some allowance, however tentative, for the completely unforeseen occurrence.

Risk response system: The on-going process put in place during the life of the project to monitor, review and update Project Risk and make the necessary adjustments. Examination of the various risks will show that some risks are greater in some stages of the Project Life Cycle than in others.

Robust technology development: An upstream activity that improves the efficiency and effectiveness of ideal function. The engineering activity focuses on developing flexible and reproducible technology before program definition.

Robustness: The ability of a product or process to perform its intended function under a variety of environmental and other uncontrollable conditions throughout the life cycle at the lowest possible cost. Thus a robust design is insensitive to "noise." In other words, it is the condition of a product or process design that remains relatively stable with a minimum of variation even though factors that influence operations or usage, such as environment and wear, are constantly changing.

ROCOF: Rate of change of failure, or rate of change of occurrence of failure.

Role-playing: A training technique whereby participants spontaneously perform in an assigned scenario taking specific roles.

Rolled through-put-yield: The cumulative calculation of defects through multiple steps in a process; total input units, less the number of errors in the first process step number of items "rolled through" that step; to get a percentage, take the number of items coming through the process correctly divided by the number of total units going into the process; repeat this for each step of the process to get an overall rolled-throughput percentage. See also *Yield.*

Root cause analysis: A quality tool used to distinguish the source of defects or problems. It is a structured approach that focuses on the decisive or original cause of a problem or condition.

Route: For inspection, inventory, or other in-plant data collection. the sequence or path that the operator follows in the data collection process.

Row-centering standardization: See *Within-case standardization.*

Run: A set of consecutive units, i.e., sequential in time. In SPC a signal condition of seven consecutive points above or below the center line.

Run chart, or time plot: Measurement display tool showing variation in a factor over time; indicates trends, patterns, and instances of special causes of variation. See Memory jogger and Volume 3 for construction/use tips; see also *Control chart; Special cause; Variation.*

Run chart: A line graph showing data collected during a run or an uninterrupted sequence of events. A trend is indicated when the series of collected data points head up or down crossing the center line. When the run chart is referred to as *time plot* it is meant to be a measurement display tool showing variation in a factor over time; indicates trends, patterns, and instances of special causes of variation. *see* Control Chart; Special Cause; Variation.

s: Symbol used to represent standard deviation of a sample.

Sales leveling: A strategy of establishing a long-term relationship with customers to lead to contracts for fixed amounts and scheduled deliveries in order to smooth the flow and eliminate surges.

S Curves: Graphical display of the accumulated costs, labor hours or quantities, plotted against time for both budgeted and actual amounts.

Sample (statistics): A representative group selected from a population. The sample is used to determine the properties of the population. (A finite number of items of a similar type taken from a population for the purpose of examination to determine whether all members of the population would conform to quality requirements or specifications.)

Sample size: The number of elements, or units, in a sample, chosen from the population.

Sample standard deviation chart (s-chart): A control chart in which the subgroup standard deviation, s, is used to evaluate the stability of the variability within a process.

Sampling: The process of selecting a sample of a population and determining the properties of the sample with the intent of projecting those conclusions to the population. The sample is chosen in such a way that its properties (even though small) are representative of the population. Sampling is the foundation of statistics; it can save time, money, and effort; allows for more meaningful data; can improve accuracy of measurement system.

Sampling bias: When data can be prejudiced in one way or another and do not represent the whole.

Sampling variation: The variation of a sample's properties from the properties of the population from which it was drawn.

Scatter diagram: A graphical technique to analyze the relationship between two variables. Two sets of data are plotted on a graph, with the *y-axis* being used for the variable to be predicted and the *x-axis* being used for the variable to make the prediction. The graph will show possible relationships (although two variables might appear to be related, they might not be: Those who know most about the variables must make that evaluation). Sometimes this is referred to as the *Scatter Plot*. The scatter diagram is one of the seven tools of quality. See also *Correlation coefficient*.

Scatter plot: Representation of the relationship between two metric variables portraying the joint values of each observation in a two-dimensional graph. It is a plot on which each unit is represented as a dot at the x, y position corresponding to the measured values for the unit. The scatter plot is a useful tool for investigating the relationship between the two variables. See *Scatter diagram.*

Scatter plot or diagram: Graph used to show relationship or correlation between two factors or variables. See also *Correlation coefficient.*

Scenario planning: A strategic planning process that generates multiple stories about possible future conditions, allowing an organization to look at the potential impact on them and different ways they could respond.

Schedule refinement: The rework, redefinition or modification of the logic or data that may have previously been developed in the Planning Process as required to property input milestones, restraints and priorities.

Schedule revision: In the context of scheduling, a change in the Network Logic or in resources which requires redrawing part or all of the network.

Schedule status: It may be viewed as affecting time constraint status. Or, as Technical Performance Status: as affecting quality. See *Scope Reporting.*

Schedule update: Revision of the schedule to reflect the most current information on the project.

Schedule variance: Any difference between the projected duration for an activity and the actual duration of the activity. Also the difference between projected start and finish dates and actual or revised start and finish dates.

Schedule work unit: A calendar time unit when work may be performed on an activity.

Schedule: A display of project time allocation.

Schedule. pictorial display: A display in the form of a still picture, slide or video, that represents scheduling information.

Scheduling: The recognition of realistic time and resource restraints which will, in some way, influence the execution of the plan.

Scientific management: Aimed at finding the one best way to perform a task so as to increase productivity and efficiency.

Scope: Defines the boundaries of the process or the Process Improvement project; clarifies specifically where opportunities for improvement reside (start- and end-points); defines where and what to measure and analyze; needs to be within the sphere of influence and control of the team working on the project - the broader the scope, the more complex and time - consuming the Process Improvement efforts will be. In other words, it is the work content and products of a project or component of a project. scope is fully described by naming all activities performed, the resources consumed and the end products which result, including quality standards. a statement of scope should be introduced by a brief background to the project, or component, and the general objective(s).

Scope change: A deviation from the originally agreed project scope.

Scope constraints: Applicable restrictions which will affect the scope.

Scope criteria: Standards or rules composed of parameters to be considered in defining the project.

Scope interfaces: Points of interaction between the project or its components and its/their respective environments.

Scope management: The function of controlling a project in terms of its goals and objectives through the processes of conceptual development, full definition or scope statement, execution and termination.

Scope of work: A narrative description of the work to be accomplished or resource to be supplied.

Scope performance/quality: Basic objective of the project. Defines the characteristics of the project's end product as required by the sponsor.

Scope reporting: A process of periodically documenting the status of basic project parameters during the course of a project. The three areas of scope reporting are: 1) Cost Status: as affecting financial status 2) Schedule Status: as affecting time constraint status, and 3) Technical Performance Status: as affecting quality.

Scope schedule: Basic time constraints.

Scope statement: A documented description of the project as to its output, approach and content.

Screening: Technique used for reviewing, analyzing, ranking and selecting the best alternative for the proposed action.

SDFBETA: See *DFBETA.*

SDS: System design specifications

Search space: Range of potential solutions defined for various learning models, such as *genetic algorithms* or *neural networks.* The solutions can be specified in terms of *dimensions* considered, data elements included, or criteria for feasible or infeasible solutions.

Selective listening: Where one hears what they are pre dispositioned to hear.

Self-directed learning: see *Learner-controlled instruction.*

Self-explicated model: *Compositional* technique for performing conjoint analysis in which the respondent provides the *part-worth estimates* directly without making choices.

Self-inspection: The process by which employees inspect their own work according to specified rules.

Self-managed team: A team that requires little supervision and manages itself and the day-to-day work it does; self-directed teams are responsible for whole work processes with each individual performing multiple tasks.

Semantics: The language used to achieve a desired effect on an audience.

Semipartial correlation: See *Part correlation coefficient.*

Sensitivity (of a measuring instrument): The smallest change in the measured quantity which the instrument is capable of detecting.

Sequential SS: The sum of squares for a model term in ANOVA, which partitions out the variation due to the term after accounting for variation due to earlier terms in the model hierarchy. Sequential SS (often referred to as Type-I SS) are suitable for orthogonal designs; non-orthogonal designs use adjusted SS.

Sequential threshold method: *Nonhierarchical clustering* procedure that begins by selecting one *cluster seed*. All *objects* within a prespecified distance are then included in that cluster. Subsequent cluster seeds are selected until all objects are grouped in a cluster.

Serviceability: The ease with which a system can be repaired. Serviceability is a characteristic of the system design, primarily considering accessibility.

Setup time: The time taken to change over a process to run a different product or service.

Seven basic tools of quality: Tools that help organizations understand their processes in order to improve them. The tools are the cause-and-effect diagram, check sheet, control chart, flowchart, histogram, Pareto chart, and scatter diagram (see Individual entries).

Seven management tools of quality: The tools used primarily for planning and managing are activity network diagram (AND) or arrow diagram, affinity diagram (KJ method), interrelationship digraph, matrix diagram, priorities matrix, process decision program chart (PDPC), and tree diagram.

Shape: Pattern or outline formed by the relative position of a large number of individual values obtained from a process.

Shewhart control chart: A graphic continuous test of hypothesis. Commonly known as Xbar and R charts. See also *Control chart*.

Shewhart cycle: See *Plan-do-check-act cycle*.

Shewhart, Walter A.: Walter A. Shewhart was a statistician at the forefront of applying statistical methods to quality management, in the late 1920s. Amongst other things, he helped to formalize the PDCA cycle.

Ship-to-stock program: An arrangement with a qualified supplier whereby the supplier ships material directly to the buyer without the buyer's incoming inspection; often a result of evaluating and approving the supplier for certification.

Short term plan: A short duration schedule, usually 4 to 8 weeks, used to show in detail the activities and responsibilities for a particular period. a management technique often used "as needed" or in a critical area of the project.

Short term schedule: See *Short term plan*.

Short-run SPC: A set of techniques used for SPC in low volume, short duration manufacturing.

Should-be process mapping: Process-mapping approach showing the design of a process the way it should be (e.g., without non-value-adding activities; with streamlined workflow and new solutions incorporated). Contrasts with the "As-Is" form of process mapping. See also *Process Redesign, Value Adding Activities; Non-Value Adding Activities.*

SI System: The metric system of units of measure. The basic units of the system are the meter, the kilogram, and the second; the system is sometimes called the MKS system for this reason.

Sigma (σ): It is the symbol for standard deviation of the population. However, it is also used in the Six Sigma methodology as a statistical unit of measure which reflects process capability. This capability may be measured from either short or long term perspective. The conversions are as follows:

	Short Term	Long Term
Short term	No action (z value)	Add 1.5σ
Long Term	Subtract 1.5σ	No action

Sigma hat (σ̂): Symbol used to represent the estimate standard deviation given by the formula Rbar/d2. The estimated standard deviation may only be used if the data is normally distributed and the process is in control.

Sigma limits: For histograms: lines marked on the histogram showing the points n standard deviations above and below the mean.

Sigmoid function: Nonlinear function with a general S-shaped distribution. One common example is the logistic function.

Signal-to-noise ratio (S/N ratio): A mathematical equation that indicates the magnitude of an experimental effect above the effect of experimental error due to chance fluctuations.

Significance level: See *Alpha.*

Significance: The strength of evidence for an effect, measured by a p-value associated with the F-ratio from analysis of variance. A significant effect has a small p-value indicating a small chance of making a Type I error. For example, $p < 0.05$ means a less than 5% chance of mistakenly rejecting a true null hypothesis. For many tests this would be considered a reasonable level of safety for rejecting the null hypothesis of no effect, in favour of the model hypothesis of a significant effect on the response. The significance of an effect is not directly informative about the size of the effect. Thus an effect may be statistically highly significant as a result of low residual variation, yet have little biological significance as a result of a small effect size in terms of the amount of variation between sample means or the slope of a regression. A nonsignificant effect should be interpreted with reference to the Type II error rate, which depends on the power of the test to detect significant effects.

Silo: (as in "functional silo") An organization where cross-functional collaboration and cooperation is minimal and where the functional "silos" tend to work towards their own goals to the detriment of the organization as a whole. Also known as chimneys.

Similarities data: Data used to determine which *objects* are the most similar to each other and which are the most dissimilar. Implicit in similarities measurement is the ability to compare all pairs of objects. Three procedures to obtain similarities data are paired comparison of objects, *confusion data*, and *derived measures.*

Similarity scale: Arbitrary scale, for example, from –5 to +5 that allows the representation of an ordered relationship between objects from the most similar (closest) to the least similar (farthest apart). This type of scale is appropriate only for representing a single dimension.

Similarity: See *Interobject Similarity and Similarities data.*

Simple regression: Regression model with a single independent variable.

Simulation: Creation of multiple data input matrices based on specified parameters that reflect variation in the distribution of the input data. Specification error Lack of model goodness-of-fit resulting from the omission of a relevant variable from the proposed model. Tests for specification error are quite complicated and involve numerous trials among alternative models. The researcher can avoid specification error to a high degree by using only theoretical bases for constructing the proposed model. In this manner, the researcher is less likely to "overlook" a relevant construct for the model.

Simulation (modeling): Using a mathematical model of a system or process to predict the performance of the real system. The model consists of a set of equations or logic rules which operate on numerical values representing the operating parameters of the system. The result of the equations is a prediction of the system's output.

Single-linkage: *Hierarchical clustering* procedure in which *similarity* is defined as the minimum distance between any *object* in one cluster and any object in another. This simply means the distance between the closest objects in two clusters. This procedure has the potential for creating less compact, or even chainlike, clusters. This differs from the *complete linkage* method, which uses the maximum distance between objects in the cluster.

Single-minute exchange of dies (SMED): A goal to be achieved in reducing the setup time required for a changeover to a new process; the methodologies employed in devising and implementing ways to reduce setup.

Single-piece flow: A method whereby the product proceeds through the process one piece at a time, rather than in large batches, eliminating queues and costly waste.

Singularity: The extreme case of *collinearity* or *multicollinearity* in which an independent variable is perfectly predicted (a correlation of ±1.0) by one or more independent variables. Regression models cannot be estimated when a singularity exists. The researcher must omit one or more of the independent variables involved to remove the singularity.

SIPOC: Acronym for Suppliers, Inputs, Process, Outputs, and Customer; enables an "at-a-glance," high-level view of a process.

SIT: Structured inventive thinking. A derivative of the Russian acronym (TRIZ).

Situational leadership: A leadership theory that maintains that leadership style should change based on the person and the situation, with the leader displaying varying degrees of directive and supportive behavior.

Six Sigma: Level of process performance equivalent to producing only 3.4 defects for every one million opportunities or operations. The term is used to describe Process Improvement initiatives using sigma-based process measures and/or striving for Six Sigma level performance. The sigma value indicates how often defects are likely to occur. The higher the sigma value, the less likely a process will produce defects. As sigma increases, costs go down, cycle time goes down, and customer satisfaction goes up.

Six-Sigma approach: A quality philosophy; a collection of techniques and tools for use in reducing variation; a program of improvement methodology.

Six-Sigma quality: A term used generally to indicate that a process is well controlled, that is, process limits ±3 sigma from the centerline in a control chart, and requirements/tolerance limits ±6 sigma from the centerline. The term was initiated by Motorola.

Skewness: Measure of the symmetry of a distribution; in most instances the comparison is made to a *normal distribution*. A positively skewed distribution has relatively few large values and tails off to the right, and a negatively skewed distribution has relatively few small values and tails off to the left. Skewness values falling outside the range of −1 to +1 indicate a substantially skewed distribution.

Skill: An ability and competence learned by *practice*.

Skip-level meeting: An evaluation technique which occurs when a member of senior management meets with persons two or more organizational levels below, without the intervening management present, to allow open expression about the effectiveness of the organization.

Slack time: The time an activity can be delayed without delaying the entire project; it is determined by calculating the difference between the latest allowable date and the earliest expected date (see Project evaluation and Review technique). smaller elements and to show the functional and physical relationships between the elements.

Smoothing: Differences between two groups are played down and the strong points of agreement are given the most attention.

SOFFIT: See *DFFIT.*

Solution statement: A clear description of the proposed solution(s); used to evaluate and select the best solution to implement.

Spaghetti chart: A before improvement chart of existing steps in a process and the many back and forth interrelationships (can resemble a bowl of spaghetti); used to see the redundancies and other wasted movements of people and material.

Span of control: Refers to how many subordinates a manager can effectively and efficiently manage.

Spatial map: See *Perceptual map.*

SPC: Statistical process control; use of data gathering and analysis to monitor processes, identify performance issues, and determine variability/capability. See also *Run charts; Control charts.*

Special cause: A special cause is an event that is a departure from the process. That is: an instance or event that impacts processes only under "special" circumstances (i.e., not part of the normal, daily operation of the process). An example is a defective cutting wheel that is installed during routine maintenance and then quickly fails. Special causes are not predictable; they come and usually go on their own. On run charts, a special cause can manifest itself in many ways (depending upon duration). The most frequent signal is a point outside of the control limits. Caution: Multiple points outside the control limits could signal a new, permanently worse level of performance (common cause shift). See also *Common cause of variation, Special causes, Common causes, Tampering,* or *Structural variation.*

Special causes: Causes of variation that arise because of special circumstances. They are not an inherent part of a process. Special causes are also referred to as assignable causes (See also Common causes).

Specification (of a product): A listing of the required properties of a product. The specifications may include the desired mean and/or tolerances for certain dimensions or other measurements; the color or texture of surface finish; or any other properties which define the product.

Specification (time mgt.): An information vehicle that provides a precise description of a specific physical item, procedure, or result for the purpose of purchase and/or implementation of the item or service. (Contract/Procurement Mgt.) Written, pictorial or graphic information which describes, defines or specifies the services or items to be procured.

Specification control: A system for assuring that project specifications are prepared in a uniform fashion and only changed with proper authorization.

Specification: The engineering requirement, used for judging the acceptability of a particular product/service based on product characteristics, such as appearance, performance, and size. In statistical analysis, specifications refer to the document that prescribes the requirements with which the product or service has to perform.

Specific variance: Variance of each variable unique to that variable and not explained or associated with other variables in the factor analysis.

Specification error: Error in predicting the dependent variable caused by excluding one or more relevant independent variables. This omission can bias (impacting the estimated effects) the estimated coefficients of the included variables as well as decrease the overall predictive power of the regression model. In addition, specification error is the lack of model goodness-of-fit resulting from the omission of a relevant variable from the proposed model. Tests for specification error are quite complicated and involve numerous trials among alternative models. The researcher can avoid specification error to a high degree by using only theoretical bases for constructing the proposed model. In this manner, the researcher is less likely to "overlook" a relevant construct for the model.

Sponsor (or champion): Person who represents team issues to senior management; gives final approval on team recommendations and supports those efforts with the Quality Council; facilitates obtaining of team resources as needed; helps Black Belt and team overcome obstacles; acts as a mentor for the Black Belt.

Sporadic problem: A sudden adverse change in the status quo that can be remedied by restoring the status quo. For example, actions such as changing a worn part or proper handling of an irate customer's complaint can restore the status quo.

Stability (of a process): A process is said to be stable if it shows no recognizable pattern of change. See also *Control and Constant cause system*.

Stabilization: The period of time between continuous operation and normal operation. This period encompasses those activities necessary to establish reliable operation at design conditions of capacity, product quality, and efficiency.

Stack chart: A chart with month of production on the X-axis and R/1000 or CPU on the Y-axis. Values for more than one time in service, e.g. 1-MIS, 3-MIS and 6-MIS, resulting in a "stack" of lines.

Stages of creativity: One model gives the following stages: generate, percolate, illuminate, and verify.

Stages of experimental design: The process of conducting the experiment. The typical stages in conducting a typical experimental design experiment are: 1) Set Objective 2) Select Team 3) Define Characteristic 4) Define Characteristic of Interest 5) Determine Capability 6) Select Factors 7) Select Levels 8) Plan the experiment 9) Select the Appropriate and Applicable Experimental Plan 10) Conduct the Analysis 11) Take Action based on the data.

Stages of team growth: Refers to the four development stages through which groups typically progress: forming, storming, norming, and performing. Knowledge of the stages help team members accept the normal problems that occur on the path from forming a group to becoming a team.

Stakeholders: People, departments, and organizations that have an invest-ment or interest in the success or actions taken by the organization.

Standard: A basis for the uniformity of measuring performance. Also, a doc-ument that prescribes a specific consensus solution to a repetitive design, operating, or maintenance problem. In other words, a state-ment, specification, or quantity of material against which measured outputs from a process may be judged as acceptable or unacceptable.

Standard (measurement): A reference item providing a known value of a quantity to be measured. Standards may be primary (i.e., the stan-dard essentially defines the unit of measure) or secondary (trans-fer) standards, which have been compared to the primary standard (directly or by way of an intermediate transfer standard). Standards are used to calibrate instruments which are then employed to make routine measurements.

Standard deviation: A measure of the spread of the process output or the spread of the sampling statistic from the process (e.g., of subgroup averages); denoted by the Greek letter, sigma (σ for population and s for sample).

Standard error of the estimate (SEE): Measure of the variation in the pre-dicted values that can be used to develop confidence intervals around any predicted value. It is similar to the standard deviation of a variable around its mean.

Standard error: Expected distribution of an estimated *regression coefficient.* The standard error is similar to the standard deviation of original data values. It denotes the expected range of the coefficient across multiple samples of the data. This is useful in statistical tests of sig-nificance that test to see if the coefficient is significantly different from zero (i.e., whether the expected range of the coefficient con-tains the value of zero at a given level of confidence). The *t* value of a *regression coefficient* is the coefficient divided by its standard error.

Standard procedure: Prescribes that a certain kind of work be done in the same way wherever it is performed.

Standardization: Process whereby raw data are transformed into new measurement variables with a mean of 0 and a standard deviation of 1. When data are transformed in this manner, the b_0 term (the intercept) assumes a value of 0. When using standardized data, the regression coefficients are known as *beta coefficients,* which allow the researcher to compare directly the relative effect of each indepen-dent variable on the dependent variable.

Standardized residual: Rescaling of the *residual* to a common basis by divid-ing each residual by the standard deviation of the residuals. Thus, standardized residuals have a mean of 0 and standard deviation of

1. Each standardized residual value can now be viewed in terms of standard errors in middle to large sample sizes. This provides a direct means of identifying outliers as those with values above 1 or 2 for confidence levels of .10 and .05, respectively.

Starting value: Initial parameter estimate used for incremental or iterative estimation processes, such as Linear Structural Relations (LISREL).

Start-up: That period after the date of initial operation, during which the unit is brought up to acceptable production capacity and quality. Start-up is the activity that is often confused (used interchangeably) with date of initial operation.

Statement of work (SOW): A description of the actual work to be accomplished. It is derived from the work breakdown structure and, when combined with the project specifications, becomes the basis for the contractual agreement on the project (also referred to as scope of work).

Statistic: A summary value calculated from the data, such as an average or standard deviation, used to describe process performance. It can also be an estimate of a population parameter using a value calculated from a random sample.

Statistical confidence: (Also called "statistical significance"). The level of accuracy expected of an analysis of data. Most frequently it is expressed as either a "95% level of significance," or "5% confidence level."

Statistical control (of a process): A process is said to be in a state of statistical control when it exhibits only random variations.

Statistical control: The condition describing a process from which all special causes of variation have been eliminated and only common causes remain; evidenced on a control chart by the absence of nonrandom patterns or trends and all points within the control limits.

Statistical inference: The process of drawing conclusions on the basis of statistics.

Statistical power: The probability of a statistical test detecting an effect if it truly occurs. A test with low probability of mistakenly accepting a false null hypothesis (i.e., a low 'Type-II error' rate, β) has a correspondingly high power $(1 - \beta)$. Power increases with more replication. It should therefore be estimated prospectively, as part of the process of planning the design of data collection. For any balanced model, power is completely described by the following six variables:

- The threshold Type-I error rate, α;
- Numerator degrees of freedom for the model, p;
- Denominator degrees of freedom, q;
- Replication, n, which is given by the product of all the variables contributing to the denominator degrees of freedom that do not also contribute to the numerator degrees of freedom;

- Treatment effect size, $\theta = \sqrt{\sum d_i^2 / p}$, where d_i is the expected deviation of treatment level i from the average treatment effect across all levels, in the absence of residual variation (so θ is the standard deviation of the treatment variability);
- Error effect size, σ (the standard deviation of the random unmeasured variation).

Power increases with α, p, q, n, θ, and decreases with σ: Note that power is only raised by more replication if this is applied at an appropriate scale. For example a response measured per leaf for a treatment applied across replicate trees includes trees as a random factor nested in the treatment levels. The power of the design depends on the number of replicate trees per treatment level, and not on the number of replicate leaves per tree.

Power estimation may require prior estimation of θ and/or σ from a pilot study. Values of the treatment and error mean squares, (TMS and EMS), from pilot samples of size n will yield unbiased estimates of the treatment effect, $\theta = [(\text{TMS} - \text{EMS})/n]^{1/2}$, and the random error effect, $\sigma = (\text{EMS})^{1/2}$. Data collection can then be planned to ensure sufficient replication to achieve a high power (e.g., $1 - \beta = 0.8$) for distinguishing a real treatment effect ($\theta > 0$) from the error effect (σ), or for detecting some specified minimum θ or θ/σ. Specifying a threshold effect size of interest has the desirable consequence that a nonsignificant effect can be deemed an uninteresting effect. A nonsignificant effect is otherwise difficult to interpret, even from a design planned for high power. It could result from there being no true effect ($\theta = 0$); alternatively, it could result from θ having been overestimated in the power calculation used to plan the experimental design, which is consequently underpowered for detecting a small but real treatment effect.

The calculation of β, and hence power, is rather involved, being the integral to critical $F_{[a]}$ of the density function for the noncentral F-distribution:

$$\beta = \int_0^{F_{[a].p,q}} \left[\sum_{j=0}^{\infty} \frac{e^{-\lambda/2}}{j!} \left(\frac{\lambda}{2} \right)^j \frac{\left(\frac{p}{q} \right)^{(p+2j)/2} . F^{p+2,j-2} . \left(1 + \frac{p}{q} . F \right)^{-(p+q+2j)/2}}{B\left(\frac{p+2j}{2}, \frac{q}{2} \right)} \right] dF$$

where the noncentrality parameter $\lambda = p \cdot n \cdot \theta^2 / \sigma^2$, and the beta function $B(x, y) = \Gamma(x) \cdot \Gamma(y) / \Gamma(x + y)$, Figure 1 shows how the noncentral distribution is shifted to the right of the central distribution, with the displacement being a function of λ. Thus the power, $1 - \beta$, of a given test increases with more replication and a larger effect size, and decreases with larger error variation.

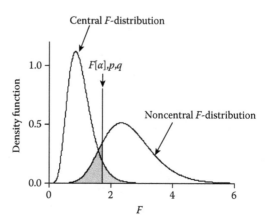

FIGURE G.2
Central and noncentral *F*-distribution.

In the absence of a treatment effect, $\theta = 0$, and $F = \text{TMS/EMS}$ follows the central *F*-distribution, with α given by the red-shaded area under its right-hand tail above the critical value $F_{[\alpha]}$. In the presence of a treatment effect, $\theta > 0$, and $F = \text{TMS/EMS}$ follows the noncentral *F*-distribution, with β given by the blue-shaded area under its left-hand tail up to the critical value $F_{[\alpha]}$. This example yields power $1 - \beta = 0.86$ for the B*A effect in cross-factored and fully replicated model S'(B|A) with $a, b = 5$ so $p = 16$, $n = 5$ so $q = 100$, $\theta/\sigma = 0.559$ so $\lambda = 25.0$, $\alpha = 0.05$ so $F_{[0.05],16,100} = 1.75$.

Calculation of power may be best left to a computer package. For example, Minitab will estimate $1 - \beta$ for any balanced ANOVA with specified α and proposed replication, n, and either an expected θ/σ or an observed *F*-value from a pilot study. In some software they go even further. For example: For a given n, it will also find the threshold θ/σ to achieve a target power. The freeware Piface by Russell V. Lenth allows further explorations of the relationships between sample size, θ, σ and power for specified designs. (Source: http://www.soton.ac.uk/~cpd/anovas/datasets/Power.htm).

Statistical process control: The use of statistical techniques such as control charts to analyze variation of the process or its outputs so as to take appropriate actions to achieve predictable performance. [Remember, that with warranty (R/1000 and CPU), the target is 0.]

Statistical quality control (SQC): The application of statistical methods for measuring, controlling and improving the quality of a processes. SPC is one method included in SQC. Often, however, the term "statistical process control" is used interchangeably with "statistical quality control" although statistical quality control includes acceptance sampling as well as statistical process control.

Statistical relationship: Relationship based on the correlation of one or more independent variables with the dependent variable. Measures of association, typically correlations, represent the degree of relationship because there is more than one value of the dependent variable for each value of the independent variable.

Statistical thinking: A philosophy of learning and action based on three fundamental principles: 1) all work occurs in a system of interconnected processes 2) variation exists in all processes and 3) understanding and reducing variation are vital to improvement

Statistics: A science dealing with decision making based on data collected from processes operating under variable conditions. The mathematical methods used to determine the best range of probable values for a project and to assess the degree of accuracy or allowance for unpredictable future events such as accidents, technological innovations, strikes, etc. that can occur during the project life. The techniques that can be used are Risk Analysis with Monte Carlo simulation, confidence levels, range analysis, etc.

Status system: System for tracking status at lowest level of detail.

Status: The condition of the project at a specified point in time.

Statusing: Indicating most current Project Status.

Stem and leaf diagram: A variant of the *histogram* which provides a visual depiction of the variable's distribution as well as an enumeration of the actual data values.

Stepdown analysis: Test for the incremental discriminatory power of a dependent variable after the effects of other dependent variables have been taken into account. Similar to stepwise regression or discriminant analysis, this procedure, which relies on a specified order of entry, determines how much an additional dependent variable adds to the explanation of the differences between the groups in the MANOVA analysis.

Stepwise estimation: Method of selecting variables for inclusion in the regression model that starts by selecting the best predictor of the dependent variable. Additional independent variables are selected in terms of the incremental explanatory power they can add to the regression model. Independent variables are added as long as their *partial correlation coefficients* are statistically significant. Independent variables may also be dropped if their predictive power drops to a nonsignificant level when another independent variable is added to the model.

Stimulus: Specific set of *levels* (one per *factor*) evaluated by respondents (also known as a *treatment*). One method of defining stimuli (*factorial design*) is achieved by taking all combinations of all levels. For example, three factors with two levels each would create eight (2 x 2 x 2) stimuli. However, in many conjoint analyses, the total number of combinations is too large for a respondent to evaluate them all. In these instances, some subsets of stimuli are created according to a systematic plan, most often a *fractional factorial design.*

Stopping rule: *Algorithm* for determining the final number of clusters to be formed. With no stopping rule inherent in cluster analysis, researchers have developed several criteria and guidelines for this determination. Two classes of rules exist that are applied *post hoc* and calculated by the researcher are (1) measures of *similarity* and (2) adapted statistical measures.

Stop work order: Request for interim stoppage of work due to non conformance, or funding or technical limitations.

Storyboard: A pictorial display of all the components in the DMAIC process, used by the team to arrive at a solution; used in presentations to Sponsor, senior management, and others.

Storyboarding: A technique that visually displays thoughts and ideas and groups them into categories, making all aspects of a process visible at once. Often used to communicate to others the activities performed by a team as they improved a process.

Strategic fit review: A process by which senior mangers assess the future of each project to a particular organization in terms of its ability to advance the mission and goals of that organization.

Strategic plan: The target plan prioritized by critical Total Float from the current schedule.

Strategic planning: A process to set an organization's long range goals and identify the actions needed to reach the goals.

Strategy: A framework guiding those choices that determine the nature and direction to attain the objective.

Stratification (of a sample): If a sample is formed by combining units from several lots having different properties, the sample distribution will show a concentration or clumping about the mean value for each lot: this is called stratification. In control charting, if there are changes between subgroups due to stratification, the R-chart points will all tend to be near the centerline. In other words, stratification is a way of looking at data in multiple layers of information such as what (types, complaints, etc.), when (month, day, year, etc.), where (region, city, state, etc.), and who (department, individual)

Stratified (random) sampling: A technique to segment (stratify) a population prior to drawing a random sample from each strata, the purpose being to increase precision when members of different strata would, if not stratified, cause an unrealistic distortion.

Stress measure: Proportion of the variance of the *disparities* (optimally scaled data) not accounted for by the MDS model. This type of measurement varies according to the type of program and the data being analyzed. The stress measure helps to determine the appropriate number of *dimensions* to include in the model.

Stress-strength interference: Identification and quantification of failures probability resulting from the probability of stress exceeding product strength.

Structural equation modeling (SEM): Multivariate technique combining aspects of multiple regression (examining dependence relationships) and factor analysis (representing unmeasured concepts—factors—with multiple variables) to estimate a series of interrelated dependence relationships simultaneously.

Structural model: Set of one or more dependence relationships linking the hypothesized model's *constructs*. The structural model is most useful in representing the interrelationships of variables between dependence relationships.

Structured query language (SQL): Method for extracting information from a database according to specified criteria. It differs from data mining in that the only results provided are those meeting the specific request conditions. SQL cannot perform any analytical processing, such as developing classification models or determining levels of association. For example, SQL can extract information for all customers between ages 35 and 55 who bought a product within the last month by credit card, but it cannot group customers into segments without specified criteria.

Structural variation: Variation caused by regular, systematic changes in output, such as seasonal patterns and long-term trends.

Structured inventive thinking (SIT): A method for applying TRIZ in a shorter time, with less reliance on external databases.

Studentized residual: The most commonly used form of standardized *residual*. It differs from other methods in how it calculates the standard deviation used in *standardization*. To minimize the effect of a single outlier, the residual standard deviation for observation i is computed from regression estimates omitting the ith observation in the calculation of the regression estimates. This is done repeatedly for each observation, each time omitting that observation from the calculations. This approach is similar to the *deleted residual,* although in this situation the observation is omitted from the calculation of the standard deviation.

Study: The methodical examination and/or analysis of a question or problem.

Subgroup: For control charts: a sample of units from a given process, all taken at or near the same time.

Subjective clustering: See *Confusion data.*

Subjective dimension: See *Perceived dimension.*

Subjective evaluation: Method of determining how many *dimensions* are represented in the MDS model. The researcher makes a "subjective inspection" of the spatial maps and asks whether the configuration looks reasonable. The objective is to obtain the best fit with the least number of dimensions.

Suboptimization: The need for each business function to consider overall organizational objectives, resulting in higher efficiency and effectiveness of the entire system, although performance of a function may be suboptimal.

Success tree analysis: Analysis that focuses on what must happen at the top-level event to be a success.

Sum of squared errors (SSE): Sum of the squared *prediction errors (residuals)* across all observations. It is used to denote the variance in the dependent variables not yet accounted for by the regression model. If no independent variables are used for prediction, this becomes the squared errors using the mean as the predicted value and thus equals the *total sum of squares.*

Sum of squares regression (SSR): Sum of the squared differences between the mean and predicted values of the dependent variable for all observations. This represents the amount of improvement in explanation of the dependent variable attributable to the independent variable(s).

Summated scales: Method of combining several variables that measure the same concept into a single variable in an attempt to increase the *reliability* of the measurement through *multivariate measurement.* In most instances, the separate variables are summed and then their total or average score is used in the analysis.

Summative quality evaluation: The process of determining what lessons have been learned after the project is completed. the objective is to document which behaviors helped determine, maintain or increase quality standards and which did not (for use in future projects).

Supervised: *Learning* process that employs a *training sample* and provides feedback to the *neural network* concerning errors at the output *nodes.*

Supplier: Any person or organization that feeds inputs (products, services, or information) into the process; in a service organization, many times the customer is also the supplier. In other words, any provider whose goods and services may be used at any stage in the production, design, delivery, and use of another company's products and services. Suppliers include businesses, such as distributors, dealers, warranty repair services, transportation contractors, and franchises, and service suppliers, such as healthcare, training, and education. Internal suppliers provide materials or services to internal customers.

Supplier audits: Reviews that are planned and carried out to verify the adequacy and effectiveness of a supplier's quality program, drive improvement, and increase value.

Supplier certification: The process of evaluating the performance of a supplier with the intent of authorizing the supplier to self-certify shipments if such authorization is justified.

Supply chain: The series of processes and/or organizations that are involved in producing and delivering a product to the final user.

Supply chain management: The process of effectively integrating and managing components of the supply chain.

Supplier default: Failure on the part of a supplier to meet technical or delivery requirements of the contract.

Supplier expediting: Actions taken to ensure that the goods/services are supplied in accordance with the schedule documented in the contract.

Supplier quality assurance: Confidence that a supplier's product or service will fulfill its customers' needs, This confidence is achieved by creating a relationship between the customer and supplier that ensures the product will be fit for use with minimal corrective action and inspection. According to J. M. Juran, there are nine primary activities needed: (1) define product and program quality requirements, (2) evaluate alternative suppliers, (3) select suppliers, (4) conduct joint quality planning, (5) cooperate with the supplier during the execution of the contract, (6) obtain proof of conformance to requirements, (7) certify qualified suppliers, (8) conduct quality improvement programs as required, and (9) create and use supplier quality ratings.

Supplier ranking: Qualitative and/or quantitative determinations of prospective suppliers' qualifications relative to the provision of the proposed goods/services.

Supplier selection strategy and criteria: Selection of new suppliers is based on the type and uniqueness of the product or service to be purchased, and the total cost. Suppliers of commodity-type items and basic supplies may be selected from directories and catalogs. For more sophisticated products and services stringent evaluation criteria may be established.

Support: Percentage of the total sample for which an *association rule* is valid. Support indicates the substantiality of the group.

Support systems: Starting with top-management commitment and visible involvement, support systems are a cascading series of interrelated practices or actions aimed at building and sustaining support for continuous quality improvement. Such practices/ actions may include: mission statement, transformation of company culture, policies, employment practices, compensation, recognition and rewards, employee involvement, rules and procedures, quality level agreements, training, empowerment, methods and tools for improving quality, tracking-measuring-evaluating-reporting systems, etc.

Surrogate variable: Selection of a single variable with the highest *factor loading* to represent a factor in the data reduction stage instead of using a *summated scale* or *factor score*.

Surveillance: Continual monitoring of a process.

Surveillance audit: The regular audits conducted by registrars to confirm that a company registered to the ISO 9001 standard still complies; usually conducted on a 6-month or 1-year basis.

Survey: An examination for some specific purpose; to inspect or consider carefully; to review in detail (survey implies the inclusion of matters not covered by agreed upon criteria.). Also, a structured series of questions designed to elicit a predetermined range of responses covering a pre selected area of interest. May be administered orally by a survey-taker, by paper and pencil, or by computer. Responses are tabulated and analyzed to surface significant areas for change.

Suspended data: Unfailed units which are still functioning at removal or end of test.

SWOT analysis: An assessment of an organization's key strengths, weaknesses, opportunities, and threats. It considers factors such as the organization's industry, the competitive position, functional areas, and management.

Symptom: An indication of a problem or opportunity.

System design specification (SDS): A specific term that refers to an Engineering Specification which defines standards and evaluation criteria for acceptable system performance

System: A system can be defined as a combination of elements that have an influence on each other and are organized to fulfill a purpose; A system can be the product itself or a process used for manufacturing or assembling that product; A methodical assembly of actions or things forming a logical and connected scheme or unit; A network of connecting processes that work together to accomplish the aim of the whole.

Systematic sampling: Sampling method in which elements are selected from the population at a uniform interval (e.g., every half-hour, every twentieth item); this is recommended for many Six Sigma measurement activities.

Systematic variation (of a process): Variations that exhibit a predictable pattern. The pattern may be cyclic (i.e., a recurring pattern) or may progress linearly (trend).

Systemic approach: A philosophy and approach that emphasizes that a product cannot be engineered solely by working on the parts. What is more important, is how the parts work together creating a whole system that performs so as to exceed customer expectations. To achieve this desired end result, (and customer symptom) level - the systemic approach seeks to answer the question, "What problems is the customer experiencing?". Once the issues have been prioritized, systems engineering/robust design practices should be followed to achieve substantial improvement.

Systems approach to management: A management theory that views the organization as a unified, purposeful combination of interrelated parts; managers must look at the organization as a whole and understand that activity in one part of the organization affects all parts of the organization (also known as systems thinking).

Systems engineering: Systems Engineering is a process to transform customer's needs into effective designs. The process enables product engineers to optimize designs within and across systems.

Tacit knowledge: Unarticulated heuristics and assumptions used by any individual or organization.

Tactical plans: Short-term plans, usually of one- to two-year duration, that describe actions the organization will take to meet its strategic business plan.

Tactics: The strategies and processes that help an organization meet its objectives.

Taguchi approach to experimentation: A special approach to parameter design using linear graphs. The Process of Taguchi Method is the following: (1) **Define the problem** (Be realistic; Size the scale of the problem; Narrow the problem to find the weakest link) (2) **Determine the objective** (Identify the measurable characteristic(s); Make it as close to the end use of the product as possible); (3) **Identify ALL factors which induce variability** (Process flow chart; Cause-and-effect-diagram); (4) **Eliminate ALL factors that cannot be measured** (or find a means of measuring the variable); (5) **Separate factors into two groups** (Controllable; Noise); (6) **Identify the controllable factors** (Establish the level; Determine the value of each level (**BE BOLD**); (7) **Identify interaction(s) between controllable factors** (Keep interactions to a minimum emphasize Main Factors); (8) **Determine the number of parts run per experiment** (Remember the cost of experimentation); (9) **Choose an OA** (Consider the cost and time for experimentation; Reduce the number of controllable factors—if necessary; Fill up an array with controllable factors—if available; (10) **Identify the noise factors** (Limit to the most important factors (no more than 3); Establish the number of levels (usually 2-3); Determine where the levels will be set (BE BOLD)); (11) **Choose an OA** (If small enough use full factorial); and (12) **RUN THE EXPERIMENT!!!**

Taguchi "Methods": Specific quality engineering philosophy and application of specific DOE (Design of Experiments) methodology to improve product designs and processes. It is based on the Loss Function.

Taguchi loss function: Pertains to where product characteristics deviate from the normal aim and losses increase according to a parabolic function; by merely attempting to produce a product within specifications doesn't prevent loss (loss is that inflicted on society after shipment of a product).

Taguchi, Genichi: Genichi Taguchi is a leading Japanese engineering specialist whose name is often taken to be synonymous with Quality and the Design of Experiments as applied to engineering. His major contributions have been the integration of engineering practice and statistical methods to achieve robust designs. He developed the Quality Loss Function and Signal to Noise Ratio concepts.

t-statistic: Test statistic that assesses the statistical significance between two groups on a single dependent variable (see *t*-Test).

t-Test: A test of the statistical hypothesis that two population means are equal. The population standard deviations are unknown, but thought to be the same. The hypothesis is rejected if the t value is outside the acceptable range listed in the t-table for a given a risk and degrees of freedom. Another way of describing it is a method for testing hypotheses about the population mean; the *t*-statistic measures the deviation between the sample and population means, in terms of the number of standard errors. The *t*-test is a special case of ANOVA for two groups or levels of a treatment variable.

Take-off: A term used for identifying and recording from drawings the material and quantities required for estimating the time and cost for the completion of an activity.

Takt time: The available production time divided by the rate of customer demand. Operating to takt time sets the production pace to customer demand.

Tally sheet: Another name for "checksheet."

Tampering: Action taken to compensate for variation within the control limits of a stable system. Tampering increases rather than decreases variation, as evidenced in the funnel experiment.

Target date: The date an activity is desired to be started or completed; accepted as the date generated by the initial CPM schedule operation and resource allocation process..

Target plan: The target plan prioritized by critical Total Float from the current schedule.

Target reporting: A method of reporting the current schedule against some established baseline schedule and the computation of variances between them.

Task types: Characterization of tasks by resource requirement, responsibility, discipline, jurisdiction, function, etc.

t-Distribution: For a sample with size n, drawn from a normally distributed population, with mean Xbar and standard deviation s. The true population parameters are unknown. The *t*-distribution is expressed as a table for a given number of degrees of freedom and a risk. As the degrees of freedom get very large, it approaches a z-distribution.

Taxonomy: Empirically derived classification of actual *objects* based on one or more characteristics. Typified by the application of cluster analysis or other grouping procedures. This classification can be contrasted to a *typology*.

Team: A set of two or more people who are equally accountable for the accomplishment of a purpose and specific performance goals; it is also defined as a small number of people with complementary skills who are committed to a common purpose. When used as an acronym it means: **T**ogether **E**veryone **A**ccomplishes **M**ore.

Team-based structure: Describes an organizational structure in which team members are organized around performing a specific function of the business, such as handling customer complaints or assembling an engine.

Team building: The process of influencing a group of diverse individuals, each with their own goals, needs, and perspectives, to work together effectively for the good of the project such that their team will accomplish more than the sum of their individual efforts could otherwise achieve. In other words, it is the process of transforming a group of people into a team and developing the team to achieve its purpose.

Team building/development: The process of transforming a group of people into a team and developing the team to achieve its purpose.

Team decision making: The process by which the Project Manager and his team determine feasible alternatives in the face of a technical, psychological, or political problem, and make a conscious selection of a course of action from among these available alternatives.

Team dynamics: The interactions which occur among team members under different conditions.

Team facilitation: Deals with both the role of the facilitator on the team and the techniques and tools for facilitating the team.

Team members: The individuals reporting either part-time or full-time to the Project Manager who are responsible for some aspect of the project's activities.

Team motivation: The process by which the Project Manager influences his Project Team to initiate effort on the Project Tasks, expend increasing amounts of effort on those tasks, and to persist in expending effort on these tasks over the period of time necessary for project goal accomplishment.

Team performance evaluation, rewards, and recognition: Special metrics are needed to evaluate the work of a team (to avoid focus on any individual on the team) and as a basis for rewards and recognition for team achievements.

Team reward system: The process by which the Project Team receives recognition for their accomplishments.

Team structure: A type of organization based on teams.

Technical quality administration: the technical process of establishing the plan for monitoring and controlling the project's satisfactory completion. this plan also includes policies and procedures to prevent or correct deviations from quality specifications/ requirements.

Technical quality specifications: The process of establishing the specific project requirements, including execution criteria and technologies, project design, Measurement Specifications, and Material Procurement and Control that satisfy the expectations of the client, shareholders and Project Team.

Technical quality support: The process of providing technical training and expertise from one or more support group(s) to a project in a timely manner. Effects of these groups could generate considerations for future client needs or warranty services.

Technical specifications: Documentation which describes, defines or specifies the goods/services to be supplied. See also *Specifications*.

Tetrachoric correlation: Measure of association used for relating two binary measures. Also see *Polychoric correlation*.

TGR: Things gone right.

TGW: Things gone wrong.

Theory: A systematic set of *causal relationships* providing a consistent and comprehensive explanation of a phenomenon. In practice, a theory is a researcher's attempt to specify the entire set of dependence relationships explaining a particular set of outcomes. A theory may be based on ideas generated from one or more of three principal sources: (1) prior empirical research; (2) past experiences and observations of actual behavior, attitudes, or other phenomena; and (3) other theories that provide a perspective for analysis. Thus, theory building is not the exclusive domain of academic researchers; it has an explicit role for practitioners as well. For any researcher, theory provides a means to address the "big picture" and assess the relative importance of various concepts in a series of relationships.

Theory of constraints (TOC): Goldratt's theory deals with techniques and tools for identifying and eliminating the constraints (bottlenecks) in a process.

Theory of inventive problem solving (TRIZ): Is a method for developing creative solutions to technical problems.

Theory of knowledge: A belief that management is about prediction, and people learn not only from experience but also from theory. When people study a process and develop a theory, they can compare their predictions with their observations; profound learning results.

Theory X and theory Y: A theory developed by Douglas McGregor that maintains that there are two contrasting assumptions about people, each of which is based on the manager's view of human nature. Theory X managers take a negative view and assume that most employees do not like work and try to avoid it. Theory Y managers take a positive view and believe that employees want to work, will seek and accept responsibility, and can offer creative solutions to organizational problems.

Theory Z: Coined by William G. Ouchi, refers to a Japanese style of management that is characterized by long-term employment, slow promotions, considerable job rotation, consensus-style decision making, and concern for the employee as a whole.

Things gone right (TGR): Product attributes or characteristics that produce a positive reaction from customers. Usually expressed as a rate; TGR/100.

Things gone wrong (TGW): Product attributes or characteristics that produce a negative reaction from customers (includes both component malfunctions and "correct" component functions that don't meet customer expectations).

Throughput time: The total time required (processing + queue) from concept to launch or from order received to delivery, or raw materials received to delivery to customer.

Time management: The function required to maintain appropriate allocation of time to the overall conduct of the project through the successive stages of its natural life-cycle, (i.e., concept, development, execution, and termination) by means of the processes of time planning, time estimating, time scheduling, and schedule control.

Time periods: Comparing calculated time versus specified time in relation to constraints and time span objectives.

Time-in-service (TIS): A concept used in the automotive industry to depict the fact that warranty R/1000 and CPU data have a length of service dimension. This dimension is normally measured by the number of months that the vehicle has been in the customer's hands, usually denoted by MIS (months-in-service). For example, a TIS matrix has rows of R/1000 or CPU data at 0 MIS (called pre-delivery), 1 MIS (which includes the 0 MIS claims as *well* as the 1 MIS claims), etc.

TIS: Time in Service.

TNI: Trouble Not Identified.

Tolerance (general): The permissible range of variation in a particular dimension of a product. Tolerances are often set by engineering requirements to ensure that components will function together properly. In other words, the variability of a parameter permitted and tolerated above or below a nominal value.

Tolerance: Commonly used measure of collinearity and multicollinearity. The tolerance of variable i (TOL_i) is $1 - R^{2*}_1$, where R^{2*}, is the coefficient of determination for the prediction of variable i by the other independent variables. As the tolerance value grows smaller, the variable is more highly predicted by the other independent variables (collinearity).

Tolerance design (Taguchi): Provides a rational grade limit for components of a system; determines which parts and processes need to be modified and to what degree it is necessary to increase their control capacity; a method for rationally determining tolerances.

Tolerance design: Determination of the tolerances which should be tightened or loosened and by how much in order to reduce the response variability of the overall system to the desired level. Tolerance design increases product/manufacturing cost and should be done when parameter design has not sufficiently reduced variation.

Top management: From the viewpoint of the project manager (Black Belt), Top Management includes the individual to whom he or she reports on project matters and other managers senior to that individual (Champion, and executive committee).

Top-management commitment: Participation of the highest-level officials in their organization's quality improvement efforts. Their participation includes establishing and serving on a quality committee, establishing quality policies and goals, deploying those goals to lower levels of the organization, providing the resources and training that the lower levels need to achieve the goals, participating in quality improvement teams, reviewing progress organization-wide, recognizing those who have performed well, and revising the current reward system to reflect the importance of achieving the quality goals. Commitment is top management's visible, personal involvement as seen by others in the organization.

Total cost (in warranty): The Material Cost plus X% allowance, plus Labor Cost of a specific repair including noncausal part(s).

Total cost: Material cost, plus labor cost plus any other cost associated with the item.

Total indicator readout (runout): Measurement of a characteristic such as concentricity or flatness for maximum or minimum value as well as range. The part is rotated or moved such that the gage measures the entire area in question. The value(s) of interest such as maximum are either read manually from the gage or determined automatically by automated data collection equipment.

Total productive maintenance (TPM): aimed at reducing and eventually eliminating equipment failure, setup and adjustment, minor stops, reduced speed, product rework, and scrap.

Total quality control (TQC): A management philosophy of integrated controls, including engineering, purchasing, financial administration, marketing and manufacturing, to ensure customer quality satisfaction and economical costs of quality.

Total quality management (TQM): A term initially coined by the Naval Air Systems Command to describe its management approach to quality improvement. Total quality management (TQM) has taken on many meanings. Simply put, TQM is a management approach to long-term success through customer satisfaction. TQM is based on the participation of all members of an organization in improving processes, products, services, and the culture they work in. TQM benefits all organization members and society. The methods for implementing this approach are found in the teachings of such quality leaders as Philip B. Crosby, W. Edwards Deming, Armand V Feigenbaum, Kaoru Ishikawa, J. M. Juran, and others.

The concept of TQM is based on (a) planning and (b) Communication. To support this concept there are six principles. (1) TQM starts on top (2) TQM requires total involvement (3) TQM focuses on the customer (4) TQM uses teams (5) TQM requires training for everybody, and (6) TQM uses tools to measure and follow progress. These principles transform the organization if applied appropriately. The fundamental steps of implementing TQM are: (a) Create a steering committee to oversee the implementation (b) Develop measures of quality and quality costs before the improvement program begins (c) Provide support to the teams, and (d) Reward success.

Total sum of squares (TSS): Total amount of variation that exists to be explained by the independent variables. This "baseline" is calculated by summing the squared differences between the mean and actual values for the dependent variable across all observations.

Trace: Represents the total amount of variance on which the factor solution is based. The trace is equal to the number of variables, based on the assumption that the variance in each variable is equal to 1.

Traceability: The ability to trace the history, application, or location of an item or activity and like items or activities by means of recorded identification.

Trade-off method: Method of presenting stimuli to respondents in which attributes are depicted two at a time and respondents rank all combinations of the levels in terms of preference.

Traditional conjoint analysis: Methodology that employs the "classic" principles of conjoint analysis, using an *additive* model of consumer preference and *pairwise* comparison or *full-profile* methods of presentation.

Traditional organizations: Those organizations not driven by customers and quality policies. Also refers to organizations managed primarily through functional units.

Training: Refers to the skills that employees need to learn in order to perform or improve the performances of their current job or tasks, or the process of providing those skills. See *Learning*.

Training evaluation: The techniques and tools used and the process of evaluating the effectiveness of training.

Training needs assessment: The techniques and tools used and the process of determining an organization's training needs.

Training sample: Observations used in the calibration of a *neural network*. It must contain actual values for the output *node* so that errors in output value prediction can be determined and used in the *learning* process.

Transaction: Action upon an *operational data element* (creation, modification or deletion) representing a single business event.

Transaction code: The financial accounts into which repairs are binned.

Transactional leadership: A style of leading whereby the leader articulates the vision and values necessary for the organization to succeed. The leader sees the work as being done through clear definitions of tasks and responsibilities and the provision of resources as needed.

Transformational leadership: A style of leading whereby the leader articulates the vision and values necessary for the organization to succeed.

Transformation: See *Data transformation*.

Transition tree: A technique used in applying Goldratt's Theory of Constraints.

Transmit: To send or convey from one person or place to another.

Treatment: Independent variable the researcher manipulates to see the effect (if any) on the dependent variable(s), such as in an experiment. See also *Stimulus*.

Tree diagram: A management and planning tool that shows the complete range of subtasks required to achieve an objective. A problem-solving method can be identified from this analysis.

Trend: A gradual, systematic change with time or other variable; Consecutive points that show a nonrandom pattern.

Trend analyses: Mathematical methods for establishing trends based on past project history allowing for adjustment, refinement or revision to predict cost. Regression analysis techniques can be used for predicting cost/schedule trends using data from historical projects

Trend analysis: A process used to analyze data from repetitive operations that encourages the coupling of product knowledge and basic statistical principles. It emphasizes the graphical display of data over time to ensure that the overall direction (better, worse, about the same) is understood, and also that individual data points are judged in context with past variability.

Trend monitoring: A system for tracking the estimated cost/schedule/resources of the project vs. those planned.

Trend reports: indicators of variations of Project Control parameters against planned objectives.

Trending budget: The review of proposed changes in resources allocation and the forecasting of their impact on budget; to be effective, trending should be regularly performed and the impacts on budget plotted graphically; used in this manner, trending supports a decision to authorize a change.

TRIZ: Theory of Inventive Problem Solving (TRIZ is a Russian acronym).

True Xbar causes: For Xbar control charts: changes in the Xbar control chart which are due to actual changes in the mean produced by the process. True Xbar changes are usually accompanied by a stable pattern in the R-chart.

Type I error (Alpha error—α): Deciding that a significant change in performance has occurred, when in fact, it has not. This means that an incorrect decision to reject something (such as a statistical hypothesis or a lot of products) when it is acceptable. Also known as "producer's risk" and "alpha risk." (In control chart analysis): concluding that a process is unstable when in fact it is stable. In a jury trial, this is analogous to convicting an innocent person. In warranty, this is often loosely described as "a witch hunt". The effect on the Company is that these mistakes consume valuable resources. There is a tradeoff between type I and type II errors; neither can be completely eliminated. The choice of control limits and run rules affects the tradeoff. Also termed *alpha (α)*. Typical levels are 5 or 1 percent, termed the 0.05 or 0.01 level, respectively.

Type II error (Beta error—β): It is also known as the Beta error. It is the probability of incorrectly failing to reject the null hypothesis—in simple terms, the chance of not finding a correlation or mean difference when it does exist. In other words: an incorrect decision to accept something when it is unacceptable. Also known as "consumer's risk" and "beta risk." (In control chart analysis): concluding that a process is stable when in fact it is unstable. Also termed *beta (β)*, it is inversely related to *Type I error*. The value 1 minus the Type II error is defined as *power*.

Typology: Conceptually based classification of *objects* based on one or more characteristics. A typology does not usually attempt to group actual observations, but instead provides the theoretical foundation for the creation of a *taxonomy*, which groups actual observations.

U-chart or u-chart: A control charting methodology developed for discrete, attribute data.It is a count per unit chart. In warranty, these charts can be used with simple systems and parts whose average R/1000 is below about 30. They don't work well for commodities above 30 R/1000, or for CPU charting. When in doubt, the I-Chart should be used.

UCL: Upper control limit.

U statistic: See *Wilks' lambda.*

Uncertainty: Lack of knowledge of future events. See also *Project risk.*

Unconditional guarantee: An organizational policy of providing customers unquestioned remedy for any product or service deficiency.

Underidentified model: *Structural model* with a negative number of *degrees of freedom*. This indicates an attempt to estimate more parameters than possible with the input matrix.

Unfolding: Representation of an individual respondent's *preferences* within a common (aggregate) stimulus space derived for all respondents as a whole. The individual's preferences are "unfolded" and portrayed as the best possible representation within the aggregate analysis.

Unidimensionality: Characteristic of a set of *indicators* that has only one underlying trait or concept in common. From the match between the chosen indicators and the theoretical definition of the unidimensional *construct*, the researcher must establish both conceptually and empirically that the indicators are reliable and valid measures of only the specified construct before establishing unidimensionality. Similar to the concept of *reliability*.

Uniform distribution: This distribution means that all outcomes are equally likely.

Unit: A discrete item (lamp, invoice, etc.) which possesses one or more CTQ (Note: "Units" must be considered with regard for the specific CTQ(s) of concern by a customer and/or for a specific process)

Unit of measure: The smallest increment a measurement system can indicate. See also *Resolution*.

Unit price (UP) contract: A fixed price contract where the supplier agrees to furnish goods/services at unit rates and the final price is dependent on the quantities needed to carry out the work.

Univariate analysis of variance (ANOVA): Statistical technique to determine, on the basis of one dependent measure, whether samples are from populations with equal means.

Universe: See *Population*.

Unobserved concept or variable: See *Latent construct* or *Variable*.

Unrestricted model: A mixed model (i.e., with random and fixed factors) is termed unrestricted if a random factor is allowed to have fixed cross factors amongst its components of variation estimated in the population. This freedom influences the choice of error Mean Square for random effects. It is generally used for unbalanced designs analysed with a General Linear Model. Balanced and orthogonal designs can be analysed with a restricted model.

Update: To revise the schedule to reflect the most current information on the project.

Upper control limit (UCL): For control charts: the upper limit below which a process remains if it is in control. The UCL just like the LCL are Process driven.

Upstream: Processes (tasks, activities) occurring prior to the task or activity in question.

Useful Life: Tolerance design increases product/manufacturing cost and should be done when parameter design has not sufficiently reduced variation. Uses arrow-diagramming techniques to demonstrate both the time and cost required to complete a project. It provides one time estimate-normal time.

USL (upper specification limit): The highest value of a product dimension or measurement which is acceptable. Customer driven.

Utility: A subjective preference judgment by an individual representing the holistic value or worth of a specific object. In conjoint analysis, utility is assumed to be formed by the combination of *part-worth estimates* for any specified set of *levels* with the use of an *additive model*, perhaps in conjunction with *interaction effects*.

Validation stimuli: Set of stimuli that are not used in the estimation of part-worths. Estimated part-worths are then used to predict preference for the validation stimuli to assess validity and reliability of the original estimates. Similar in concept to the validation sample of respondents in discriminant analysis.

Validation: Confirmation by examination of objective evidence that specific requirements and/or a specified intended use are met.

Validity: Extent to which a measure or set of measures correctly represents the concept of study—the degree to which it is free from any systematic or nonrandom error. Validity is determined to a great extent by the researcher, because the original definition of the construct or concept is proposed by the researcher and must be matched to the selected indicators or measures. Validity does not guarantee reliability, and vice versa. A measure may be accurate (valid) but not consistent (reliable). Also, it may be quite consistent but not accurate. Thus validity and reliability are two separate but interrelated conditions. Validity is concerned with how well the concept is defined by the measure(s), whereas *reliability* relates to the consistency of the measure(s).

Value-added: Refers to tasks or activities that convert resources into products or services consistent with customer requirements, The customer can be internal or external to the organization.

Value-adding activities: Steps/tasks in a process that meet all three criteria defining value as perceived by the external customer: 1) the customer cares; 2) the thing moving through the process changes; and 3) the step is done right the first time.

Value analysis, value engineering, and value research (VA, VE, VR): Value analysis assumes that a process, procedure, product, or service is of no value unless proven otherwise. In other words, it is an activity devoted to optimizing cost performance. It is the systematic use of techniques which identify the required functions of an item, establish values for those functions and provide the functions at the lowest overall cost without loss of performance (optimum overall cost). Value analysis assumes that a process, procedure, product, or service is of no value unless proven otherwise. It assigns a price to every step of a process and then computes the worth-to-cost ratio of that step. VE points the way to elimination and reengineering. Value research, related to value engineering, for given features of the service/product, helps determine the customers' strongest "likes" and "dislikes" and those for which customers are neutral. Focuses attention on strong dislikes and enables identified "neutrals" to be considered for cost reductions.

Value chain: See *Supply chain.*

Value-enabling activities: Steps/tasks in a process enabling work to move forward and add value to the customer but not meeting all three of the value-adding criteria; should still be scrutinized for time and best practices-can it be done better?

Values: Statements that clarify the behaviors that the organization expects in order to move toward its vision and mission. Values reflect an organization's personality and culture.

Value stream mapping: The technique of mapping the value stream.

Value stream: The primary actions required to bring a product from concept to placing the product in the hands of the end-user.

Variability: The property of exhibiting variation, i.e., changes or differences, in particular in the product of a process.

Variable data: Data resulting from the measurement of a parameter or a variable as opposed to attributes data. A dimensional value can be recorded and is only limited in value by the resolution of the measurement system. Generally, they are quantitative data, where measurements can be directly used for analysis. However, in warranty data are not technically variables data and are often analyzed with the I-Chart (Individuals or X-Chart). See also *Attribute*. On the other hand, control charts based on variables data include average (Xbar) chart, individuals (X) chart, range (R) chart, sample standard deviation (s) chart, and CUSUM chart.

Variable sampling plan: A plan in which a sample is taken and a measurement of a specified quality characteristic is made on each unit. The measurements are summarized into a simple statistic, and the observed value is compared with an allowable value defined in the plan.

Variables: Quantities which are subject to change or variability.

Variance: in statistics: The square of the standard deviation.

Variance: (Of a sample). Variance is a measure of the spread of the data. $[s^2 = (Y_i - Y)^2/(n - 1) = SS/df]$.

Variance: Any actual or potential deviation from an intended or budgeted figure or plan. A variance can be a difference between intended and actual time. Any difference between the projected duration for an activity and the actual duration of the activity. Also, in relation to a project, the difference between projected start and finish dates and actual or revised start and finish dates.

Variance analysis (project oriented): The analysis of the following (1) Cost Variance = BCWP − ACWP; (2) %Over/Under = [ACWP-BCWP]/ BCWPX100; (3) Unit Variance Analysis for at least the following items: (a) Labor Rate (b) Labor Hours/Units of Work Accomplished (c) Material Rate and d) Material Usage; and (4) Schedule/Performance = BCWP − BCWS

Variance reports: Documentation of project performance about a planned or measured performance parameter.

Variance extracted measure: Amount of "shared" or common variance among the *indicators* or *manifest variables* for a *construct*. Higher values represent a greater degree of shared representation of the indicators with the construct.

Variance inflation factor (VIF): Indicator of the effect that the other independent variables have on the standard error of a regression coefficient. The variance inflation factor is directly related to the *tolerance* value $(VIF_i = 1/\ TOL_i)$. Large VIF values also indicate a high degree of *collinearity* or *multicollinearity* among the independent variables.

Variate: Linear combination of variables formed in the multivariate technique by deriving empirical weights (coefficients to the independent variables) applied to a set of variables specified by the researcher.

Variation: A change in data, a characteristic, or a function that is caused by one of four factors: special causes, common causes, tampering, or structural variation. Change or fluctuation of a specific characteristic which determines how stable or predictable the process may be; affected by environment, people, machinery/equipment, methods/procedures, measurements, and materials; any Process Improvement should reduce or eliminate variation. See also *Common cause; Special cause.*

VARIMAX: One of the most popular orthogonal factor rotation methods.

Vector: Method of portraying an ideal point or attribute in a perceptual map. Involves the use of *projections* to determine an *object's* order on the vector. Set of real numbers (e.g., $X_i \dots X_n$) that can be written in either columns or rows. Column vectors are considered conventional and row vectors are considered transposed.

Verification: The act of reviewing, inspecting, testing, checking, auditing, or otherwise may establishing and documenting whether items, processes, services, or documents conform to specified requirements. In other words, the confirmation of data, application of judgment and comparison with other sources and previous monitor results.

Vertical icicle diagram: Graphical representation of clusters. The separate *objects* are shown horizontally across the top of the diagram, and the hierarchical clustering process is depicted in combinations of clusters vertically. This diagram is similar to an inverted *dendrogram* and aids in determining the appropriate number of clusters in the solution.

Vertically integrate: To bring together more of the steps involved in producing a product in order to form a continuous chain owned by the same firm; typically involves taking on activities that were previously in the external portion of the supply chain.

Vision: A statement that explains what the company wants to become and what it hopes to achieve.

Visual control: A technique of positioning all tools, parts, production activities, and performance indicators so that the status of a process can be understood at a glance by everyone; provide visual clues: to aid the performer in correctly processing a step or series of steps, to reduce cycle time, to cut costs, to smooth flow of work, to improve quality.

Vital few, useful many: A term used by J. M. Juran to describe his use of the Pareto principle, which he first defined in 1950. (The principle was used much earlier in economics and inventory control methodologies.) The principle suggests that most effects come from relatively few causes; that is, 80 percent of the effects come from 20 percent of the possible causes. The 20 percent of the possible causes are referred to as the "vital few"; the remaining causes are referred to as the "useful many." When Juran first defined this principle, he referred to the remaining causes as the "trivial many," but realizing that no problems are trivial in quality assurance, he changed it to "useful many."

Voice of the customer (VOC): An organization's efforts to understand the customers' needs and expectations ("voice") and to provide products and services that truly meet such needs and expectations. The data (complaints, surveys, comments, market research, etc.) representing the views/needs of a company's customers; should be translated into measurable requirements for the process. Generally, the VOC is identified as a functionality that the customer is seeking.

Walk the talk: Means not only talking about what one believes in but also being observed acting out those beliefs. Employees' buy in of the TOM concept is more likely when management is seen as committed and involved in the process, every day.

Walkabout: A visual, group technique used in resolving resource planning conflicts among organizational components.

Ward's method: *Hierarchical clustering* procedure in which the *similarity* used to join clusters is calculated as the sum of squares between the two clusters summed over all variables. This method has the tendency to result in clusters of approximately equal size due to its minimization of within-group variation.

Waste: Activities that consume resources but add no value; visible waste (for example, scrap, rework, downtime) and invisible waste (for example, inefficient setups, wait times of people and machines, inventory). It is customary to view waste as any variation from target

WBS: See *Work breakdown structure.*

Wearout: A failure mode characterized by a hazard rate that increases with age; i.e., old units are more likely to fail than new units. The product life-cycle phase that begins after the design's expected life.

Weibull analysis: Procedure for finding the Weibull distribution that best describes a sample of unit lifetimes in order to estimate reliability.

Weibull distribution: A distribution of continuous data that can take on many different shapes and is used to describe a variety of patterns; used to define when the "infant mortality rate" has ended and a steady state has been reached (decreasing failure rate); relates to the "bathtub" curve.

Wilks' lambda: One of the four principal statistics for testing the null hypothesis in MANOVA. Also referred to as the maximum likelihood criterion or *U-statistic.*

Wisdom: The culmination of the continuum from data to information to knowledge to wisdom.

Within-case standardization: Method of standardization in which a respondent's responses are not compared to the overall sample but instead to their own responses. Also known as ipsitizing, the respondents' average responses are used to standardize their own responses.

Work acceptance: Work is considered accepted when it is conducted, documented and verified as per acceptance criteria provided in the technical specifications and contract documents.

Work analysis: The analysis, classification and study of the way work is done. Work may be categorized as value-added work (necessary work), non-value-added (rework, unnecessary work, idle). Collected data may be summarized on a Pareto chart, showing how people within the studied population work. The need for and value of all work is then questioned and opportunities for improvement identified. A Time Use Analysis may also be included in the study.

Work breakdown structure (WBS): A project management technique by which a project is divided into tasks, subtasks, and units of work to be performed. In other words, It is a task-oriented "family tree" of activities which organizes, defines and graphically displays the total work to be accomplished in order to achieve the final objectives of a project. each descending level represents an increasingly detailed definition of the project objective. it is a system for subdividing a project into manageable work packages, components or elements to provide a common framework for scope/cost/ schedule communications, allocation of responsibility, monitoring and management.

Work group: A group composed of people from one functional area who work together on a daily basis and whose goal is to improve the processes of their function.

Work instruction: A document which answers the question: How is the work to be done? (see Procedure).

World-class quality: A term used to indicate a standard of excellence: best of the best.

WRA: Warranty Responsible Activity.

WRT: Warranty Reduction Team.

X: Variable used to signify factors or measures in the Input or Process segments of a business process or system. It appears as part of the $Y = F(X)$. It is the independent factor that can control and or predict the F(Y). The X must be directly correlated with the customer's needs, wants or expectations. Is also known as: Cause, Control, and Problem. It is the basis for identifying the projects for the Black Belts.

Xbar and R-charts: For variables data: control charts for the average and range of subgroups of data. See also *Control chart*.

Xbar and Sigma charts: For variables data: control charts for the average and standard deviation (sigma) of subgroups of data. This chart is much more effective than the Xbar and R-chart. See also *Control chart*.

X bar chart: Average chart.

Y: Variable used to signify factors or measures at the Output of a business process or system. Equivalent to "results." A key principle of Six Sigma is that Y is a function of upstream factors; or $Y = f(x)$. It is the depended variable that the predictors of $f(x)$ trying to define. It is also known as output, effect, symptom and monitor.

Yield: Total number of units handled correctly through the process step(s). Mathematically it may be shown as the ratio between salable goods produced and the quantity of raw materials and/or components put in at the beginning of the process.

YTD report: This report contains a commodity's YTD R/1000 or CPU values across selected models years for selected TIS values. The summaries are grouped by product and or repair entities.

Youden square: is a reduction from randomized complete blocks to a balanced incomplete design in which one row or column has been removed from a Latin square (so it is no longer square). It is a useful design for balancing out the effects of treatment order in a repeated-measures sequence. The example on these web pages could pertain to a predator odour treatment (A with $a = 4$ levels) tainting the food offered to each of $a = 4$ mice (random factor B). Each mouse is tested with $a - 1 = 3$ odour types in order (random factor C) assigned by the Youden square.

z-Distribution: For a sample size of n drawn from a normal distribution with mean μ and standard deviation σ. Used to determine the area under the normal curve.

Zero defects: A performance standard popularized by Philip B. Crosby to address a dual attitude in the workplace. People are willing to accept imperfection in some areas, while, in other areas they expect the number of defects to be zero. This dual attitude had developed because of the conditioning that people are human and humans make mistakes. However, the zero-defects methodology states that if people commit themselves to watching details and avoiding errors, they can move closer to the goal of zero.

Zero investment improvement: Another name for a kaizen blitz.

$Z_{max}/3$: The greater result of the formula when calculating C_{pk}. Shows the distance from the tail of the distribution to the specification which shows the greatest capability. See C_{pk}.

$Z_{min}/3$: The smaller result of the formula when calculating C_{pk}. Shows the distance from the tail of the distribution to the specification which is the smallest capability. See C_{pk}.

z-test: A test of a statistical hypothesis that the population mean (μ) is equal to the sample mean ($\bar{X}$) when the population standard deviation is known.

Selected Bibliography

Anderson, T. W. and Darling, D. A. (1952). Asymptotic theory of certain "goodness of fit" criteria based on stochastic processes. *Annals of Mathematical Statistics* 23:193–212.

Bartholomew, D. J., Steele, F., Mustaki, I. and Galbrainth, J. (2008). *Analysis of Multivariate Social Science Data.* 2nd ed., CRC Press, Boca Raton, FL.

Brown, H. and Prescott, R. (2000). *Applied Mixed Models in Medicine*, John Wiley & Sons, New York.

Buonaccorsi, J. P. (2010). *Measurement Error: Models, Methods and Applications.* CRC Press. Boca Raton, FL.

Chatterjee, S., Hadi, A. S. and Price, B. (1999). *Regression Analysis by Example,* John Wiley & Sons, New York.

Choulakian, V., Lockhart, R. A. and Stephens, M. A. (1994). Cramér–von Mises statistics for discrete distributions. *The Canadian Journal of Statistics* 22:125–137.

Cook, R. D. and Weisberg, S. (1999). *Applied Regression Including Computing and Graphics,* John Wiley & Sons, New York.

Cook, R.D. (1998). *Regression Graphics: Ideas for Studying Regressions through Graphics.* John Wiley & Sons, New York.

Cramér, H. (1928). On the composition of elementary errors. *Skandinavisk Aktuarietidskrift.* 11:13–74, 141–180.

Doncaster, C. P. and Davey, A. J. H. (2007). *Analysis of Variance and Covariance: How to Choose and Construct Models for the Life Sciences.* Cambridge University Press, Cambridge.

Draper, N. R. and Smith, H. (1998). *Applied Regression Analysis,* 3rd ed., John Wiley & Sons, New York.

Einmahl, J. and McKeague, I. (2003). Empirical likelihood based hypothesis testing. *Bernoulli* 9:267–290.

Freund, R. J. and Littell, R. C. (2000). *SAS System for Regression,* John Wiley & Sons, New York.

Hettmansperger, T. P. and Mckeen, J. W. (2010). *Robust Nonparametric statistical methods.* 2nd ed, CRC Press, Boca Raton, FL.

Hosmer, D.W. and Lemeshow, S. (2000). *Applied Logistic Regression,* 2nd ed., John Wiley & Sons, New York.

From, S. G. (1996). A new goodness of fit test for the equality of multinomial cell probabilities verses trend alternatives. *Communications in Statistics–Theory and Methods.* 25:3167–3183.

Hoenig, J. M. and Heisey, D. M. (2001). `The Abuse of Power: The Pervasive Fallacy of Power Calculations for Data Analysis. *The American Statistician* 55:19–24.

Khattree, R. and Naik, D.N. (1999). *Applied Multivariate Statistics with SAS Software,* John Wiley & Sons, New York.

Khattree, R., Naik, D. N. and SAS Institute Inc. (2000). *Multivariate Data Reduction and Discrimination with SAS Software,* John Wiley & Sons, New York.

Khuri, A., Mathew, T. and Sinha, B. K. (1998). *Statistical Tests for Mixed Linear Models,* John Wiley & Sons, New York.

McCulloch, C. E. and Searle, S. R. (2000). *Generalized, Linear, and Mixed Models,* John Wiley & Sons, New York.

McLachlan, G. and Peel, D. (2000). *Finite Mixture Models,* John Wiley & Sons, New York.

Mulaik, S. A. (2010). *Foundations of Factor Analysis.* 2nd ed. CRC Press, Boca Raton, FL.

Pearson, K. (1900). On the criterion that a given system of deviations from the probable in the case of a correlated system of variables is such that it can be reasonably supposed to have arisen from random sampling. *Philosophical Magazine* 5:157–175.

Pettitt, A. N. and Stephens, M. A. (1977). The Kolmogorov–Smirnov goodness-of-fit statistic with discrete and grouped data. *Technometrics* 19:205–210.

Read, T. R. C. and Cressie, N. A. C. (1988). *Goodness-of-Fit Statistics for Discrete Multivariate Data,* Springer-Verlag, New York.

Rencher, A. C. (1999). *Linear Models in Statistics,* John Wiley & Sons, New York.

Ryan, T. P. (1996). *Modern Regression Methods,* John Wiley & Sons, New York.

Schimek, M. G. (2000). *Smoothing and Regression: Approaches, Computation and Application,* John Wiley & Sons, New York.

Seber, G. A. F. and Wild, C. J. (1989). *Nonlinear Regression,* John Wiley & Sons, New York.

Smirnov, N. V. (1939). Estimate of deviation between empirical distributions (Russian). *Bulletin Moscow University* 2:3–16.

Stamatis, D. H. (2002). *Six Sigma and Beyond: Design of Experiments.* St. Lucie Press, Boca Raton, FL.

Zar, J. H. (1999). *Biostatistical Analysis.* 2nd ed. Prentice-Hall, Englewood Cliffs, NJ.

Zhang, J. (2002). Powerful goodness-of-fit tests based on the likelihood ratio. *Journal of the Royal Statistical Society, Series B* 64:281–294.

Index